**Resonance Enhancement
in Laser-Produced Plasmas**

Resonance Enhancement in Laser-Produced Plasmas

Concepts and Applications

Rashid A. Ganeev
Changchun Institute of Optics, Fine Mechanics and Physics, Changchun, China

This edition first published 2018
© 2018 John Wiley & Sons Inc.

The right of Rashid A. Ganeev to be identified as the author of this work has been asserted in accordance with law.

Registered Office
John Wiley & Sons, Inc., 111 River Street, Hoboken, NJ 07030, USA

Editorial Office
111 River Street, Hoboken, NJ 07030, USA

For details of our global editorial offices, customer services, and more information about Wiley products visit us at www.wiley.com.

Wiley also publishes its books in a variety of electronic formats and by print-on-demand. Some content that appears in standard print versions of this book may not be available in other formats.

Library of Congress Cataloging-in-Publication Data

Names: Ganeev, Rashid A., author.
Title: Resonance enhancement in laser-produced plasmas : concepts and applications / by Rashid A. Ganeev.
Description: First edition. | Hoboken, NJ : John Wiley & Sons, 2018. | Includes index. |
Identifiers: LCCN 2018013083 (print) | LCCN 2018029388 (ebook) | ISBN 9781119472353 (pdf) | ISBN 9781119472261 (epub) | ISBN 9781119472247 (cloth)
Subjects: LCSH: Laser plasmas. | Harmonics (Electric waves)
Classification: LCC QC718.5.L3 (ebook) | LCC QC718.5.L3 G375 2018 (print) | DDC 530.4/43—dc23
LC record available at https://lccn.loc.gov/2018013083

Cover design: Wiley
Cover image: © sakkmesterke/iStockphoto

Set in 10/12pt WarnockPro by SPi Global, Chennai, India

Printed in the United States of America

V10002507_071418

To my parents, wife, son, and daughter

Contents

Preface

The motivation in writing this book is to show the most recent findings of newly emerged field of resonance-enhanced high-order harmonic generation (HHG) using the laser pulses propagating through the narrow and extended laser-produced plasma plumes. It becomes obvious that the developments in this field are aimed for improvement of harmonic yield through precise study of resonance effects during fine-tuning of driving pulses to the resonances. The purpose of writing this book is to acquaint the readers with the most advanced, recently developed methods of plasma harmonics generation at the conditions of coincidence of some harmonics, autoionizing states, and some ionic transitions possessing strong oscillator strengths. This book demonstrates how one can improve the plasma harmonic technique using this approach.

There is a classical book relevant with the resonance processes in gaseous media [1]. Another book [2] is related with the spectroscopic details of nonlinear optical studies. Separate chapters of the nonlinear optical properties of matter [3] were related with the resonance processes. Some details of plasma properties relevant to those at which the resonance processes play an important role are discussed in Ref. [4]. Meanwhile, though some separate details of plasma harmonic studies were published in different edited books as the chapters, there is no collection of the various aspects of resonance-enhanced harmonic generation processes in a single book.

The dissemination of information presented in this book will help to understand the peculiarities of laser–plasma interaction, which can be used for the amendment of harmonic yield in the extreme ultraviolet (XUV) region. The book also demonstrates the limitations of this method of harmonic generation, especially in the case of gas HHG. The development of plasma harmonic spectroscopy using this approach would be useful for material science. It may help in the next steps of the development of this interaction, which lead to generation of attosecond pulses. The basics of resonance plasma harmonic studies will help the reader to acquaint with novel methods of XUV coherent sources formation.

Among most attractive key features, which the reader may find in this book, are the demonstration of novel approaches in the resonance-based amendments of harmonic generation in the laser-produced plasmas using fixed and tunable long-wavelength pulses, methods of the application of tunable laser sources of parametric waves for resonance enhancement of single harmonic, the application of proposed method for the nonlinear optical spectroscopic studies of various organic materials, and the implementation of theoretical and experimental consideration of the usefulness of mid-infrared driving pulses and two-color technique for the potential shortening of harmonic pulses toward the attosecond region.

The novelty in laser–plasma resonance interaction shown in this book may attract the interest in various groups of researchers, particularly those involved in the applications of lasers and development of short-wavelength coherent sources. Most relevant audience include the researchers, specialists, and engineers in the fields of optics and laser physics. This book will also be useful for the students of high education in the physical departments of universities and institutes. It may serve as a tutorial for the optical and nonlinear optical interactions of ultrashort pulses and low-dense plasmas produced on the surfaces of various solids.

This book would be interesting to the academic community. The researchers in laser physics and optics are the main audience, who can find interesting information regarding state-of-art developments in the field of frequency conversion of laser sources toward the short wavelength spectral range. Those involved in optics, nonlinear optics, atomic physics, resonance processes, HHG, and plasma physics are the specific potential readers of this book. Graduate students can also find plenty of novelties in this rapidly developing field of nonlinear optics and atomic physics. Any professionals interested in material science could be also interested in updating their knowledge of the new methods of material studies using nonlinear spectroscopy, developed using the resonance-induced high-order harmonic enhancement in the vicinity of autoionizing states and ionic transitions possessing strong oscillator strengths.

All these amendments of the plasma harmonic resonance studies could not be realized without the collaboration of different groups involved in the studies of the nonlinear optical properties of ablated species. Among the numerous colleagues I would like to thank are H. Kuroda, M. Suzuki, S. Yoneya, M. Baba (Saitama Medical University, Japan), T. Ozaki, L. B. Elouga Bom (Institut National de la Recherche Scientifique, Canada), P. D. Gupta, P. A. Naik, H. Singhal, J. A. Chakera (Raja Ramanna Centre for Advanced Technology, India), J. P. Marangos, J. W. G. Tisch, C. Hutchison, T. Witting, F. Frank, A. Zaïr (Imperial College, United Kingdom), M. Castillejo, M. Oujja, M. Sanz, I. López-Quintás, M. Martin (Instituto de Química Física Roca-solano, Spain), H. Zacharias, J. Zheng, M. Woestmann, H. Witte (Westfäliche

Wilhelms-Universität, Germany), T. Usmanov, G. S. Boltaev, I. A. Kulagin, V. I. Redkorechev, V. V. Gorbushin, R. I. Tugushev, N. K. Satlikov (Institute of Ion-Plasma and Laser Technologies, Uzbekistan), M. Danailov (ELETTRA, Italy), B. A. Zon, O. V. Ovchinnikov (Voronezh State University, Russia), D. B. Milošević (University of Sarajevo, Bosnia and Herzegovina), M. Lein, M. Tudorovskaya (Leibniz Universität Hannover, Germany), E. Fiordilino (University of Palermo, Italy), V. Tosa (National Institute of R&D Isotropic and Molecular Technologies, Romania), V. V. Strelkov (General Physics Institute, Russia), M. K. Kodirov, P. V. Redkin (Samarkand State University, Uzbekistan), A. V. Andreev and S. Y. Stremoukhov (Moscow State University, Russia), and Pengfei Lan (Huazhong University of Science and Technology, China).

The inspirations for all these new findings are my wife Lidiya, son Timur, and daughter Dina. Now, becoming a grandfather, I would like to include Timur's wife Anna and our beloved grandson Timofey and granddaughter Valeria in the list of most important people, who helped me to overcome various obstacles of the life of scientific tramp.

This book is organized as follows. In Chapter 1, high-order harmonic studies showing the role of resonances on the temporal and efficiency characteristics of converted coherent pulses, as well as different approaches, are discussed. Particularly, resonance harmonic generation in gases (theory and experiment), as well as the role of resonances in plasma harmonic experiments are analyzed. In Chapter 2, different theoretical approaches in plasma HHG studies at resonance conditions are described. Comparative analysis of the HHG in laser-ablation plasmas prepared on the surfaces of complex and atomic targets, nonperturbative HHG in indium plasma (theory of resonant recombination), and important consequences of different theories are analyzed. We show the simulations of resonant HHG in three-dimensional fullerene-like system by means of multiconfigurational time-dependent Hartree–Fock approach, describe the basics of the nonlinear optical studies of fullerenes, as well as endohedral fullerenes, present the model of resonant high harmonic generation in multi-electron systems. We also analyze the drawbacks of different theories of resonant HHG. Chapter 3 is dedicated to the comparison of resonance harmonics through experimental and theoretical studies. We discuss the experimental and theoretical studies of two-color pump resonance-induced enhancement of odd and even harmonics from different plasmas, provide the comparative studies of resonance enhancement of harmonic radiation in indium plasma using multicycle and few-cycle pulses and tunable near-infrared (NIR) pulses. Resonance enhancement of harmonics in laser-produced Zn II- and Zn III-containing plasmas using tunable NIR pulses, single- and two-color pumps of plasma, and applications of tunable NIR radiation for resonance enhancement of harmonics in tin, antimony, and chromium plasmas are also discussed, alongside the model of resonant high harmonic generation in multi-electron systems. In Chapter 4, early studies of resonance enhancement of harmonics in metal-ablated plasmas are analyzed. Strong resonance enhancement of single harmonic generated in XUV range is described through the chirp-induced enhancement of harmonic generation from metal-containing plasmas, such as chromium, manganese, tin, and antimony plasma plumes. Here, we also discuss the enhancement of high harmonics from plasmas using two-color pump and chirp variation of 1 ;kHz Ti:sapphire laser pulses, and show the advances in using high pulse repetition source for HHG in plasmas. Other topics discussed here are the spatial coherence measurements of nonresonant and resonant high-order harmonics generated in different plasmas, demonstration of the 101st harmonic generation from laser-produced manganese plasma, and isolated sub-fs XUV pulse generation in Mn plasma ablation. Chapter 5 is dedicated to the resonance processes occurring in ablated semiconductors. We discuss the quasi-phase-matching and properties of semiconductor plasmas, such as GaAs, Te, Sb, and others. In Chapter 6, the resonance processes at different conditions of harmonic generation are discussed. Particularly, we analyze the application of picosecond pulses for HHG in gases and plasmas, show the resonance processes in lead and carbon plasmas using 1064 nm pulses. Size-related resonance processes influencing harmonic generation in plasmas are discussed as well. We also discuss the resonance-enhanced harmonic generation in nanoparticle-containing plasmas and fullerenes. Chapter 7 is dedicated to the comparison of the resonance-, nanoparticle-, and quasi-phase-matching-induced processes leading to the growth of high-order harmonic yield. Particularly, we describe the quasi-phase-matched HHG in laser-produced plasmas and influence of few-atomic silver molecules on HHG in the laser-produced plasmas. We compare the experimental conditions for observation of the control of harmonic enhancement in the cases of featureless and resonance-enhanced harmonic distributions, and compare plasma and harmonic spectra allowing generation of resonantly enhanced harmonics in zinc, cadmium, and manganese plasmas. Finally, we provide the analysis of the comparison of micro- and macroprocesses during HHG in laser-produced plasmas. We conclude with a summary section and underline the most important findings analyzed alongside this book.

References

1 Reintjes, J. (1984). *Nonlinear Optical Parametric Processes in Liquids and Gases*. Academic Press.
2 Mukamel, S. (1999). *Principles of Nonlinear Optical Spectroscopy*. Oxford University Press.

3 Palpant, B. (2006). Third-order nonlinear optical response of metal nanoparticles. In: *Nonlinear Optical Properties of Matter*, vol. 1 (ed. M.G. Papadopoulos, A.J. Sadlej and J. Leszczynski), 461–508. Dordrecht: Springer.
4 Hippler, R., Kersten, H., and Schmidt, M. (2008). *Low Temperature Plasmas: Fundamentals, Technologies and Techniques*. Wiley-VCH.

January, 2018

Rashid A. Ganeev
Changchun, China

1

High-Order Harmonic Studies of the Role of Resonances on the Temporal and Efficiency Characteristics of Converted Coherent Pulses: Different Approaches

1.1 Resonance Harmonic Generation in Gases: Theory and Experiment

Excitation of atomic resonances exhibits a simple way to enhance high-order harmonic conversion efficiencies. The basic idea is straightforward: the driving laser is tuned to an atomic resonance (usually a multiphoton resonance, e.g. with n photons from the driving laser involved). The resonance enhances the nonlinear susceptibility $\chi^{(n)}$ of order n. If permitted by selection rules, this supports generation of the nth harmonics of the driving laser or frequency mixing processes with m additional photons from the same laser field, e.g. generating harmonics of order $(n + m)$. Such resonantly enhanced frequency conversion is well known from low-order frequency conversion processes, driven by lasers of moderate intensities. As a simple example, one can note resonantly enhanced four-wave mixing in atomic gases. Here, a first laser pulse drives a two-photon transition, which serves to resonantly enhance a sum or difference frequency mixing process with a second laser pulse. However, during initial studies of this process it was not obvious, that resonance enhancement occurs for generation of high-order harmonics driven by high-intensity ultrashort laser pulses.

The strong electric field of the laser significantly perturbs the level structure of the medium and may destroy any resonance effect in conversion processes. From this simple consideration it becomes clear, that resonantly enhanced harmonic generation with ultra short pulses may be efficient, if the driving radiation field is, on one hand, sufficiently strong to drive high-order harmonic generation (HHG), and on the other hand the field is still not too strong to destroy the resonance structure of the medium. In the terminology of high intensity laser–matter interactions and photoionization, one can consider operation in the regime of "multi-photon ionization" rather than "tunneling ionization" (see Ref. [1] and references therein). This choice provides appropriate conditions to observe pronounced resonance effect. The restriction toward not-too-strong laser intensities still enables a large range of applications and the possibility to exploit resonances for efficient harmonic generation. One can

Resonance Enhancement in Laser-Produced Plasmas: Concepts and Applications,
First Edition. Rashid A. Ganeev.
© 2018 John Wiley & Sons, Inc. Published 2018 by John Wiley & Sons, Inc.

also note that proper investigation and application of resonance effects also requires tunable lasers and moderate frequency bandwidth (i.e. not-too-short pulse durations) to properly address isolated atomic and ionic resonances.

There are only few studies of resonance enhancement in harmonic generation in gases via bound atomic states (for example, [2, 3]). The enhancement of particular harmonics in those studies was observed at specific laser intensities, which allowed an increase in the yield of the nth harmonic by exciting a dynamically shifted n-photon resonance. Another consideration of this enhancement was suggested in Ref. [4], where the phase-matching effects or multiphoton resonances were attributed for the harmonic enhancement. Contrary to gases, a sequence of experiments on pronounced enhancement of single harmonics in laser-driven plasmas [5] has shown that this effect is attributed to dynamically shifted ionic resonances close to specific harmonics. The details of latter studies will be discussed in the following chapters. Notice that in most cases only single harmonics were enhanced, while the theoretical predictions show that excitation of n-photon resonances should also affect harmonics with order larger than n [6–8]. The resonance HHG can also be realized through atomic Fano resonances [9].

In Ref. [1], the coupling scheme and relevant energy levels in a jet of argon atoms were considered (see Figure 1.1). The intense picosecond laser pulses were in the vicinity of the five-photon resonance $3p^{61}S_0 \rightarrow 4s$, $[1/2]_1$ at $95\,400$ cm^{-1}, corresponding to a fundamental laser wavelength of 524 nm. In the experiment, the resonantly enhanced 5th harmonic generation of the driving laser radiation as well as harmonics of higher order (indicated by dashed arrows in the figure) were observed. The authors investigated harmonic generation in a dense Ar gas jet driven in the vicinity of above resonance by intense tunable picosecond radiation pulses from dye amplifier system. The laser system combined sufficient intensity (i.e. up to 100 TW cm^{-2}) to approach the regime of HHG with still fine spectral resolution to address and exploit single atomic resonances. In a first experiment on resonantly enhanced 5th harmonic generation, they determined pronounced AC-Stark shifts and line broadenings of the five-photon resonance. Moreover, they found evidence for an additional difference frequency mixing process "six minus one photon" via a set of highly excited states in Ar, which also generates radiation at the 5th harmonic of the driving laser. In a second experiment, they investigated the effect of resonant multiphoton excitation on the generation of harmonics (i.e. with higher order than the involved multiphoton transition). When the laser frequency was tuned to the Stark-shifted five-photon resonance, a pronounced resonance enhancement was found not only of the 5th, but also of the 7th and 9th harmonic. They pointed out that, as an important feature of resonance enhancement, the laser wavelength must be matched to the position of the Stark-shifted atomic resonance, which depends upon the applied laser

Figure 1.1 Coupling scheme in argon atoms with relevant energy levels. The short designation (5p/5p′) indicates the manifold of closely spaced states 5p′ [3/2]$_2$, 5p [1/2]$_0$, 5p [3/2]$_2$, and 5p [5/2]$_2$. Full arrows depict the driving laser at 524 nm, dashed arrows indicate the 5th, 7th, and 9th harmonics, as investigated in the experiment. *Source:* Ackermann et al. 2012 [1]. Reproduced with permission from Optical Society of America.

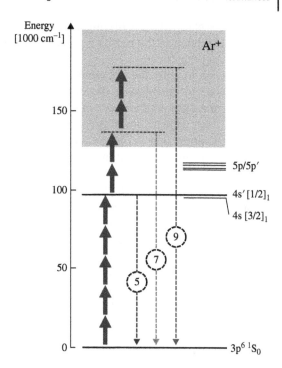

intensity. The experimental data clearly demonstrate the effect of resonant multiphoton excitation to enhance harmonic generation.

Resonantly enhanced harmonic generation is a particular example of the more general HHG technique. Although resonantly enhanced HHG has been shown to increase the harmonic yield in a limited range of settings, it has more recently been explored for its potential to reveal the dynamics of bound and quasi-bound states in the presence of a strong driving field [10–16]. Several mechanisms for resonant enhancement have been discussed in the literature, generally all involving an intermediate, resonant step in the semiclassical model. The resonant step may occur either in the ionization process, via a multiphoton resonance between the ground state and the Stark-shifted excited state, or in the rescattering process via enhanced recombination, or by capture into an excited bound state that subsequently decays via spontaneous emission of light. The capture and spontaneous emission process has been explored in detail for short-lived quasi-bound states embedded in a continuum, for which it can give rise to very large enhancements [10, 13]. For bound-state resonances with long lifetimes, the capture and spontaneous emission process can generally be distinguished from the coherently driven resonant enhancement (via multiphoton ionization or enhanced recombination), since it will give rise to narrow-band radiation at the field-free resonance frequency, given that it largely takes place after the driving laser pulse is over [13]. More details on the theoretical models of the resonant enhancement of HHG will be discussed in the following chapter.

In contrast to this, the coherently driven resonantly enhanced response will give rise to emission at the difference frequency between the ground state and the Stark-shifted excited state since this process only takes place while the laser field is on. The coherently driven resonance-enhanced harmonic generation and investigation of the interplay between the resonant enhancement and the quantum path dynamics of the harmonic generation process is of utmost interest to understand the peculiarities of this process. In particular, authors of Ref. [17] have studied how the amplitude and phase of the different quantum path contributions to the harmonic yield in helium are changed in the vicinity of a bound-state resonance. They presented a study of the interplay between resonant enhancement and quantum path dynamics in near-threshold harmonic generation in helium and analyzed the driven harmonic generation response by time-filtering the harmonic signal so as to suppress the long-lasting radiation that would result from population left in excited states at the end of the pulse. By varying the wavelength and intensity of the near-visible driving laser field, they identified a number of direct and indirect enhancements of H7, H9, and H11 via the Stark-shifted 2p–5p states. For H9, the Autler–Townes-like splitting of the enhancement was observed due to the 3p state, when the wavelength and intensity are such that the driving field strongly couples the 3p state to the nearby dark 2s state.

In was found, in terms of the quantum path dynamics, that both the short- and the long-trajectory contributions to the harmonic emission can be easily identified for harmonics that are resonantly enhanced via the Stark-shifted np states. The authors of Ref. [17] found that both contributions are enhanced on resonance and that the maximum of the envelope of the resonant harmonic is delayed by approximately 0.5 optical cycles. It was interpreted to mean that the enhancement happens via a multiphoton resonance between the ground state and the Stark-shifted excited state and that the electron is then trapped for a while in the excited state before entering the continuum. Furthermore, they found that only the long-trajectory contribution acquires a phase shift, which leads to a delay in emission time of approximately 0.125 optical cycles, suggesting that the phase shift is acquired in the interaction between the returning electron wave packet and the ion core, for which there is a large difference in the short- and long-trajectory dynamics. Finally, they have shown that both the enhancement and the phase shift are still visible in the macroscopic response. This means that these effects could potentially be explored experimentally, especially considering that the calculations predict that the macroscopic response is dominated by the long-trajectory contribution, which exhibits the on-resonance phase shift.

Some experiments have considered an interesting alternative for lower intensities below 10^{14} W cm^{-2}. Authors of Ref. [3] note atomic resonance effects on the 13th harmonic for argon in a Ti:sapphire field. Theoretically, there have been investigations of model systems in which resonant enhancement of harmonic generation has been noted. Plaja and Roso [18] have demonstrated that HHG resonant structures occur in a two-level atom when AC-Stark-shifted level energies (the real parts of the complex Floquet quasi-energies discussed in Refs. [19, 20]) undergo avoided crossings. Gaarde and Schafer [7] have performed one-electron pseudo-potential calculations modeling potassium in a Gaussian laser pulse, which shows resonant enhancement of harmonics. In Ref. [20], theoretical predictions of resonance enhanced harmonic generation for argon in a KrF laser field, which take full account of atomic structure effects, are presented. These are ab initio calculations with no empirical adjustable parameters. They have described how the resonant behavior of the harmonics arises from the interaction between quasi-energies. The mechanisms and general features of their results could be applied to general atomic resonances in laser fields.

Another approach is related with the influence of plasmonic resonances on the harmonic efficiency. Particularly, the study [21] was devoted to numerical simulations of the laser–cluster interaction with emphasis on the nonlinear collective electron dynamics and the generation of low-order harmonics in a comparatively small cluster consisting of $\sim 10^3$ atoms. The classical molecular-dynamics simulation model taken into account almost all of the ingredients of the laser–cluster interaction. In full qualitative agreement with the phenomenological picture of nonlinear oscillations of a cloud of electrons trapped inside a charged cluster in a strong laser field [22–24], they have observed a strong 3rd harmonic excitation when the tripled laser frequency became close to the Mie frequency that is, near the third-order resonance with the Mie frequency. This corresponds to nonlinear excitation of the dipole surface plasmon. The resonant behavior can be seen both in the total electron acceleration (which is responsible for 3rd harmonic generation by the cluster) and in the internal electric field of the cluster (which is responsible for ionization of the cluster constituents). Varying the laser wavelength produces a pronounced resonance curve whose width affords an estimate of the width of the Mie resonance in a cluster irradiated by a strong laser field. The time-dependent envelopes of the total electron acceleration and of the inner electric field at the position of the central cluster ion have been calculated at the fundamental frequency and at the 3rd harmonic. They have confirmed the presence of the 3rd harmonic (and of higher harmonics, especially the 5th) in the total electron acceleration as well as in the inner electric field at the position of the central ion and its resonant behavior when the frequency of the incident laser sweeps through the corresponding nonlinear resonance with the time-dependent Mie frequency. The dependence of the time-dependent yield of the 3rd and 5th harmonics on the various parameters of the laser–cluster interaction was analyzed.

Basically, the described approaches require one-photon resonance for fundamental and generated radiations. The claimed primary implementation of such studies is to extend the generated radiation to the extreme ultraviolet (XUV) range where the efficiency to be achieved exceeds that in crystals. Therefore, the coupling of the generated XUV radiation and fundamental radiation becomes almost inevitable for achieving large enhancement of harmonic yield.

Although HHG via interaction of intense laser pulses with matter provides a unique source of coherent femtosecond and attosecond pulses in the XUV, the low efficiency of the process is a serious limit to its wide application. Using the resonances of the generating medium is a natural way to boost the efficiency, as was already suggested in early HHG experimental [3] and theoretical [18, 25] studies. Generation of high harmonics with frequencies close to that of the transition from the ground state to an autoionizing state (AIS) of the generating particle was experimentally investigated in plasma media [26] and in noble gases [27]. A number of theories describing HHG enhancement based on the specific properties of AIS were developed [10, 28–30]. These theories involve the rescattering model in which the HHG is described as a result of tunneling ionization, classical free electronic motion in the laser field, and recombination accompanied by the XUV emission upon the electron's return to the parent ion. In particular, in Ref. [10] a four-step resonant HHG model was suggested. The first two steps are the same as in the three-step model, but instead of the last step (radiative recombination from the continuum to the ground state) the free electron is trapped by the parent ion, so that the system (parent ion + electron) lands in the AIS, and then it relaxes to the ground state emitting XUV. In addition, there are several theoretical studies in which the HHG efficiency was calculated using the recombination cross-section. It was done heuristically [31] and analytically [32] for the Coulomb interaction and by generalizing the numerical results for the molecules [33].

In Ref. [34], the HHG theory considering an AIS in addition to the ground state and the free continuum treated in the theory for the nonresonant case was suggested. They have shown that such accurate consideration verifies the model [10]. Moreover, they have shown that the intensity of the resonant HHG is described with a Fano-like factor that includes the scattering cross-section. However, in contrast to previously suggested theories, this approach also allows calculating the resonant harmonic's phase. Their theory generalizes the strong-field approximation approach for HHG to the resonant case, considering an AIS in addition to the ground state and the free continuum state; the latter two states are treated in the same way as in the theories developed for the nonresonant case. This theory allows calculating not only the resonant harmonic intensity but also its phase. It was shown that there is a rapid variation of the phase in the vicinity of the resonance. These calculations reasonably agree with the RABBIT harmonic phase measurements. The theory predicts that in the case of a resonance covering a group of harmonics the resonance-induced phase variation can compensate for the attochirp in a certain spectral region. In the following we discuss one of their results.

The calculated spectrum of the resonant 17th harmonic from Sn plasma calculated using the numerical time-dependent Schrödinger equation (TDSE) solution as described in Ref. [35] averaged for laser intensities up to 0.8×10^{14} W cm^{-2} is shown in Figure 2.1. The laser pulse duration is 50 fs. One can see that different detuning from the resonance led to different peak harmonic intensities and, even more interesting, to different harmonic line shapes: for the 793, 796, and 808 nm fundamentals the harmonic line consists of two peaks. It is known for the nonresonant HHG that these peaks can be attributed to the contributions of the short and the long electronic trajectories. In Figure 1.2 one can see that the long trajectory's contribution is, in general, weak, but it becomes more pronounced when its frequency is closer to the exact resonance, as is the case for the 793 nm fundamental. These results illustrate the fact that the harmonic line shape can be well understood via the factorization of the harmonic signal. This straightforward factorization is a remarkable fact, considering the complexity of the dynamics of both the free electronic wave packet and the AIS, which determine the harmonic line shape.

Attosecond pulse production using high-order harmonics generated by an intense laser field [36, 37] is essentially based on the phase locking of the

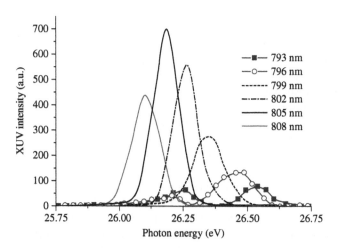

Figure 1.2 The calculated harmonic spectrum in the vicinity of the resonance for different fundamental wavelengths, leading to different detunings from the resonance. The resonant transition is the $4d^{10}5s^25p\ ^2P_{3/2} \leftrightarrow 4d^95s^25p^2\ (^1D)_2\ D_{5/2}$ transition in Sn$^+$; the transition frequency is 26.27 eV, which is close to the 17th harmonic of a Ti:sapphire laser. *Source:* Strelkov et al. 2014 [34]. Reproduced with permission from American Physical Society.

harmonics. This phase locking is well understood [38] for the case when there are no resonances affecting the process. However, recently much attention has been paid to the role of resonances in HHG in gases and plasma plumes. It was shown that when the high-harmonic frequency is close to the transition to an excited quasi-stable state of the generating particle, the harmonic can be much more intense than the off-resonant ones. For the HHG in plasma plumes such an enhancement can be as high as an order of magnitude greater.

An enhancement of the efficiency of XUV generation due to giant resonance in Xe was predicted in Ref. [29] and observed in Refs. [12, 39]. Namely, the XUV in the spectral region of about 20 eV is more intense than the lower frequency XUV, and the enhancement near the center of the resonance is approximately an order of magnitude.

Broadband resonant enhancement potentially allows generating attosecond pulses using resonant harmonics. This approach is interesting not only because of the higher generation efficiency of the resonant HHG, but also because it essentially reduces the requirements for harmonic filtering (the resonant region naturally stands out). However, phase locking of resonant HHG differs from the one of nonresonant HHG, so attosecond pulse production in the former case is not straightforward. In Ref. [40], this aspect of resonant HHG was investigated both numerically and analytically. The effect of resonance on the phase difference between the neighboring harmonics was studied, which allowed calculating the duration of the attosecond pulses produced by resonant harmonics. The conditions were found for which the free-motion-induced attochirp can be compensated by the resonantly induced attochirp, leading to phase synchronization of a group of resonant harmonics. It was shown that attopulses with a duration of 165 as can be obtained using resonantly enhanced harmonics generated in Xe. This duration is smaller than the minimal duration of the attosecond pulse formed by the off-resonant harmonics; it can be further reduced down to almost a hundred attoseconds using the two-color driver. Resonant HHG enhancement leads to an increase of the attopulse intensity by more than an order of magnitude and relaxes the requirements for XUV filtering: only harmonics much lower than the resonance should be suppressed by the filter.

Conversion efficiency is the most important parameter in HHG, and many other schemes have been proposed to improve the conversion efficiency. In theory, using excited atoms with Rydberg states was proved to be an effective way to generate harmonics with both large cutoff energy and high conversion efficiency [41–44]. In experiment, Paul et al. [45] reported on the observation of enhanced HHG from the excited Rb vapor with resonance excitation by using a weak diode laser. However, there were almost no experimental reports on the enhanced HHG from the excited atoms of rare gas with Rydberg states. The experimental demonstration of the enhanced HHG from optically prepared excited atoms with Rydberg states, which is different from resonance excitation and can be created by tunneling ionization [46–48], was recently demonstrated in Ref. [49]. They used an effective pump-probe scheme to experimentally demonstrate the enhanced HHG from a superposition state of ground state and excited states in argon. In their experiment, an obvious enhancement plateau of HHG with a half lifetime of dozens of picoseconds is observed when controlling the time delay between the pump and probe laser pulse, and the harmonic intensity is enhanced by nearly 1 order of magnitude compared to the case without pre-excitation. Then, the gradual enhancement process is demonstrated with the increasing intensity of pump pulse for improving the population of excited states. A theoretical simulation with excited populations by solving TDSE well explained those experimental results.

1.2 Role of Resonances in Plasma Harmonic Experiments: Intensity and Temporal Characterization of Harmonics

A breakthrough in this area of studies was reported in Ref. [50] where strong resonant enhancement in HHG from low-ionized indium plasma was experimentally demonstrated, which was attributed to the superposition of ground state and excited states [28, 51, 52]. The results, which constitute the first temporal characterization of the femtosecond envelope of the resonant high-order-harmonic emission from ablation plasma plumes, were discussed in Ref. [53]. The complex nature of this medium containing different kinds of ions and a rather high free electron density does not allow relying on straightforward analogies with the well-known HHG in neutral gases. The confirmation, found in their results, that the XUV emission from indium plasmas, both resonant and nonresonant, has a femtosecond envelope thus constituting an important advance. While the determined harmonic pulse durations bear significant relative uncertainties, they consistently found XUV pulse durations that are shorter than the driving laser pulse for all plasma targets.

While those results for the resonance-enhanced harmonics are not yet conclusive with regard to Strelkov's four-step model, they give new experimental input to the theoretical effort of modeling. The surprising similarity of the pulse envelope for resonant and nonresonant harmonics is very good news which, in view of the very high conversion efficiencies of 10^{-4} observed in other experiments with indium plasma, opens the perspective of a very high peak flux tabletop XUV source. The central wavelength of these pulses could be selected by choosing a target material with a strong transition at the desired energy. These measurements could be complemented by more advanced temporal characterization. For instance, in so-called XFROG measurements, using a shorter infrared (IR) probe pulse, such that $\tau_{XUV} > \tau_{IRprobe}$, would allow a more precise determination of the XUV pulse duration as well as a measurement of the harmonic chirp rate, as demonstrated in Ref. [54]. Alternatively, a high-order harmonic SPIDER measurement would give direct access to both temporal envelope and phase. Finally, a complete characterization of the harmonic emission including its attosecond (sub)structure with the FROG–CRAB method [55] would of course be the most desirable progress, albeit certainly also the most challenging one. All of these would give information on the intensity dependence of the harmonic phase, which may be different for resonant and nonresonant harmonics, thus shedding more light on the mechanism behind resonant enhancement.

Major efforts up to now have been devoted to increasing the photon flux of resonant harmonics. However, there have only been a few experiments that were aimed for better understanding of the physics involved behind this resonant harmonics generation. These experiments have only shown that the resonant harmonics intensity depends strongly on the ellipticity of the driving laser [56], and follows phase-matching conditions similar to other conventional gas harmonics [57].

Experimentally observed intensity enhancements of resonance harmonics have been compared with theory for several materials [10], and the relative phase between resonant harmonics and nonresonant harmonics have been experimentally measured and compared with theory [51]. However, intensity and phase measurements of the harmonics are still indirect evidence of the four-step model, and thus there currently lacks direct and concrete evidence that the AIS is involved in resonant harmonics generation. In Ref. [58], the electron quantum paths in the vicinity of AIS with mid-IR tunable driving fields were studied. Those results allow to unambiguously clarify the mechanism involved in resonance harmonics generation. The study has shown that resonant harmonics generation process involves the AIS for coherent harmonic emission at resonant energy via the microscopic response. Further, it has been revealed that the resonant harmonics generation process involves the dressed AIS for coherent harmonic emission at frequencies $\pm 2\Omega$ from the resonant harmonic frequency (Ω represents laser frequency). As resonant harmonic is an excellent candidate as a source for intense harmonics, the involvement of dressed states in HHG opens the perspective to expand the bandwidth of those harmonics (for example, one can generate harmonics at $\pm 2\Omega$, $\pm 4\Omega$, and above frequencies), thus opening the possibility to generate intense and ultrashort attosecond pulses, useful for the applications of attosecond science [59, 60]. The direct involvement of AIS in resonant harmonics generation via microscopic response will also provide opportunities to study emission dynamics of AIS with ultrafast lasers. For example, one can extract information about the absolute time that the electron stays in the AIS before emission, study electron–electron interaction, and interference of resonant harmonics and direct harmonic at attosecond timescale by driving harmonics using pulses with certain duration, as the electron stays for a short time in the AIS. Further, harmonic emission from virtual quantum states will provide opportunities to understand the nature of virtual states and their influence on the physical systems both in physics and chemistry, hence providing better control on the systems and roots to find new applications.

Previous studies of resonant harmonics generation were focused at a single state (particularly, the AIS for the case of Sn II), which is in the multiphoton resonance with the driving laser [51, 61]. However, experimental observations show that resonant harmonics generation can involve three different states: the actual AIS and the two dressed states located at $\pm 2\Omega$ around the resonant AIS. At the third step of the four-step model the system finds itself in one of these states and then, at the fourth step, it recombines to the ground state and emits one of the three resonant harmonics. It is obvious that when the driving wavelength corresponds to the exact multiphoton resonance, the harmonics from these dressed states overlap with the neighboring nonresonant harmonics, and thus the two are difficult to differentiate. However, these satellite harmonics can be distinguished from each other when the driving wavelength is detuned. Thus, this observation shows that the driving wavelength detuning from the resonance allows to observe the very pronounced separation of resonance harmonic and driving harmonic, as well as the spectral peaks generated due to the dressed AIS.

Note that very pronounced enhancement of the harmonics neighbor to the resonant one was found theoretically in Ref. [28]. This theory considers a coherent superposition of ground and excited states formed by pumping by the resonance harmonic, and thus assumes quite high intensity of this harmonic. The moderate intensity of resonant harmonic in the discussed experiments [58] (in general, comparable to the one of the dressed harmonics) prevents the direct application of this theory for their interpretation. Those observations provide the accurate information of the quantum path that the electron follows in the vicinity of the AIS for resonant harmonics generation. It was shown that for resonance harmonic generation the electron quantum path is perturbed not only from the AIS but also from the dressed states, resulting in three coherent harmonic generations. In high-order harmonic attosecond science, pump-probe absorption spectroscopy has been recently started to study the light–matter interaction with dressed AIS [62]. However, the propagation and emission dynamics of the electron from dressed AIS have been highly ignored. The observation of satellite harmonics from dressed AIS shows that the AIS dressing by the laser field should be taken into account to describe also the emission processes during the light–matter interaction for elements containing AIS. As AIS exist in different materials, the discussed work will be applicable in general to study the influence of dressed states on electron quantum path in all these materials. As dressed states respond within the driving laser pulse, the nonlinear properties of the dressed AIS will provide the opportunity to study and control electronic motion at fast timescales, which will be an advantage as electrons remain in the AIS for only a few hundred attoseconds to a few femtoseconds.

The study [61] gave an affirmative answer to the practical question of whether resonance-enhanced HHG is indeed a source of intense ultrashort XUV pulses. They reported that enhanced harmonic order has the same femtosecond duration as the nonresonant ones. On the attosecond timescale however, they found significant distortions of the phase of the near-resonant harmonic. Those results suggest the detuning from the resonance as an effective handle controlling the resonant harmonic phase. From a more fundamental viewpoint, previous studies of the HHG phase properties focused mainly on the phase accumulated by the quasi-free electron in the continuum, or on the recombination step as a probe of molecular structure and dynamics [63, 64]. The discussed results present experimental evidence of the dramatic influence of the recombination step on the phase of resonant harmonics from the plasma. It was shown that the recombination dipole matrix element alone can describe the origin of the phase distortions observed in the experiments. This is the basis for the so-called "self-probing" schemes to extract structural and dynamic information about the generating system from intensity, phase, and polarization measurements of high-order harmonics. Those results thus suggest the possibility of devising "self-probing" schemes for atomic resonances based on the advanced characterization of resonance-enhanced high harmonics. In particular, the rapid phase variation responsible for the intriguing suppression of the RABBIT oscillations may encode characteristic features of the involved AIS.

Recently, alongside the plasma medium, resonance HHG has been discovered in other systems, for example, in atomic Fano resonances [9] and in SF_6 molecule [65]. Resonance-enhanced HHG introduces the possibility of increasing the conversion efficiency of a specific harmonic by more than an order of magnitude compared to nonresonant HHG in noble gases. If this effect could be combined with phase-matching effects and/or coherent control of HHG, then an intense XUV source could be generated with only a single harmonic in the spectrum. Such a unique radiation source will truly be ideal for accelerating its various applications in physics, chemistry, biology, and for exploring new fields such as nonlinear X-ray optics and in attosecond physics. It would have excellent spatial coherence [66], the possibility of high repetition rate (of the order of kilohertz to hundreds of kilohertz), and improved conversion efficiency.

There are two possible explanations for the mechanism of this resonant harmonic. The most common is based on modified three-step model. However, a second explanation relies on the collective response of the medium instead of the single-atom response, i.e. on the phase-matching conditions [67]. The results reported in Ref. [57] clearly indicate the existence of a

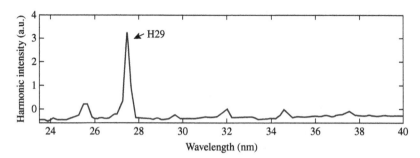

Figure 1.3 Harmonic spectra from chromium plasma including the resonance harmonic (H29). *Source:* Rosenthal and Marcus [57]. Reproduced with permission from American Physical Society.

resonance enhanced harmonic in harmonic spectrum of chromium plasma (Figure 1.3). Nevertheless, analysis of coherence length does not show any dramatic change in the coherence length of that harmonic with respect to the neighboring harmonics. Those results do not support the phase-matching enhancement hypothesis. Rather, they favor the viewpoint that the single atom is responsible. However, the single-atom response is not fully understood, and further investigation must be carried on.

References

1 Ackermann, P., Münch, H., and Halfmann, T. (2012). *Opt. Express* 20: 13824.

2 L'Huillier, A., Balcou, P., and Lompré, L.A. (1992). *Phys. Rev. Lett.* 68: 166.

3 Toma, E.S., Antoine, P., Bohan, A., and Muller, H.G. (1999). *J. Phys. B* 32: 5843.

4 Barkauskas, M., Brandi, F., Giammanco, F. et al. (2005). *J. Electron. Spectrosc. Relat. Phenom.* 144–147: 1151.

5 Ganeev, R.A. (2009). *Phys. Usp.* 52: 55.

6 Taïeb, R., Véniard, V., Wassaf, J., and Maquet, A. (2003). *Phys. Rev. A* 68: 033403.

7 Gaarde, M. and Schafer, K. (2001). *Phys. Rev. A* 64: 013820.

8 de Morisson Faria, C., Kopold, R., Becker, W., and Rost, J. (2002). *Phys. Rev. A* 65: 023404.

9 Rothhardt, J., Hädrich, S., Demmler, S. et al. (2014). *Phys. Rev. Lett.* 112: 233002.

10 Strelkov, V. (2010). *Phys. Rev. Lett.* 104: 123901.

11 Redkin, P.V. and Ganeev, R.A. (2017). *J. Phys. B* 50: 185602.

12 Shiner, A., Schmidt, B., Trallero-Herrero, C. et al. (2011). *Nat. Phys.* 7: 464.

13 Tudorovskaya, M. and Lein, M. (2011). *Phys. Rev. A* 84: 013430.

14 Jin, C., Bertrand, J.B., Lucchese, R.R. et al. (2012). *Phys. Rev. A* 85: 013405.

15 Chu, X. and Groenenboom, G.C. (2013). *Phys. Rev. A* 87: 013434.

16 Xiong, W.-H., Geng, J.-W., Tang, J.-Y. et al. (2014). *Phys. Rev. Lett.* 112: 233001.

17 Camp, S., Schafer, K.J., and Gaarde, M.B. (2015). *Phys. Rev. A* 92: 013404.

18 Plaja, L. and Roso, L. (1993). *J. Mod. Opt.* 40: 793.

19 Plummer, M. and Noble, C.J. (2000). *J. Phys. B* 33: L807.

20 Plummer, M. and Noble, C.J. (2002). *J. Phys. B* 35: L51.

21 Fomichev, S.V., Zaretsky, D.F., Bauer, D., and Becker, W. (2005). *Phys. Rev. A* 71: 013201.

22 Fomichev, S.V., Popruzhenko, S.V., Zaretsky, D.F., and Becker, W. (2003). *J. Phys. B* 36: 3817.

23 Fomichev, S.V., Popruzhenko, S.V., Zaretsky, D.F., and Becker, W. (2003). *Opt. Express* 11: 2433.

24 Fomichev, S.V., Popruzhenko, S.V., and Zaretsky, D.F. (2003). *Laser Phys.* 13: 1188.

25 Oleinikov, P.A., Platonenko, V.T., and Ferrante, G. (1994). *J. Exp. Theor. Phys. Lett.* 60: 246.

26 Ganeev, R.A. (2012). *J. Mod. Opt.* 59: 409.

27 Gilbertson, S., Mashiko, H., Li, C. et al. (2008). *Appl. Phys. Lett.* 93: 111105.

28 Milošević, D. (2007). *J. Phys. B* 40: 3367.

29 Frolov, M.V., Manakov, N.L., Sarantseva, T.S. et al. (2009). *Phys. Rev. Lett.* 102: 243901.

30 Ivanov, A. and Kheifets, A.S. (2008). *Phys. Rev. A* 78: 053406.

31 Platonenko, V.T. (2001). *Quantum Electron.* 31: 55.

32 Frolov, M.V., Manakov, N.L., Sarantseva, T.S., and Starace, A.F. (2011). *Phys. Rev. A* 83: 043416.

33 Morishita, T., Le, A.-T., Chen, Z., and Lin, C.D. (2008). *Phys. Rev. Lett.* 100: 013903.

34 Strelkov, V.V., Khokhlova, M.A., and Shubin, N.Y. (2014). *Phys. Rev. A* 89: 053833.

35 Ganeev, R.A., Strelkov, V.V., Hutchison, C. et al. (2012). *Phys. Rev. A* 85: 023832.

36 Paul, P.M., Toma, E.S., Breger, P. et al. (2001). *Science* 292: 1689.

37 Tzallas, P., Charalambidis, D., Papadogiannis, N.A. et al. (2003). *Nature* 426: 267.

38 Antoine, P., L'Huillier, A., and Lewenstein, M. (1996). *Phys. Rev. Lett.* 77: 1234.

39 Facciala, D., Pabst, S., Bruner, B.D. et al. (2016). *Phys. Rev. Lett.* 117: 093902.

40 Strelkov, V.V. (2016). *Phys. Rev. A* 94: 063420.

41 Watson, J.B., Sanpera, A., Chen, X., and Burnett, K. (1996). *Phys. Rev. A* 53: R1962.

42 Zhai, Z., Zhu, Q., Chen, J. et al. (2011). *Phys. Rev. A* 83: 043409.

43 Chen, J., Wang, R., Zhai, Z. et al. (2012). *Phys. Rev. A* 86: 033417.

44 Zhao, D., Jiang, C., and Li, F. (2015). *Phys. Rev. A* 91: 033414.

45 Paul, P.M., Clatterbuck, T.O., Lynga, C. et al. (2005). *Phys. Rev. Lett.* 94: 113906.

46 Nubbemeyer, T., Gorling, K., Saenz, A. et al. (2008). *Phys. Rev. Lett.* 101: 233001.

47 Shvetsov-Shilovski, N.I., Goreslavski, S.P., Popruzhenko, S.V., and Becker, W. (2009). *Laser Phys.* 19: 1550.

48 Landsman, A.S., Pfeiffer, A.N., Hofmann, C. et al. (2013). *New J. Phys.* 15: 013001.

49 Yuan, X., Wei, P., Liu, C. et al. (2015). *Appl. Phys. Lett.* 107: 041110.

50 Ganeev, R.A., Suzuki, M., Baba, M. et al. (2006). *Opt. Lett.* 31: 1699.

51 Milošević, D.B. (2006). *J. Opt. Soc. Am. B* 23: 308.

52 Redkin, P.V., Kodirov, M.K., and Ganeev, R.A. (2011). *J. Opt. Soc. Am. B* 28: 165.

53 Haessler, S., Elouga Bom, L.B., Gobert, O. et al. (2012). *J. Phys. B* 45: 074012.

54 Mauritsson, J., Johnsson, P., Martens, R.L. et al. (2004). *Phys. Rev. A* 70: 021801.

55 Mairesse, Y. and Quéré, F. (2005). *Phys. Rev. A* 71: 011401.

56 Suzuki, M., Baba, M., Ganeev, R.A. et al. (2006). *Opt. Lett.* 31: 3306.

57 Rosenthal, N. and Marcus, G. (2015). *Phys. Rev. Lett.* 115: 133901.

58 Fareed, M.A., Strelkov, V.V., Thiré, N. et al. (2017). *Nat. Commun.* 8: 16061.

59 Krausz, F. and Ivanov, M. (2009). *Rev. Mod. Phys.* 81: 163.

60 Corkum, P. and Krausz, F. (2007). *Nat. Phys.* 3: 381.

61 Haessler, S., Strelkov, V., Elouga Bom, L.B. et al. (2013). *New J. Phys.* 15: 013051.

62 Wu, M., Chen, S., Camp, S. et al. (2016). *J. Phys. B* 49: 062003.

63 Haessler, S., Caillat, J., and Salières, P. (2011). *J. Phys. B* 44: 203001.

64 Salières, P., Maquet, A., Haessler, S. et al. (2012). *Rep. Prog. Phys.* 75: 062401.

65 Ferré, A., Boguslavskiy, A.E., Dagan, M. et al. (2015). *Nat. Commun.* 6: 5952.

66 Ganeev, R.A., Witting, T., Hutchison, C. et al. (2013). *Phys. Rev. A* 88: 033838.

67 Kulagin, A., Kim, V.V., and Usmanov, T. (2011). *Quantum Electron.* 41: 801.

2

Different Theoretical Approaches in Plasma HHG Studies at Resonance Conditions

In this chapter, we analyze simulations of resonant high-order harmonic generation (HHG) by means of multiconfigurational time-dependent Hartree–Fock (MCTDHF) approach for three-dimensional fullerene-like systems. The results proved the theory of resonant recombination proposed in the article and showed the ways of resonant HHG optimization. The results of MCTDHF calculation of the HHG for C_{60} were found in good qualitative coincidence with reported experimental data. We also discuss the HHG in ordinary buckyball C_{60}, endohedral $In@C_{60}$ fullerene, and endohedral $Sb@C_{60}$ fullerene by means of time-dependent density functional theory (TDDFT). In all cases, the resonant enhancement of several high-order harmonics in the plateau region up to 2 orders of magnitude was predicted. Theoretical investigation of absorption spectra of C_{60} showed absorption peaks in the regions of the resonant HHG enhancement. In both endohedral fullerenes, peaks of plasmon absorption and resonance enhancement were shifted toward the corresponding atomic resonant transitions, which may be used to control the spectral distribution of groups of resonant harmonics and obtain bright attosecond pulse trains.

We also describe different models describing the resonance enhancement of harmonics in laser-produced plasmas (LPPs). The theoretical calculations show the enhancement of specific harmonics for the Sb, Te, and Cr plasmas in the double-target configurations. Another study was devoted to finding an explanation of the observed phenomena of resonance enhancement of the single high-order harmonic in indium plasma. The computations were performed on the base of TDDFT. The results of TDDFT calculation of the HHG for indium ion were found in good qualitative coincidence with experimental data. This allowed us for the first time to introduce a theory of resonant recombination for the HHG. It can also be used to predict the most promising targets for resonant HHG and to increase its efficiency by control of the pump radiation's parameters.

Resonance Enhancement in Laser-Produced Plasmas: Concepts and Applications,
First Edition. Rashid A. Ganeev.
© 2018 John Wiley & Sons, Inc. Published 2018 by John Wiley & Sons, Inc.

2.1 Comparative Analysis of the High-Order Harmonic Generation in the Laser Ablation Plasmas Prepared on the Surfaces of Complex and Atomic Targets

Here we apply the model of resonant HHG to some of the targets showing resonance enhancement. The analysis of single harmonic enhancement is crucial for understanding the processes influencing the nonlinear optical response of the complex targets. In this section, we analyze the results of application of the model of resonant HHG [1, 2] to the Sb II, Cr II, and Te II plasma ions used in double target configuration [3]. This model assumes that, due to large absorption strength of a particular transition in these ions, the population of a metastable excited state, having the excitation energy $\Delta\omega = E_2 - E_1$, is increased, so that a coherent superposition of the ground energy (E_1) and excited (E_2) states is formed. If the wavelength of the applied linearly polarized laser field is resonant with the mentioned transition, then, beside the standard single-state harmonics having the frequency $(2k + 1)\omega$, the resonant odd harmonics of frequency $\Omega = \Delta\omega \pm 2n\omega = (2n_R + 1 \pm 2n)\omega$, $n = 0$, 1, 2, ..., are emitted.

Numerical results for the Sn II ion presented in Ref. [1] show that the 17th harmonic intensity is strongly enhanced, which is in agreement with the experiment [4]. Other harmonics form a plateau followed by a cutoff, in agreement with the three-step model of the HHG [5]. In discussed model the neighboring harmonics (15th and 19th orders) are also slightly enhanced, while in the experiment only the single (17th) harmonic is enhanced. The possible explanation of this discrepancy is that in this model the strong-field approximation was used, which may fail for low-order processes (near resonance) and which does not take into account the Coulomb effects. However, there is a study of HHG from He⁺ ions by a combined laser and harmonic pulse in which the HHG spectra were obtained by solving the time-dependent Schrödinger equation (TDSE) [6]. In the case when the 27th harmonic photon energy was close to the 1s–2p transition in He II, the strong enhancement of HHG was observed. In particular, the neighboring harmonics to the resonant one were also strongly enhanced. Obtained numerical results indicate that fine-tuning of the fundamental wavelength is not necessary for the enhancement. Since the method used in Ref. [6] is based on ab initio solutions of the TDSE, and since it gives the results qualitatively similar to results [2], the inadequacy of the strong-field approximation is a less probable explanation of the mentioned discrepancy. From the microscopic point of view, more possible explanation is that the Sn II ions are much more complex media than the He II ions.

The macroscopic effects due to the propagation of these harmonics in the plasma medium were not taken into account neither in Ref. [2] nor in Ref. [6]. However, some experimental observations (in particular, the dependences of the harmonic yield on the beam waist position, plasma sizes, and laser radiation intensity) point out the effects related to a collective character of the HHG from the laser plumes. Among the factors enhancing harmonic output are the effects related to the difference in the phase conditions for different harmonics. The phase mismatch ($\Delta k = nk_1 - k_i$, where k_1 and k_i are the wave numbers of the laser radiation and ith harmonic) changes due to the ionization caused by propagation of the driving pulse through the plume. According to calculations [7], the phase mismatch caused by the influence of free electrons is about one to two orders higher than those caused by the influence of the atoms and ions. A relevant example where the phase mismatch induced by the presence of free electrons is properly accounted for in a complete phase-matching analysis was presented in Ref. [8]. At the resonance conditions, when the harmonic frequency is close to the frequency of the atomic or ionic transition, the variation of the wave number of a single harmonic could be considerable [9], and the influence of free-electron-induced mismatch can be compensated for by the atomic dispersion for specific harmonic order. In that case, improvement of the phase conditions for single-harmonic generation can be achieved.

The suppression during propagation can be another reason why the neighboring harmonics are not enhanced in the experiment. There is a recent study of the influence of the propagation effects on the enhancement of HHG in He, Ar, and Xe atoms [10]. It shows that the enhancement of the whole plateau survives the propagation, but no conclusions were given about the relative yield of the resonant harmonic and its neighbors. This still remains to be explored. Taking all this into account, it is worthwhile to explore how the discussed model works in the case of other plasma ions, which possess the large absorption strength [3].

In Figure 2.1, three such examples for the laser intensity 4×10^{14} W cm^{-2} and the wavelengths, such that the resonance with the corresponding transition is achieved, are presented. One can see that for Sb II (lower panel; $E_1 = -16.63$ eV, $\Delta\omega = 32.79$ eV) the 21st harmonic is enhanced, which is in agreement with Ref. [3]. For Cr II (middle panel; $E_1 = -16.48$ eV, $\Delta\omega = 45.23$ eV), the 29th harmonic is also enhanced in accordance with the experimental data. Finally, the 27th harmonic is enhanced in Te II plasma (top panel; $E_1 = -18.6$ eV, $\Delta\omega = 40.18$ eV), which is again in accordance with the experiment. All three presented curves finish with a cutoff. The enhancement of the closest harmonics to the resonantly enhanced one, still seen in Figure 2.1, can be explained by the above-presented reasons.

The aim of those studies was to analyze the HHG from the plasma formed by the complex targets. Since there is no information about the existence of strong transition lines in the plasmas preformed from the molecular targets, one can only speculate that the oscillator strengths of the corresponding resonant transitions in molecular ions are much smaller than that of the molecule constituents. Therefore, one can expect that the resonant enhancement of single harmonics from molecular targets is suppressed in comparison with the

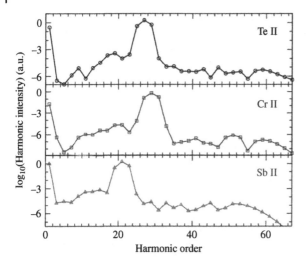

Figure 2.1 Harmonic intensities as functions of the harmonic order for the HHG from the Sb II (bottom panel), Cr II (middle panel), and Te II (top panel). The intensity of applied linearly polarized laser field is 4×10^{14} W cm^{-2}. *Source:* Ganeev and Milošević 2008 [3]. Reproduced with permission from Optical Society of America.

enhancement observed in plasma formed by pure atomic targets. This agrees with the earlier experimental findings.

Let us now analyze the HHG from the double targets. The plasma preformed from such a target will be a mixture of two ions M_1 II and M_2 II. One can suppose that both ions have a strong resonant transition at some frequencies $n_1\omega$ and $n_2\omega$ where ω is the fundamental laser frequency. The field of the nth harmonic emitted at the time t is: $E_{jn}(t) = n^2 D_j(n) \exp(-in\omega t)$, where $D_j(n)$ is the nth harmonic strength, defined in Ref. [2], and the index $j = 1, 2$ stands for the jth component of the target. Let us denote the ratio of the amplitudes of these strengths by $R_n = |D_2(n)| / |D_1(n)|$. Then the nth harmonic intensity is $n^4 |D_1(n)|^2 |1 + R_n \exp(i\varphi_n)|^2$, where φ_n is the relative phase between the nth harmonic strengths $D_1(n)$ and $D_2(n)$.

The phase of the resonantly enhanced harmonic is approximately equal to $\pi/4$, while the nonresonant phases strongly depend on the laser intensity. In general, this phase includes the contribution of the microscopic strengths and the macroscopic contribution, which comes from the fact that the plasma ions M_1 II and M_2 II may be at different places in the laser focus. In principle, the interference caused by this relative phase can be constructive or destructive. The final result can be obtained only after the averaging over the laser focus intensity distribution and after taking into account the propagation effects. Let us consider two examples. In the case when the difference between the wavelengths of the single resonantly enhanced harmonics n_1 and n_2 is large the interference is absent. In the bottom panel of Figure 2.2 we show the harmonic spectra for the Sb and Cr ions and for the double target SbCr. Since the single-resonant harmonics 21 (for Sb) and 29 (for Cr) are well separated and since the strength of nonresonant harmonics is much smaller than that

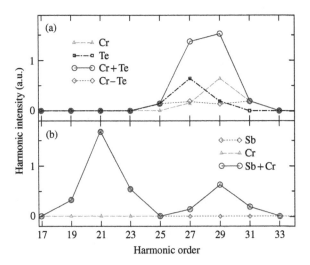

Figure 2.2 Harmonic intensities as functions of the harmonic order for the HHG by a linearly polarized laser field having the intensity 4×10^{14} W cm^{-2}. (a) Cr, Te, and double target Cr $\pm$ Te with constructive (+) and destructive (−) interferences of the resonant harmonic contributions. (b) Sb, Cr, and double target SbCr. *Source*: Ganeev and Milošević 2008 [3]. Reproduced with permission from Optical Society of America.

of resonant harmonics, we have $R_n \ll 1$ and the interference is absent. In this case, the harmonic intensity from the double target is a simple sum of the intensities of harmonics generated from Sb and Cr. A different situation appears for the case presented in the top panel of Figure 2.2: the resonantly enhanced harmonics of Cr (H29) and of Te (H27) are the neighbor harmonics and the interference is possible. Furthermore, harmonics 27, 29, and 31 for Cr and harmonics 25, 27, and 29 for Te are in phase, so that the interference for the harmonics 27 and 29 generated from the double target is constructive (see the solid line denoted by Cr + Te and circles). Since the relative harmonic phase depends also on the position of the Cr II and Te II ions in laser focus, in Figure 2.2 we have also presented the case of destructive interference (dotted curve Cr−Te with diamonds). In order to obtain the destructive interference the relative harmonic phase was changed by π. In the real experiment, the cases between these two extreme constructive and destructive interferences can be presented. However, due to the separation between the species in double-target scheme, the appearance of destructive interference is unlikely.

We have to underline that in most cases the comparison of experimental and theoretical results shows the qualitative character. The discussed results clearly point out the obstacles of the used approach. The theoretical model predicts an enhancement, which regards a band of harmonic peaks (three consecutive orders, with the stronger central peak), whereas, often, the experimental counterpart evidences a single harmonic peak enhancement. The theoretically predicted enhancement is always higher than actually observed in the experiments.

2.2 Nonperturbative HHG in Indium Plasma: Theory of Resonant Recombination

As we already mentioned, first observation of resonant HHG in LPP was reported in Ref. [11] for low-charged indium plasma prepared by laser abla- tion. In those and other experiments with indium plasma [12–14], the 13th harmonic of 800 nm Ti:sapphire laser radiation was almost 10^2 times more intense than neighboring harmonics. This was attributed to a strong multipho- ton resonance with exceptionally strong transition of single-charged indium ion, which can easily be Stark-shifted toward the 13th harmonic of 800 nm pump radiation. However, it could not be explained by a direct lasing effect because conversion efficiency dropped rapidly with the interaction length. The study reported in Ref. [15] is devoted to finding such an explanation of the observed phenomena. It can also be used to predict the most promising targets for resonant HHG and to increase its efficiency by control of the pump radiation's parameters. In the following, we discuss the results of a theoretical study of resonant HHG from low-charged indium vapors. As the driving electric field amplitudes needed for this process are close to intra-atomic ones, resonant HHG becomes nonperturbative and can no more be treated with time-independent methods. This means, the pump laser field can no longer be treated as a small perturbation of the system. A promising way to investigate this process is to solve numerically a TDSE in some approximation for a given system.

2.2.1 Principles of Theory

The computations were performed on the base of TDDFT [16] with the aid of real-space real-time code [17]. This means that the wavefunctions of the sys- tem and their evolution are computed only in certain points in the real space, called the grid points, and they are extrapolated to the other regions of space. In the TDDFT approach itself the many-body time-dependent wave-function is replaced by the time-dependent density $n(r,t)$, which is a simple function of the three-dimensional vector r. $n(r,t)$ is obtained with the help of a fictitious system of noninteracting electrons by solving the time-dependent Kohn–Sham equations [18]:

$$n(r,t) = \sum_i^{occ} |\varphi_i(r,t)|^2 \qquad (2.1)$$

These are one-particle equations, so most many-electron problems can be solved with only a linear increase of computational effort with the number of electrons.

The density of the interacting system is obtained from the time-dependent Kohn–Sham orbitals:

$$i\frac{\partial}{\partial t}\varphi_i(r, t) = \left[-\frac{\nabla^2}{2} + v_{KS}(r, t)\right]\varphi_i(r, t) \tag{2.2}$$

$$v_{KS}(r, t) = v_{ext}(r, t) + v_{Hartree}(r, t) + v_{xc}(r, t) \tag{2.3}$$

where $v_{ext}(r, t)$ is the external potential (laser field), and $v_{Hartree}(r, t)$ accounts for the classical electrostatic interaction between the electrons.

$$v_{Hartree}(r, t) = \int d^3r' \frac{n(r, t)}{|r - r'|} \tag{2.4}$$

The interaction of a strong laser field with a model indium ion in a single-atom response and within the so-called adiabatic local density approximation was studied, which assumes that the potential is the time-independent exchange-correlation potential evaluated at the time-dependent density.

$$v_{xc}^{adiabatic}(r, t) = \tilde{v}_{xc}[n](r)|_{n=n(t)} \tag{2.5}$$

The indium ion was represented by Hartwigsen–Goedecker–Hutter effective core pseudopotentials [19] with addition of Slater exchange and Perdew and Zunger correlation functionals [20]. The need for additional functionals arises from the fact that, by definition, TDDFT is a single-electron method. Implementation of these functionals reproduces the essential features of the exchange and correlation processes correctly at least in the vicinity of the nucleus. As a result, one can go well beyond the single active electron approximation.

For taking into account the chirp-induced spectral broadening, the laser pulse was chosen as linear polarized continuous wave of frequency 0.057 atomic units (au), which corresponds to $\lambda = 800$ nm, multiplied by a Gaussian temporal envelope with the maximum at 827 au (20 fs). Peak electric field amplitude of the pump laser was set to 0.1166 au (4.8×10^{14} W cm^{-2}). The time-propagation was performed for the first 48 fs. To obtain the photoabsorption spectrum of the studied systems the delta-kick method [21] was used with the same simulation region. This means, a strong short delta-shaped pulse was applied in the beginning of the simulation, after which no external field was applied to the system so that wavefunction density relaxed freely to the ground state.

The harmonic spectra presented in Figure 2.3 were obtained by least-squares approximation of the time-dependent dipole expectation value to a set of harmonics of the fundamental radiation frequency. To take into account slight deviations from monochromaticity of harmonics induced by the pump

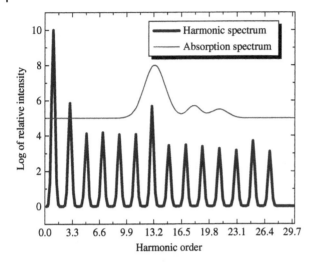

Figure 2.3 Intensity distribution of harmonics for singly charged indium ion and absorption spectrum of the singly charged indium ion. *Source*: Redkin et al. 2011 [15]. Reproduced with permission from Optical Society of America.

intensity modulation and by possible computational inaccuracies these harmonics were considered Gaussian-broadened by frequency. Certainly, one cannot approximate a complicated function by a single set of quasi-sinusoidal functions. That's why this approximation was made piecewise and simply summed the indices of corresponding harmonics for all approximation intervals. One should note that least-square approximation can be used only for sampled (discrete and equidistantly spaced) data. As this method is free from noise induced by signal length finiteness, it is thus superior to widely used for these purposes discrete Fourier transforms. It is well known that for analytically given expressions (such as in semiclassical approaches) the Fourier transform gives exact results, but it turned out not to hold true for discrete Fourier transforms. That's why the piecewise least-squares approximation can be considered a novel and efficient method for analysis of HHG on the base of sampled time-dependent quantities.

These results describe the enhancement of a single 13th harmonic in indium plasma. Analysis of absorption spectrum by means of delta-kick method also revealed a broad region of absorption, which is shown in Figure 2.3 for convenience.

2.2.2 Discussion

The resonance-induced enhancement can be achieved by changing the harmonic wavelength, which enabled the overlap of the 13th harmonics with the ionic transition with strong oscillator strength. Another approach for the tuning of harmonic wavelength based on the chirp variations of driving radiation was demonstrated during gas HHG studies [22–24]. Moreover, the tuning of harmonics and their "sharpness" were demonstrated using a combination of the external control of laser chirp and the intensity-induced variation of laser chirp inside a nonlinear medium [23, 24].

The very few demonstrations of resonance enhancement of the high-order harmonics in previous laser-gas HHG experiments can be understood by considering the difference between the excited spectra of atoms and ions of available solid targets and few rare gases. In the case of plasma from various targets, there is a higher probability to find a proper target for which the fulfillment of multiresonance conditions in extreme ultraviolet (XUV) range can lead to the enhancement of the single harmonic yield, while neighboring harmonics do not enhance. The suppression during propagation can be another reason why the neighboring harmonics are not enhanced in the experiment. A study of the influence of the propagation effects on the enhancement of HHG in He, Ar, and Xe atoms was reported in Ref. [10]. It shows that the enhancement of the whole plateau survives the propagation, but no conclusions were given about the relative yield of the resonant harmonic and its neighbors. This still remains to be explored.

The discussed computations were found in good qualitative agreement with the experimental results [11]. This allows to apply the theory of resonant recombination, where the probability of electron's recombination into the ground state is greatly multiplied when the sum of its kinetic energy acquired from the laser field and its ionization energy equals the energy difference between the resonant and ground-state levels.

Let us compare this theory of resonant recombination with the theory of harmonic generation from coherent superposition of states having different parity [2]. They are presented in Figure 2.4, where the nonresonant case (Figure 2.4a) is given for comparison. As well as in the nonresonant case (Figure 2.4a), in discussed theory of resonant recombination (Figure 2.4c), the electron is first ejected from the ground state by tunnel ionization process (dashed arrow), whereas in theory [2] (Figure 2.4b) the electron leaves the ground state due to resonant multiphoton excitation (solid arrow) to the excited state. In all three cases, the electron accelerates in the field of laser radiation (solid curved arrow) and acquires the kinetic energy E_k. Then in all three cases the recombination of electron into the ground state with a HHG photon emission occurs. Only in discussed theory this resonant HHG process (Figure 2.4c) takes place when the energy of HHG photon equals (within the spectral width of the pump radiation) kinetic energy of an electron plus ionization energy, otherwise there is an ordinary nonresonant HHG (see Figure 2.4a).

In theory [2] (Figure 2.4b), the resonant HHG originates from electrons starting from the excited state, which was populated by a multiphoton resonance, and recombining into the ground state. One can easily see that this process

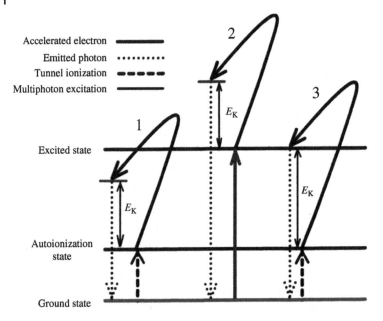

Figure 2.4 Comparative plots of three possible harmonic generation models.
1 – Nonresonant HHG case (ordinary three-step process); 2 – HHG from coherent
superposition of states having different parity; 3 – HHG from resonant recombination.
Source: Redkin et al. 2011 [15]. Reproduced with permission from Optical Society of America.

can lead to simultaneous resonant enhancement of several neighboring harmonics. On the contrary, in experiments [11], the harmonics closest to a resonant one are even suppressed compared to nonresonant case. It can easily
be explained by means of resonant recombination theory that the neighboring
harmonics should be decreased. Indeed when the electron starts from ground
state and recombines into it again, the increase in probability of recombination
with emission of a certain harmonic reduces the probability of other recombination. So, the main advantage of discussed theory of resonant recombination
lies in the fact that it allows explanation of resonant enhancement of a single
harmonic.

Another fascinating peculiarity of this theory of resonant recombination is
the fact that all equations given in Ref. [2] can describe discussed theory as well
without any changes. The only difference lies in the fact that additional resonant
term

$$\sum_{j,j'} a_j^* a_{j'} \langle j|r|j' \rangle \exp[-i(E_{j'} - E_j)t] \tag{2.6}$$

in discussed theory not merely increases the effectiveness of ionization, as considered in Ref. [2], but plays the decisive role in the resonant recombination

process. So the discussed theory can be easily explained semiclassically as well, giving full effectiveness of semiclassical approach.

At the same time, resonant recombination is not a simple lasing process, as nonresonant nonperturbative HHG lies in its origin. In addition, the resonant recombination does not require a not-so-probable exact multiphoton resonance at all. It is proved by the facts that resonant HHG is most efficient in case of tunneling, not multiphoton, ionization regime. As an important consequence of the theory of resonant recombination, the most promising targets for the resonant HHG should have high-lying excited states, which have a compromise between spontaneous radiation probability and resonance lifetime. This is fulfilled best of all in the systems having delocalized electrons, such as atoms with weakly screened d-electrons, metal nanoparticles, and molecules of fullerenes.

2.2.3 Important Consequences

The important consequence of this theory is in the necessity to get best resonant conditions in the region of pulse having sufficiently high intensity, that is, in the tunnel ionization regime. At the same time, theory [2] would require best resonant conditions in the low-intensity regions of the pulse, because only at relatively low intensities the multiphoton excitation is most probable. Such difference allows to check the correctness of discussed theory and theory [2], both theoretically and experimentally by investigating the spectral distribution of energy within the pulse.

As it was mentioned in the introduction, the indium plasma emission observed in the vicinity of the 13th harmonic wavelength is possibly due to the $4d^{10}5s^2 \, ^1S_0 \rightarrow 4d^95s^25p \, (^2D) \, ^1P_1$ transition of In II (19.92 eV, 62.24 nm), which is exceptionally strong. The absorption oscillator strength (gf) of this transition has previously been calculated to be 1.11, which is more than 12 times larger than that of other transitions from the ground state of the In II [25]. This transition can be driven into the resonance with the 13th harmonic (61 nm, 20.33 eV) thereby resonantly enhancing its intensity.

It was shown in Ref. [14] that a similar resonance effect can occur for the 15th harmonic. For a chirp-free laser, the 15th harmonic is weak and compared with the neighboring harmonics in the plateau. However, for the negatively chirped pulses, which tune the harmonics toward blue, the 15th harmonic shows enhancement, with its intensity comparable to that of the lower 13th harmonic. This behavior can be attributed to the tuning of 15th harmonic toward the $4d^{10}5s5p \, ^3P_2 \rightarrow 4d^95s5p^2 \, (^3P) \, ^3F_3$ transition of In II at 23.85 eV (51.99 nm), which has a gf of 0.30, and detuning of the 13th harmonic out of the resonance.

Such intensity enhancement can also be attributed to the existence of oscillating electron trajectories that revisit the ionic core twice per laser cycle [26]. Since such trajectories start from the resonantly populated excited state, with a nonzero initial kinetic energy, they still have nonzero instantaneous kinetic energies when they return to the origin. As usual, recombination results in the emission of harmonics, but due to the relatively low probabilities, the population in the laser-driven wave packets increases continuously and the probability for harmonic emission grows with the number of allowed recollisions. This multiple recollision is predicted to enhance harmonics in the spectral ranges close to the atomic and ionic resonances.

The question arises as to why a multiphoton resonance with some excited state of ions leads to a pronounced resonance in the harmonic spectrum while a multiphoton resonance with other excited states does not? The numerical simulations of this problem were reported in Ref. [27]. One can understand that the competition between the reabsorption, phase matching (mismatching), and growth of harmonic intensity, as well as the population of the excited states and the transition life-time have to be at "optimal" conditions to show the resonance-induced growth of single harmonic in the plateau region. A few types of plumes can satisfy these conditions at a given condition of the pump laser. Further search of plumes demonstrating high enhancement of single harmonic seems to be very important for extending the highest enhanced harmonic to the shorter wavelength range. The estimations show that the presented theory can predict some promising laser plumes, which can demonstrate such enhancement.

The absorption line shown in Figure 2.3 follows naturally from the model for the indium ion and not artificially introduced for purposes of computing. The absorption is computed for the same indium ion model, which is later used to compute the harmonic spectrum. In both cases the TDSE was solved. In fact, a system of Kohn–Sham equations is solved instead, according to the method. To obtain the absorption spectrum a strong and short delta-shaped pulse (0.01 au) at the beginning of computation was used to excite the system, after which the system's evolution was analyzed in time for 48 fs. The time interval was chosen taking into account the pulse duration used in the C_{60} HHG experiments. At the same time, full Gaussian-shaped sine pulse lasting for all 48 fs instead of the short delta-kick was used to obtain the harmonic spectrum by our least-squares-based routine. Nevertheless, the absorption properties were obtained by the built-in analysis tools of the program. To sum it up, the investigated system was the same both for absorption and HHG computations (i.e. the same initial electron density, potentials, and functionals). That's why the absorption of the given system can be related directly to HHG spectrum. The broadening of absorption line is only determined by the delta-kick method simulation time (in femtoseconds).

To sum it up, one needs only well-proven resonant properties of a given system, not the exact indium ion. However, the main absorption properties (triplet near 18 eV) of the real indium system could be reproduced even in these conditions within the used approximation and thus the resonant HHG in artificial indium-like system can be directly related to experimental data.

2.3 Simulation of Resonant High-Order Harmonic Generation in Three-Dimensional Fullerenelike System by Means of Multiconfigurational Time-Dependent Hartree–Fock Approach

The insignificant resonant HHG in gases can be explained by much narrower range of available resonances in UV for gases, which can be used for HHG. Plasma ablation is a way to create gaseous media from most solid targets. This determines a greater possibility of finding a certain transition favorable for resonant HHG. Regarding this, we should say that plasma ablation gives better probability of finding optimal transitions for resonant HHG compared to gases. Meanwhile, no one can say for sure that plasma ablation should give the largest possible harmonic intensity. Plasma ablation of various targets carried out by picosecond pulse, which can be optimized by delay with regard to main femtosecond pulse. In this approach, the application of heating pulse allows the creation of monoparticle- and nanoparticle-containing plasma, resonance-induced enhancement of harmonics, generation of extended plasma, etc., i.e. gives the additional freedom in variation of the nonlinear optical characteristics of medium. One of such interesting examples is the fullerene-containing medium.

2.3.1 Basics of the Nonlinear Optical Studies of Fullerenes

Theoretical studies on HHG from C_{60} involved extending the three-step model [28], analyzing an electron constrained over the surface of a rigid sphere, with geometrical parameters similar to those of the C_{60} fullerene [29], and using the dynamical simulations [30]. In the latter, high-order harmonics were shown to be due to multiple excitations and could be easily generated even with a weak laser field. Both studies reveal how HHG can be used to probe the electronic and molecular structure of C_{60}. At the same time, theoretical investigation of such systems is hampered by the fact that the Hamiltonian of HHG is time-dependent and the systems consist of many electrons. The investigation of influence of the fundamental properties of electrons on resonant HHG can be performed by means of MCTDHF approach, which has the accuracy of direct numerical solution of Schrödinger equation and is almost as simple as ordinary time-dependent Hartree–Fock approach. The discussed computations were based on the Heidelberg MCTDH software packages [31, 32]. It can easily handle MCTDHF problems as well by setting all particles identical and the **A** of the wavefunction of the system fully antisymmetric in the initially unsymmetric MCTDH approximation.

In this subsection, the simulations of resonant HHG by means of MCTDHF approach for three-dimensional fullerene-like systems are discussed. We analyze the influence of the surface plasmon resonance (SPR) of C_{60} on harmonic efficiency in the range of 60 nm ($E = 20$ eV). These results showed the ways of resonant HHG optimization and, most important, attosecond pulse train generation. The MCTDHF calculations of the HHG for C_{60} were found in good qualitative coincidence with experimental data reported in previous studies of the harmonic generation in fullerene-containing laser plumes [33].

The MCTDHF approach treats the wavefunction of multi-electronic system as

$$\Psi(Q_1, \ldots, Q_f, t) = \sum_{j_1=1}^{n_1} \cdots \sum_{j_f=1}^{n_f} A_{j_1 \cdots j_f}(t) \prod_{\kappa=1}^{f} \varphi_{j_\kappa}^{(\kappa)}(Q_\kappa, t) \qquad (2.7)$$

where $Q_1, \ldots, Q_f$ are the coordinates of electrons, $A_{j_1 \cdots j_f}$ are antisymmetrized **A** for all n_κ time-dependent expansion functions $\varphi_{j_\kappa}^{(\kappa)}$ for every degree of freedom κ. Setting $n_\kappa = n_1$ describes the direct solution of TDSE and $n_\kappa = 1$ simplifies the wavefunction to ordinary time-dependent Hartree–Fock approximation.

The equations of motion in the MCTDHF approach are derived from modified variational principle

$$\langle \delta\Psi_{\text{MCHF}}(t) | i\frac{d}{dt} - H(t) | \Psi_{\text{MCHF}}(t) \rangle = 0 \quad \forall t \qquad (2.8)$$

The MCTDHF method was applied to simulate a three-dimensional fullerene-like system represented by the so called jelly-like sphere approximation. A jellium sphere was used as a potential surface for the representation of fullerenes. Then, two electrons were considered moving in this potential, that is, the remaining electrons are considered frozen. The system under investigation was represented by spherically symmetric potential of the form [34] ($R_0 = 8.1$, $R_i = 5.3$, and $v_0 = 0.78$):

$$V(r) = \begin{cases} -3\left(\dfrac{250}{R_0^3 - R_i^3}\right)\left(\dfrac{R_0^2 - R_i^2}{2}\right), & r \le R_i \\[12pt] -\left(\dfrac{250}{R_0^3 - R_i^3}\right)\left(\dfrac{3R_0^2}{2} - \left[\dfrac{r^2}{2} + \dfrac{R_i^3}{r}\right]\right) - v_0, & R_i < r < R_0 \\[12pt] -250/r, & r \ge R_0 \end{cases}$$

$$(2.9)$$

where $r_n = x_n^2 + y_n^2 + z_n^n$ for nth electron ($n = 1, 2$).

The Coulomb repulsion between electrons was:

$$V_{ee} = 1/\sqrt{(x_1 - x_2)^2 + (y_1 - y_2)^2 + (z_1 - z_2)^2 + 2} \tag{2.10}$$

Below we present the results of the study of the interaction of the fullerene-like system with the Gaussian femtosecond electric pulse

$$E(t) = \exp\left(\frac{(t - t_0)^2}{\tau^2}\right) E_0 \sin(\omega t) \tag{2.11}$$

where $\omega = 0.046$ au ($\lambda = 991$ nm) or $\omega = 0.057$ au ($\lambda = 800$ nm). The first frequency was chosen as a source of even (16th) harmonic, which coincides with the central region of the SPR of C_{60} ($\lambda = 60$ nm), and the second frequency coincides with the frequency of most frequently used laser (Ti:sapphire) in such experiments.

The simulation box size was 100 au in each direction for each degree of freedom. The intergrid spacing was 0.1 au, which is quite sufficient (the particles are indistinguishable, so the main contribution comes from the **A**, if more than one configuration is considered). No further investigations of influence of mesh size on the results were performed.

From the solution of MCTDHF equations for $\Psi(Q_1, \ldots, Q_f, t)$, the time-dependent dipole $d(t) = \sum_f \int \Psi(Q_1, \ldots, Q_f, t) Q_f \Psi^*(Q_1, \ldots, Q_f, t)$ was obtained for the estimation of the power spectrum of HHG. It is well known that for analytically given expressions (such as in semiclassical approaches) the Fourier transform gives exact results as it treats all signals as quasi-infinite. But it turned out not to hold true for discrete Fourier transforms when analyzing extremely short signals due to well-known fact of spectral leakage, which is present for any possible window-function, including rectangular one. That's why the piecewise least-squares approximation can be considered a novel and efficient method for analysis of HHG on the base of sampled time-dependent quantities. It consists of least-squares approximation of $d(t)$ to a sum of harmonics on every sufficiently small time interval and summation of the resulting indices to get the resulting HHG spectrum $d(t) = \sum_{i=1}^{N} a_i \sin(b_i \omega t)$, $N \leq T/dt$ (T – pulse length, dt – timestep). Usually N is two or three times smaller than T/dt and b_i is $1 + 2i$. The analytical Fourier transform is simply evaluation of an indefinite integral $S(\Omega) = \int d(t) e^{i\Omega t} dt$. However, discrete Fourier transforms do not directly evaluate the corresponding definite integral $S(\Omega) = \int_0^T d(t) e^{i\Omega t} dt$. Instead, it tries to make a fit $d(t) = \sum_{i=1}^{T/dt} a_i \sin(a_i \omega t)$. Thus, discrete Fourier transforms, being another way of fitting the sampled data to sinusoids, can also be done piecewise, but the resulting set of fixed sampling frequencies will be determined only by the time-step (which may be far from harmonics) and, most important, the fit is not guaranteed to be optimal and not unique (although it' is relatively exact then). At the same time, the fit obtained by means of least-squares approximation is always optimal by construction and unique. Its only disadvantage is that frequencies of harmonics should be guessed, although it is not too difficult if one knows that the resulting spectrum should consist of harmonics only. Another advantage of this method is the fact that functions for approximation should not necessarily be pure sinusoids. Actually, the fitting functions were equally Gaussian-broadened sinusoids to take into account slight deviations from monochromaticity of harmonics induced by the pump intensity modulation and by possible computational inaccuracies. The advantage of this method is not only its full mathematic correctness and accuracy, but also the fact that the harmonic spectrum becomes easily viewable by definition.

2.3.2 Simulations and Discussion

The absorption spectrum of this system was obtained by a procedure similar to delta-kick method [21]. It consists of applying to the whole system a strong rectangular pulse ($E = 0.01$ au) in the beginning of propagation and then evaluating the free propagation of the system, after which the spectrum is obtained via Fourier transform of the $x_1(t)$. This approach is not implemented into Heidelberg MCTDH package directly, so a user-defined field was used. The spectrum had absorption maxima near 0.741 au, which is the 13th harmonic of $\omega = 0.057$ au and 16th harmonic of $\omega = 0.046$ au radiations, although absorption band is rather wide (see inset in Figure 2.5). One should mention that a simple jelly-like system was considered, not the C_{60} molecule itself, so the spectrum may deviate from the experimental one.

Figure 2.5 presents the results of HHG simulation within exact MCTDHF approximation (six expansion functions) for carrier wave frequencies of 0.046 and 0.057 au. The 13th harmonic was approximately 10 times enhanced relative to the plateau harmonics. Note that experimentally observed enhancement of this harmonic was approximately same and depended on the excitation of fullerene-containing targets [35]. One can see that harmonics neighboring to 13th are not so enhanced, although they are still close to the broad absorption band of C_{60} (50–70 nm). At the same time, they are not suppressed, so it is highly possible that a competition between enhancement and absorption takes place. The pulse in general is not monochromatic, so its spectral properties can also have an influence on the simulations.

The even harmonics, being 2 orders of magnitude smaller than the neighboring odd harmonics, were also observed. This artifact can be attributed to symmetry breaking introduced by numerical grid, which is perhaps still too sparse and introduces some kind of rectangular integration box as well. Further reducing of the grid spacing may remove such unphysical result. Note, that in contrast to TDDFT method, no spherically symmetric integration box can be introduced in the MCTDHF (at least, by means of Heidelberg MCTDH package). However, these harmonics are almost two orders smaller than neighboring ones and can be thus disregarded as some kind of numerical inaccuracies.

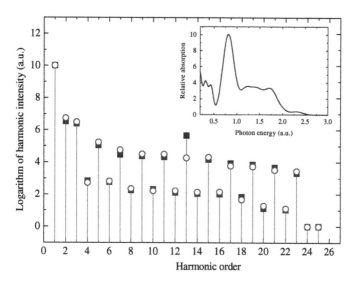

Figure 2.5 The influence of resonance on the HHG spectrum generating in fullerene-like medium in the cases of radiation of carrier wave frequencies 0.046 (open circles) and 0.057 au (filled squares). Inset: Absorption spectrum of the fullerene-like system obtained by delta-kick method. *Source*: Redkin and Ganeev 2010 [33]. Reproduced with permission from American Physical Society.

The resonant 13th harmonic of the radiation with carrier frequency $\omega = 0.057$ au was enhanced. On the contrary, the 16th harmonic of the radiation with carrier frequency $\omega = 0.046$ au was suppressed due to symmetry effects, which are still strong in this system. In both cases the maximum observed harmonic order was 23, which is close to reported experimental results at moderate excitation of fullerene-containing targets. All these results point out that MCTDHF approximation indeed allows to describe both resonant HHG and harmonic cutoff in the fullerene-like medium.

Then more simplified systems were studied at the same conditions for the Gaussian field with carrier wave frequency 0.057 au The results of resonant HHG simulations in these approximations compared to results of MCTDHF calculations with six expansion functions are shown in Figure 2.6. It is seen that reducing the number of configurations up to time-dependent Hartree–Fock (TDHF) approximation (one expansion function) did not lead to complete vanishing of resonant HHG, although the conversion efficiency of resonant harmonic was reduced. In the one-dimensional case, when only a single coordinate of each particle was taken into account, conversion efficiency of resonant harmonic decreases as well. To sum it up, only a large number of expansion functions is enough to observe resonant HHG in the one-dimensional case. However, the resonant nature of HHG for 13th harmonic is unchanged by the number of configurations.

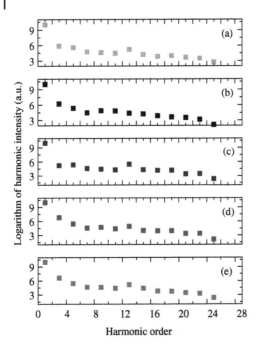

Figure 2.6 The influence of approximations on the observation of resonant HHG. (a) Without the influence of exchange, (b) without the influence of interaction, (c) six expansion functions, (d) three expansion functions, and (e) one expansion function. *Source:* Redkin and Ganeev 2010 [33]. Reproduced with permission from American Physical Society.

Two-electron interaction is a Coulomb repulsion (Eq. (2.10)) between two electrons. Neglecting the two-electron interaction resulted in nonresonant HHG without any significant suppression of other harmonics. At the same time, the representation of quasi-electrons as distinguishable particles had almost no influence on the observation of resonant HHG. The most important application of this phenomenon is the necessity of exact description of two-electron interaction, while exchange processes have almost no effect on resonant HHG simulations. Probably some enhancement of other harmonics observed in these calculations in case of six exchange functions is determined by other resonant transitions in the system regarding dynamical modifications of plasmon absorption spectrum.

The main difference between this and previous theories is the following. In theory [1, 2] resonances originate from the recombination into low-order even harmonics (0 or 2) with respect to the excited state, which stands in place of ground state. Taking into account extremely high conversion efficiency for lower orders, such resonant enhancement is mainly determined by the efficiency of multiphoton transition. At the same time, direct resonant recombination gives the emission of a single-enhanced harmonic for any existing strong transition resonant with it and has almost no dependence on the excited state's population caused by multiphoton resonance.

In order to check the correctness of these theories by means of numeric simulation, the time-dependent electric field was changed so that in the region of

Figure 2.7 Results of simulation of the HHG for various deviations of carrier wave frequency from the resonant one for regimes favorable for coherent superposition or resonant recombination. *Source*: Redkin and Ganeev 2010 [33]. Reproduced with permission from American Physical Society.

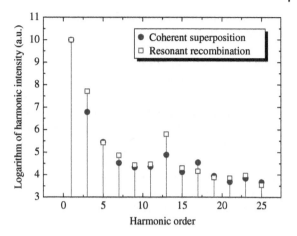

relatively low intensities (the beginning and the end of the pulse) the carrier wave had frequency 0.046 au, while in the region of higher intensities it had frequency 0.057 au (see Figure 2.7, open squares). For comparison the opposite case was considered in Figure 2.7 (closed circles) where carrier wave frequency was 0.057 au for low and 0.046 au for higher intensities, respectively. No harmonic enhancement was observed in the former case. The results of HHG simulation showed that resonant HHG requires exact resonance at higher intensities first of all. The results of numerical and analytical calculations based on this model are in quantitative agreement with the experiments, which will be discussed in the following sections.

2.4 Endohedral Fullerenes: A Way to Control Resonant HHG

Giant-resonant HHG related with plasmonic properties of fullerenes has previously been reported to increase the efficiency of ordinary HHG up to 20 times for the 11th–15th harmonics in the C_{60} plasma ([35], see also the following chapters). Resonant HHG can also lead to a better control of the properties of harmonic radiation by adjusting the laser wavelength and corresponding resonant transitions. For example, in Sb plasma, a 10-fold enhancement of resonant HHG has been reported [36] using the appropriate adjustment of experimental parameters. On the other hand, HHG in fullerenes [35] has been shown to allow the simultaneous enhancement of several neighboring harmonic orders. Thus it seems timely to combine the superior resonant HHG properties observed in atomic plasmas with multiple harmonics enhancement in fullerene plasmas inside a single system. This motivates to investigate the HHG in endohedral C_{60}

fullerenes with In and Sb implanted and compare these studies with those from ordinary C_{60} fullerenes.

Endohedral fullerenes, also called endofullerenes, are fullerenes that have additional atoms, ions, or clusters enclosed within their inner spheres. The first lanthanum C_{60} complex was synthesized in 1985 and called La@C_{60}. The sign "@" in the name reflects the notion of a small molecule trapped inside a shell. Two types of endohedral complexes exist: endohedral metallofullerenes and nonmetal-doped fullerenes. Endohedral fullerenes [37] are characterized by the fact that electrons will transfer from the metal atom to the fullerene cage and that the metal atom takes a position off-center in the cage. On the contrary, the nonmetals, for example nitrogen [38], inserted in fullerenes have almost no charge transfer in the center and represent an atomic trap that is stable at room temperature and for an arbitrarily long time.

Atomic or ion traps are of great interest since particles are present free from significant interaction with their environment, allowing unique quantum mechanical phenomena to be explored, especially those that arise from quantum confinement. It is thus interesting to investigate the behavior of fullerenes doped with semiconductors, because properties of semiconductors change greatly with confinement. The endohedral C_{60} fullerene with implanted indium has not yet been synthesized; however, various stable endohedral fullerenes have been already experimentally obtained, not only La@C_{60}. In Ref. [39], the formation of heavier atoms (Sb, Te)-incorporated fullerenes has been investigated by using radionuclides produced by nuclear reactions. From the trace of radioactivities of ^{120}Sb (^{122}Sb) or ^{121}Te after high performance liquid chromatography, it was found that the formation of endohedral fullerenes or heterofullerenes in atoms of Sb or Te is possible by a recoil process following the nuclear reactions. Results of molecular dynamics simulation in Ref. [39] can also easily describe ion implantation process, which can substitute nuclear recoil previously used to implant neutral atoms. Similar experimental and theoretical results were presented in Ref. [40] for a broader range of targets.

Endohedral fullerenes can also be produced by arc-discharge vaporization of composite rods made of graphite and the metal compounds [40, 41], although the output is relatively low. There are also techniques for insertion of foreign atoms into the C_{60} shell [42–44]. But taking into account conditions of typical HHG experiments, optimal and inexpensive technique to produce them could be ion implantation during laser ablation of C_{60}-containing superglue, because expensive purification as well as long-term stability of endohedral fullerenes would not be needed at all. Although the optimal conditions of the experiments should be investigated, production of In@C_{60} and Sb@C_{60} in quantities sufficient for HHG experiments is, in principle, possible. The macroscopic stability of C_{60} with some of implanted ions is unlikely. However, this is actually not required, because time of any HHG experiment is shorter than decay time of endohedral fullerenes obtained by means of ion implantation into vaporized fullerene molecules, so there is a high possibility of experimental verification of this effect by adding an ion beam into the standard HHG setup.

2.4.1 Theoretical Approach and Details of Computation

Spectroscopic data for endohedral fullerenes in XUV range are also not yet available. The investigation analyzed below is devoted to spectroscopic properties of C_{60} doped with indium and antimony and their effect on HHG in the case of 800 nm fundamental laser radiation. We discussed the results of theoretical studies of absorption and HHG spectra of C_{60}, In@C_{60}, and Sb@C_{60} and analyzed their possible applications [45].

Since HHG requires field strength of pump laser radiation close to intra-atomic one, the pump laser field can no longer be treated as a small perturbation, and the time-independent methods, such as perturbation theory, are not applicable. To model the HHG, TDSE was solved in approximation of TDDFT [16] with the aid of real-space real-time code [46–49]. Detailed description of TDDFT formalism can be found in Ref. [18]. Here we give only few basic equations.

In the TDDFT approach, the many-body time-dependent wavefunction is replaced by the time-dependent density $n(r, t)$, which is a simple function of the three-dimensional vector r. $n(r, t)$ is obtained with the help of a fictitious system of noninteracting electrons as

$$\rho(r, t) = \sum_{i}^{occ} |\varphi_i(r, t)|^2 \tag{2.12}$$

$\rho(r, t)$ can be found from $\varphi_i(r, t)$ by solving the time-dependent Kohn–Sham equations:

$$i\frac{\partial}{\partial t}\varphi_i(r, t) = \left[-\frac{\nabla^2}{2} + v_{KS}(r, t)\right]\varphi_i(r, t) \tag{2.13}$$

These are the one-particle equations, so most many-electron problems can be solved with only a linear increase of computational effort with the number of electrons. According to Runge–Gross theorem [42], for a given initial state, there is a one-to-one correspondence between time-dependent one-body densities $n(r, t)$ and time-dependent one-body potentials $v_{KS}(r, t)$:

$$v_{KS}(r, t) = v_{ext}(r, t) + v_{Hartree}(r, t) + v_{xc}(r, t) \tag{2.14}$$

where $v_{ext}(r, t)$ is the external potential (laser field), and $v_{Hartree}(r, t)$ accounts for the classical electrostatic interaction between the electrons and the nucleus.

$$v_{Hartree}(r, t) = \int d^3r' \frac{n(r, t)}{|r - r'|} \tag{2.15}$$

The exchange-correlation potential $v_{xc}(r, t)$, which describes the interaction between electrons, is the only approximation in TDDFT. In general, it depends on the whole history of density and orbitals and its exact form would give us an exact solution of Eq. (2.13). The need for additional functionals arises from the fact that, by definition, TDDFT is a single-electron method.

The most widely utilized in literature adiabatic local density approximation is not very useful when describing the HHG process, which is strongly nonadiabatic. For all the calculations, the Krieger–Li–Iafrate method [50] applicable for time-dependent problems was used. Here the exchange-correlation potential V_{xc}^{KLI} is expressed as a set of explicit functionals of the orbitals $\varphi_{j\sigma}$, where σ stands for the spin of the electrons. In the exchange-only case V_x^{KLI}, if correlation effects are neglected, the potential reads

$$V_x^{KLI}(r,t) = w_{x\sigma}(r,t) + \frac{1}{\rho_\sigma(r,t)} \sum_j^{n_\sigma} \rho_{j\sigma}(r,t) \times \int d^3r' \rho(r',t) V_x^{KLI}(r',t)$$

(2.16)

where $\rho_{j\sigma}(r,t) = |\varphi_{j\sigma}(r,t)|^2$ and

$$
\begin{aligned}
w_{x\sigma}(r,t) = & -\frac{1}{\rho_\sigma(r,t)} \sum_{j,k}^{n_\sigma} \varphi_{j\sigma}(r,t)\varphi_{k\sigma}^*(r,t) \int d^3r' \\
& \times \frac{\varphi_{k\sigma}(r,t)\varphi_{j\sigma}^*(r,t)}{|r-r'|} - \rho_{j\sigma}(r,t) \int d^3r'' \int d^3r' \\
& \times \frac{\varphi_{k\sigma}(r,t)\varphi_{j\sigma}^*(r,t)\varphi_{j\sigma}(r,t)\varphi_{k\sigma}^*(r,t)}{|r''-r'|}
\end{aligned}
$$

(2.17)

Before the time-dependent runs, self-consistent ground states were obtained for all the investigated systems by minimization of the overall energy of electronic orbitals $\varphi_i(r,t)|_{t=0}$ to some convergence criteria. C_{60} geometry was obtained from freely available results of geometry optimization by means of molecular dynamics. By the construction of a full wavefunction of our endohedral fullerenes, there is no atom-fullerene wavefunction overlap at all, because the intension is to model endohedral fullerenes as being produced by means of ion implantation during laser ablation. So, the self-consistent wavefunctions of fullerene shells and endohedral doping are simply added to each other and then renormalized. The numerical results show that under such circumstances the influence of screening is insignificant, that's why plasmon resonance tunes toward stronger atomic resonances.

Pseudopotentials [19] were used for carbon, indium, and antimony in endohedral fullerenes to reduce the total number of electrons. In@C_{60} and Sb@C_{60} are open shell systems, so additional electronic orbitals were added to account for the lowest excited states (in the computations three excited states were obtained from the properties of the potential without additional assumptions) and achieve convergence. They were also added for ordinary C_{60} for comparison. To decrease numerical effort without a loss of sense, the simulation box was a parallelepiped, 80 au along the propagation axis and 20 au in other two axes. Grid spacing was 0.5 au One should note that the nonspherical integration box violates the symmetry of the system, so even harmonics may appear in calculations, while they should not exist in the experiment under such conditions.

The harmonic generation was analyzed in three media (C_{60}, In@C_{60}, and Sb@C_{60}) by applying an electromagnetic field corresponding to the commonly used laser source (Ti:sapphire laser) with a central wavelength of 800 nm and a pulse duration of 48 fs. It is well known for HHG in real fullerenes that electrons from lowest energy levels are unlikely to ionize, so it was quite justified to consider them "frozen." Only the evolution of 8 of the 125 electronic orbitals was computed, although the action of Hartree and exchange potentials of these frozen orbitals on the investigated ones was fully considered. In the time-dependent runs density reaching the boundary of the integration box was removed by the mask technique similar to that proposed in [51].

It is also a good chance to check the correspondence of the studied model system to the real one. This means, a strong short delta-shaped pulse was applied in the beginning of the simulation, after which no external field was applied to the system so that density propagated freely to the ground state during the first 48 fs. Then, the propagation under the influence of strong pulse with an amplitude of 0.5 au and a frequency of 0.057 au ($\lambda = 800$ nm) broadened by a cosinoidal envelope $\cos(\pi(t - 2\tau_0 - t_0)/\tau_0)$, where $\tau_0 = 0.5$ and $t_0 = 1200$ (peak of the contour), was investigated under the same conditions for the first 48 fs.

Introduction of pseudopotentials for a multi-electronic system with corresponding exchange functionals eliminates the need of forced specification of artificial autoionizing resonances, and the time propagation methods of spectrum investigation do not require the addition of unoccupied states. So the spectrum is derived ab initio in a natural way for a given model system.

2.4.2 Results of Simulations and Discussion

The square of the absorption cross-sections of all species as a function of the photon energy in units of 0.057 au, which corresponds to the harmonic orders of 800 nm laser radiation, is presented in Figure 2.8a–c. There are signs of separate resonant peaks whose experimental observations were previously attributed to strong transitions in indium and antimony [25, 52]. This can be because C_{60} shell does not prevent the corresponding dipole excitations of implanted atoms. More important is the fact that broad plasmon peaks are shifted from the energy of 17th harmonic of 800 nm radiation in ordinary C_{60} toward the 13th and 19th–21st harmonics in In@C_{60} and Sb@C_{60}, respectively.

Owing to the nature of TDDFT formalism, it is impossible to distinguish which electrons are responsible for the observed spectral shifts. However, it is now evident that even in the absence of charge transfer, the valence electrons

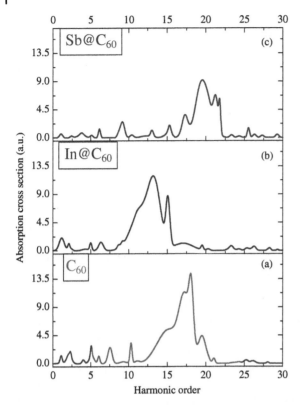

Figure 2.8 Absorption spectra of (a) original C_{60} fullerene, (b) endohedral In@C_{60} fullerene, and (c) endohedral Sb@C_{60} fullerene. Energy is given in harmonics of fundamental radiation with frequency 0.057 au *Source*: Redkin et al. 2011 [45]. Reproduced with permission from American Physical Society.

of the implanted atoms are not screened from the pump radiation, because even a slight excitation of other electrons is equal to perturbations similar to a delta-kick, and the system of delocalized electrons moves as a whole. So any noticeable resonance in atoms is actually amplified by the collective oscillation modes of C_{60} shell, but the resulting oscillation is in turn a seed for collective oscillations because its energy is mostly absorbed by collective oscillation modes that are closer to the corresponding resonance in implanted media. That's why the collective oscillations of delocalized electrons can tune to such resonances. As far as the equilibrium conditions are determined by the stability of collective oscillation modes, the initial strength of the resonance plays almost no role on the strength of resulting spectral properties.

The results of simulation of the HHG process in the three studied fullerenes are presented in Figure 2.9a–c. It should be noted that the results are plotted on a nonlogarithmic scale where the intensity of the nonshown first harmonic is 10 000, so the resonantly enhanced harmonics had intensity up to 10^{-4} of the main pulse. Real conversion efficiency is usually less by an order of magnitude due to self-defocusing on free electrons during propagation. One can easily see a good correspondence between spectral properties (Figure 2.8a–c) and

Figure 2.9 HHG spectra from (a) original C_{60} fullerene, (b) endohedral In@C_{60} fullerene, and (c) endohedral Sb@C_{60} fullerene. Intensities are normalized so that intensity of the 1st harmonic (i.e. fundamental intensity) is 10 000 (not shown). *Source*: Redkin et al. 2011 [45]. Reproduced with permission from American Physical Society.

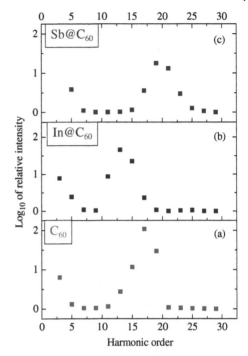

positions of enhanced groups of harmonics presented in Figure 2.9. A very important observation comes from the fact that the shifted peaks of HHG enhancement in endohedral In@C_{60} and Sb@C_{60} are very close to the corresponding shifted absorption peaks in them. It may be noted that, in the case of pure ionized fullerenes, maximum intensity of harmonics was observed in the vicinity of the broad SPR of C_{60} with the central wavelength near 50 nm (i.e. close to the 15th harmonic of an 800 nm pump).

Earlier, the enhancement of groups of harmonics was reported during HHG in fullerene-containing plasma plumes [35, 53–55]. The enhanced harmonics were observed in the range of 13th harmonic, which was attributed to the influence of the giant SPR of neutral C_{60} ($\lambda_{SPR} \cong 60$ nm with 10 nm full width at half maximum). The single harmonic enhancement using the indium and antimony plumes has also been experimentally observed. Using the indium laser-ablation, the enhancement of the13th harmonic at the wavelength of 61.26 nm has been reported [11]. The $4d^{10}5s^2\,^1S_0 - 4d^95s^25p\,(^2D)\,^1P_1$ transition of In II at the wavelength of 62.2 nm, which has an oscillator strength (*gf* value) of 1.11 [25], can be driven into the resonance with the 13th harmonic by the AC-Stark shift. The *gf* value of this transition is 10 times higher compared with the other transitions of In II in this spectral range. In the case of Sb plasma [36], the enhancement of the 21st harmonic of the 791 nm radiation was reported, although the 21st harmonic ($\lambda = 37.67$ nm) was slightly away from the $4d^{10}5s^22p$ $^3P_2$ – $4d^95s^25p^3$ (^2D) $^3D_3$ transition ($\lambda = 37.82$ nm) possessing highest gf value [52].

Thus the experimentally observed enhanced harmonics in the cases of C_{60} (13th harmonic), In (13th harmonic), and Sb (21st harmonic) were close to the enhanced harmonics calculated in the present study (17th harmonic in the case of pure fullerenes, 13th harmonic in the case of In@C_{60}, and 19th harmonic in the case of Sb@C_{60}). The difference between the experiment and theory in the case of pure fullerenes could be related with conditions of experiment, when the plasma contained mostly neutral fullerenes, which SPR lays in the range of 60 nm, (i.e. 13th harmonic of used 800 nm laser radiation).

There are also low-energy absorption peaks in absorption spectra of all systems. So, it is interesting to find out why resonant harmonic generation was observed only for those in the high-energy domain. A possible explanation of this phenomenon can be given by approaches, which attribute resonant HHG to resonant properties of the media at the moment of recombination [33, 56]. Namely, resonant recombination can proceed through every resonance, but when the electron moves back to the nucleus, the recombination into the first available resonant channel (with the highest energy) is extremely probable due to either its sufficiently large matrix dipole element [33] or inelastic collision cross-section [56]. So, a strong resonant transition in XUV range is a better choice for resonant HHG than the same transition in lower energy range. Phase-matching conditions are in general worse for higher harmonics because of self-defocusing on free electrons. So in the experiments lower intensity of higher harmonics is a result of the propagation effect, which is not included in discussed computations.

If these results will be experimentally verified, they can open a new way to control resonant HHG and this is only a matter of time because of experimental success in resonant HHG in In and Sb plasmas as well as enhancement of groups of harmonics near SPR of C_{60}. There is also no counter-evidence to the possibility of formation of stable In@C_{60}. In addition, Sb@C_{60} is stable and available in quantities sufficient for experiments. The approach proposed here can be used to increase the strength of groups of harmonics near SPR of C_{60}. Further advances in this area of research include usage of ionized C_{60} shells, which itself can promote the peak of HHG enhancement deeper into far ultraviolet spectral range, from 20 eV for neutral C_{60} to 25 eV for single-ionized C_{60}.

In the following, we briefly address the possibility of generating attosecond pulses in the fullerenes under consideration. It has been discussed in Ref. [56] that a resonant enhancement of a single harmonic is insufficient for the production of attosecond pulse trains. The situation is even worse in a mixture of different species, because resonant conditions for certain type of atoms differ greatly because of their narrow absorption peaks. The suggestion to use a mixture of endohedral fullerenes as a media for attosecond pulse train generation is only a development of the suggestions first proposed in Ref. [56] where atomic targets with close resonances were predicted to be a useful media for attosecond pulse train generation. In fact, it was shown in Ref. [56] that even a single resonant HHG is sufficient to generate a bright attosecond pulse train, but its shape is too bad due to domination of a single-resonant harmonic. In C_{60}, neighboring harmonics present a better case, since their intensities in the range of surface plasmon have relatively equal values. As far as resonant harmonics from endohedral fullerenes have the same plasmon-enhanced origin, one can propose that a simple mix of ordinary C_{60} and endohedral fullerenes can produce the required intensity distribution.

Several resonantly enhanced harmonics in C_{60} and endohedral fullerenes can be a possible solution to this problem, because the corresponding phases of resonant harmonics are linked to each other [53, 57]. For the best attosecond pulse train production these neighboring harmonics should be of the comparable intensity. In Ref. [56], an atom with several transitions at suitable frequencies and similar oscillator strengths, or a mixture of atoms was suggested for this purpose. Unfortunately, it turns out to be impossible to find corresponding atoms with such transitions for which resonant conditions could be fulfilled by a single pump pulse.

References

1 Milošević, D.B. (2006). *J. Opt. Soc. Am. B* 23: 308.

2 Milošević, D.B. (2007). *J. Phys.* B40: 3367.

3 Ganeev, R.A. and Milošević, D.B. (2008). *J. Opt. Soc. Am. B* 25: 1127.

4 Suzuki, M., Baba, M., Ganeev, R. et al. (2006). *Opt. Lett.* 31: 3306.

5 Lewenstein, M., Balcou, P., Ivanov, M.Y. et al. (1994). *Phys. Rev. A* 49: 2117.

6 Ishikawa, K. (2004). *Phys. Rev. A* 70: 013412.

7 Kubodera, S., Nagata, Y., Akiyama, Y. et al. (1993). *Phys. Rev. A* 48: 4576.

8 Altucci, C., Bruzzese, R., de Lisio, C. et al. (2003). *Phys. Rev. A* 68: 033806.

9 Reintjes, J.F. (1984). *Nonlinear Optical Parametric Processes in Liquids and Gases*. New York, NY: Academic Press.

10 Schiessl, K., Persson, E., Scrinzi, A., and Burgdörfer, J. (2006). *Phys. Rev. A* 74: 053412.

11 Ganeev, R.A., Suzuki, M., Ozaki, T. et al. (2006). *Opt. Lett.* 31: 1699.

12 Ganeev, R.A., Singhal, H., Naik, P.A. et al. (2006). *Phys. Rev. A* 74: 063824.

13 Ganeev, R.A., Elouga Bom, L.B., Kieffer, J.-C., and Ozaki, T. (2007). *Phys. Rev. A* 75: 063806.

14 Ganeev, R.A., Singhal, H., Naik, P.A. et al. (2009). *J. Opt. Soc. Am. B* 26: 2143.

15 Redkin, P.V., Kodirov, M.K., and Ganeev, R.A. (2011). *J. Opt. Soc. Am. B* 28: 165.

16 Runge, E. and Gross, E.K.U. (1984). *Phys. Rev. Lett.* 52: 997.

17 Castro, A., Appel, H., Oliveira, M. et al. (2006). *Phys. Status Solidi B* 243: 2465.

18 Marques, M.A.L., Ullrich, C.A., Nogueria, F. et al. (2003). *Time-Dependent Density Functional Theory*. Berlin: Springer.

19 Hartwigsen, A., Goedecker, S., and Hutter, J. (1998). *Phys. Rev. B* 58: 3641.

20 Perdew, J.P. and Zunger, A. (1981). *Phys. Rev. B* 23: 5048.

21 Bertsch, G.F. and Yabana, K. (1996). *Phys. Rev. B* 54: 4484.

22 Chang, Z., Rundquist, A., Wang, H. et al. (1998). *Phys. Rev. A* 58: R30.

23 Kim, H.T., Lee, D.G., Hong, K.-H. et al. (2003). *Phys. Rev. A* 67: 051801.

24 Kim, H.T., Kim, J.-H., Lee, D.H. et al. (2004). *Phys. Rev. A* 69: 031805.

25 Duffy, G. and Dunne, P. (2001). *J. Phys. B* 34: L173.

26 Taieb, R., Veniard, V., Wassaf, J., and Maquet, A. (2003). *Phys. Rev. A* 68: 033403.

27 Faria, C.F.M., Kopold, R., Becker, W., and Rost, J.M. (2002). *Phys. Rev. A* 65: 023404.

28 Ciappina, M.F., Becker, A., and Jaroń-Becker, A. (2007). *Phys. Rev. A* 76: 063406.

29 Ruggenthaler, M., Popruzhenko, S.V., and Bauer, D. (2008). *Phys. Rev. A* 78: 033413.

30 Zhang, G.P. (2005). *Phys. Rev. Lett.* 95: 047401.

31 Meyer, H.-D., Manthe, U., and Cederbaum, L.S. (1990). *Chem. Phys. Lett.* 165: 73.

32 Beck, M.H., Jäckle, A., Worth, G.A., and Meyer, H.-D. (2000). *Phys. Rep.* 324: 1.

33 Redkin, P.V. and Ganeev, R.A. (2010). *Phys. Rev. A* 81: 063825.

34 Bauer, D., Ceccherini, F., Maccani, A., and Cornolti, F. (2001). *Phys. Rev. A* 64: 063203.

35 Ganeev, R.A., Elouga Bom, L.B., Abdul-Hadi, J. et al. (2009). *Phys. Rev. Lett.* 102: 013903.

36 Suzuki, M., Baba, M., Ganeev, R.A. et al. (2007). *Opt. Express* 15: 1161.

37 Shinohara, H. (2000). *Rep. Prog. Phys.* 63: 843.

38 Saunders, M., Jiménez-Vázquez, H.A., Cross, R.J., and Poreda, R.J. (1993). *Science* 259: 1428.

39 Ohtsuki, T., Ohno, K., Shiga, K. et al. (2001). *Phys. Rev. B* 64: 125402.

40 Ohtsuki, T. and Ohno, K. (2004). *Sci. Technol. Adv. Mater.* 5: 621.

41 Chai, Y., Guo, T., Jin, C. et al. (1991). *J. Phys. Chem.* 95: 7564.

42 Saunders, M., Cross, R.J., Jiménez-Vázquez, H.A. et al. (1996). *Science* 271: 1693.

43 Braun, T. and Rausch, H. (1998). *Chem. Phys. Lett.* 288: 179.

44 Gadd, G.E., Schmidt, P., Bowles, C. et al. (1998). *J. Am. Chem. Soc.* 120: 10322.

45 Redkin, P.V., Danailov, M., and Ganeev, R.A. (2011). *Phys. Rev. A* 84: 013407.

46 Ullrich, C.A., Gossmann, U.J., and Gross, E.K.U. (1995). *Phys. Rev. Lett.* 74: 872.

47 Rubio, A., Alonso, J.A., Blase, X. et al. (1996). *Phys. Rev. Lett.* 77: 247.

48 Bertsch, G.F., Iwata, J.I., Rubio, A., and Yabana, K. (2000). *Phys. Rev. B* 62: 7998.

49 Marques, M.A.L., Castro, A., Bertsch, G.F., and Rubio, A. (2003). *Comp. Phys. Comm.* 151: 60.

50 Krieger, J.B., Li, Y., and Iafrate, G.J. (1992). *Phys. Rev. A* 45: 101.

51 Tafipolsky, M. and Schmid, R. (2006). *J. Chem. Phys.* 124: 174102.

52 D'Arcy, R., Costello, J.T., McGuinnes, C., and O'Sullivan, G. (1999). *J. Phys. B* 32: 4859.

53 Ganeev, R.A., Elouga Bom, L.B., Wong, M.C.H. et al. (2009). *Phys. Rev. A* 80: 043808.

54 Ganeev, R.A., Singhal, H., Naik, P.A. et al. (2009). *J. Appl. Phys.* 106: 103103.

55 Ganeev, R.A., Singhal, H., Naik, P.A. et al. (2010). *Appl. Phys. B* 100: 581.

56 Strelkov, V. (2010). *Phys. Rev. Lett.* 104: 123901.

57 Antoine, P., L'Huillier, A., and Lewenstein, M. (1996). *Phys. Rev. Lett.* 77: 1234.

3

Comparison of Resonance Harmonics: Experiment and Theory

3.1 Experimental and Theoretical Studies of Two-Color Pump Resonance-Induced Enhancement of Odd and Even Harmonics from a Tin Plasma

Among a few laser-produced plasmas (LPPs) demonstrating enhanced harmonics [1–11], tin represents an interesting sample of single harmonic generation, when strong transitions of singly and doubly charged ions can considerably influence this process depending on the experimental conditions (wavelength of driving radiation, laser chirp, single- or two-color pump, spectral width of driving radiation, pulse duration, etc.), which was confirmed during experimental and theoretical studies of the high harmonic generation (HHG) in a Sn plasma [5, 12–14]. In the meantime, a further search of superior properties of this plasma, together with consideration of the role of different plasma species (neutrals, singly-, and doubly charged ions) in the optimization of efficient harmonic generation, can improve the understanding of the role of various ionic transitions in the efficiency of this nonlinear optical process. Application of high pulse repetition rate lasers generating broadband ultrashort pulses can also enhance the output power of generating single harmonics from a tin plasma.

In the following, we analyze the HHG studies in a tin plasma using the 780 nm, 40 fs, 1 kHz laser by exploring different approaches (chirp variation, two-color pump). Strong enhancement of single even and odd harmonics was obtained in those studies, which was attributed to the involvement of different transitions of singly and doubly charged ions in the growth of harmonic efficiency. We also present theoretical studies of the photoabsorption spectra of different ions in the tin plasma and calculations of the harmonic output for odd and even harmonics at variable experimental conditions [15].

Resonance Enhancement in Laser-Produced Plasmas: Concepts and Applications,
First Edition. Rashid A. Ganeev.
© 2018 John Wiley & Sons, Inc. Published 2018 by John Wiley & Sons, Inc.

3.1.1 Experimental Studies

A 1 kHz chirped pulse amplification Ti:sapphire laser source was used in these studies. A part of the uncompressed radiation (780 nm central wavelength, 1.5 mJ, 20 ps, 1 kHz) was split from the beam line prior to the laser compressor stage and was directed into the vacuum chamber to create a plasma on the tin target (Figure 3.1). This picosecond radiation was focused using a 400 mm focal length lens and created a plasma plume with a diameter of 0.4 mm using an intensity on the target surface of $I_{ps} = 5 \times 10^9$ to 3×10^{10} W cm^{-2}. Broadband femtosecond pulses (780 nm, 40 fs, 1 mJ, 1 kHz, 40 nm spectral width at half maximum) were focused in a direction orthogonal to that of the heating picosecond pulse into the laser plasma using a 200 mm focal length reflective mirror. The position of the focus with respect to the plasma area was chosen to maximize the harmonic signal. The intensity of femtosecond radiation at the focal range was estimated to be $I_{fs} = 5 \times 10^{14}$ W cm^{-2}. The delay between plasma initiation and femtosecond pulse propagation was varied in the range of 6–57 ns using an optical delay line. The harmonic radiation was analyzed by an extreme ultraviolet spectrometer (XUVS) consisting of a flat field grating and a microchannel plate coupled to a phosphor screen. The images of harmonics were detected by a charge coupled device (CCD) camera.

Insertion of a nonlinear crystal, beta barium borate (BBO), in the path of the 780 nm laser beam led to the appearance of a second wave, 2ω, which enabled us to study the joint action of two orthogonally polarized waves (ω and 2ω) during the harmonic generation in the plasma plume. Group velocity dispersion between the ω and 2ω pulses in 0.5 mm thick nonlinear crystal was compensated for using the calcite [16]. The BBO crystal was inserted in front of the focusing mirror, so that no impeding processes were observed after propagation of the laser radiation through the crystal, while the

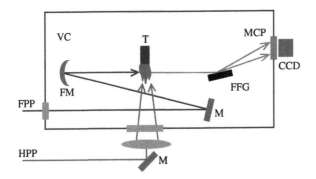

Figure 3.1 Experimental setup: HPP, heating pump pulse; FPP, femtosecond probe pulse; M, mirrors; VC, vacuum chamber; T, target; FM, focusing mirror; FFG, flat field grating; MCP, microchannel plate; and CCD, charge-coupled device. *Source:* Ganeev et al. 2012 [15]. Reproduced with permission from American Physical Society.

second harmonic conversion efficiency did not exceed 4%. The odd and even harmonic generation was achieved despite a significant difference between the two pump intensities (25 : 1).

The tuning of harmonic wavelengths was accomplished using the chirp-induced variation of the distribution of spectral components along the laser pulse. The chirp of the driving radiation was varied by adjusting the separation between the two gratings of the pulse compressor. Reducing the separation between the gratings from the chirp-free conditions resulted in the appearance of positively chirped pulses, and an increase of the distance between the gratings provided negatively chirped pulses. The artificially induced chirp was calibrated by analyzing the phase and pulse duration of the laser pulses propagating through the compressor stage at different distances between gratings by using the SPIDER [17] and measuring the pulse duration of various positively and negatively chirped pulses. At these conditions, no new frequencies are introduced, but the "ionization gating" changes the effective drive laser wavelength in the medium through chirping [18–20].

The wavelength shift of the harmonics can be explained by the redistribution of spectral components in the leading edge of the chirped pump laser. The harmonics produced with positively chirped laser pulses were red-shifted because the harmonics produced in the leading edge of the laser pulse come from the red component of the laser spectrum, with the blue component not contributing due to the accumulation of ionization over the pulse. The same can be said about the blue-shifted harmonics produced by negatively chirped pulses. Note that effective tuning can be achieved only in the case of a broadband radiation. In that case, only the leading part of the fundamental pulse consisting of either blue or red components participates in the HHG. It is worth noting that the broadband radiation is associated with a relatively short laser pulse. In discussed case, the relatively broadband radiation (40 nm) was used, which allowed a considerable change in harmonic wavelength with variation of laser chirp, which is crucial for achieving the appropriate adjustment between the wavelengths of some harmonics and ionic resonances possessing strong oscillator strengths.

Figure 3.2a shows the spectral distribution of harmonic intensity from the tin plasma using the 780 nm pump. The harmonics up to the 21st order were routinely observed, with the enhanced 17th harmonic ($E = 27.02$ eV, $\lambda = 45.88$ nm) dominating the spectrum of generated extreme ultraviolet (XUV) radiation. The harmonic cutoff from this plasma was considerably less compared with many other plasma plumes from different metal targets used in separate experiments. Variation of laser chirp allowed a considerable tuning of harmonic wavelengths (Figure 3.3), while the relative intensities of the 15th and 17th harmonics changed from the 15th harmonic being considerably suppressed in the case of positively chirped pulses to considerable enhancement of this harmonic for negatively chirped pulses. This variation of harmonic

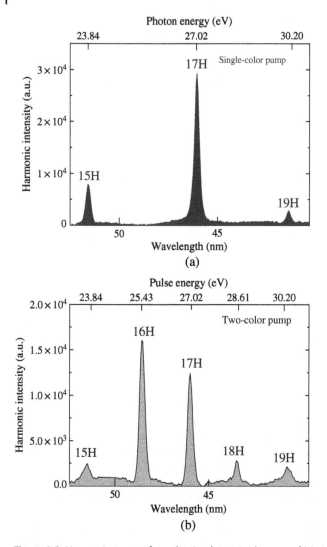

Figure 3.2 Harmonic spectra from the tin plasma in the case of (a) single-color and (b) two-color pump schemes. *Source:* Ganeev et al. 2012 [15]. Reproduced with permission from American Physical Society.

intensities can be attributed to the tuning of the 15th harmonic ($\lambda = 52$ nm, $E = 23.84$ eV at chirp-free conditions) toward the ionic resonance possessing strong oscillator strength.

One can note that, in this figure, the 17th harmonic was normalized to see the variation of the 15th harmonic with regard to the 17th harmonic over a broad range of spectral phase modulation (from +135 to −94 fs; the sign

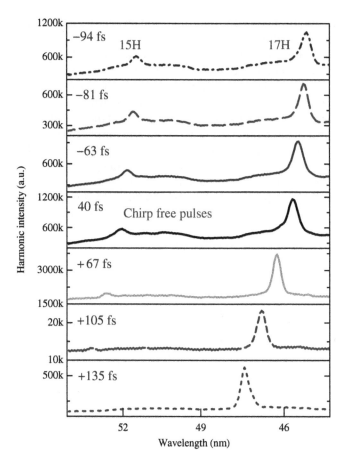

Figure 3.3 Tuning of 15th and 17th harmonics by changing the distance between the gratings in the compressor stage. Positive and negative values of pulse duration correspond to positively and negatively chirped pulses. *Source*: Ganeev et al. 2012 [15]. Reproduced with permission from American Physical Society.

of pulse duration denotes negative or positive chirp of laser radiation). In the meantime, the intensity of the 17th harmonic varied with tuning of the harmonic wavelength along the region of strong ionic resonances of singly- and doubly charged Sn ions. More discussion about the ionic transitions involved in harmonic intensity variations will be carried out in the following subsection.

The divergences of enhanced and unenhanced harmonics were analyzed and it was found that they are equal to each other. A considerable decrease of the efficiency of all harmonic orders was observed once the polarization of the driving pulse changed from linear to elliptical. These comparative experiments confirmed the same origin of resonant and nonresonant harmonics hinting strongly at recollision-induced processes.

The introduction of orthogonally polarized second-harmonic field increased the conversion efficiency of odd harmonics, which has been reported both in gaseous [21] and plasma [22] HHG studies. In the case of a tin plasma, strong enhancement of the even (16th) harmonic was also observed, which was of the same range of intensity as the odd ones (Figure 3.2b) in spite of the considerable difference in the number of interacting ω and 2ω photons in the plasma volume. The presence of even harmonics is a proof that sufficient overlap between the pulses occurs and a weak 2ω wave influences the dynamics of ejected electron.

The Sn II resonance, which influences the enhancement of harmonics, lies at the wavelength of 47.28 nm ($E = 26.22$ eV). The 780 nm radiation was used, with corresponding wavelengths of 16th (48.75 nm), 17th (45.88 nm), and 18th (43.33 nm) harmonics in the case of chirp-free radiation. In that case the 16th harmonic wavelength is close to the resonance transition, and the latter considerably influences the intensity of the former. The 17th harmonic also experiences enhancement. As for the 18th harmonic, it lies far from the Sn II and Sn III transitions.

Another case was observed in the case of the HHG studies in tin plasmas reported in Ref. [14]. The corresponding wavelengths of harmonics from the driving 800 nm radiation were 50 nm (16th harmonic), 47.07 nm (17th harmonic), and 44.44 nm (18th harmonic). The 16th harmonic was far from the 47.30 nm resonance of Sn II and did not enhance. At the same time, the influence of Sn III on the 18th harmonic intensity became stronger due to both the closeness with its resonance line and taking into account the AC-Stark shift resulting from the greater intensity of the driving radiation used (1×10^{15} W cm^{-2} in the plasma area).

3.1.2 Theoretical Approach

One can find from the published data on the Sn II transitions in the studied spectral region [23] that the observed enhancement of the 16th harmonic ($E = 25.43$ eV) and the 17th harmonic ($E = 27.02$ eV) of the chirp-free 780 nm radiation (see Figures 3.2 and 3.3) can be attributed to the transitions $4d^{10}5s^25p$ $^2P_{3/2} \rightarrow 4d^95s^25p^2$. The frequencies of these transitions, some of which possess reasonably large oscillator strengths, lie in the photon energy range of $24.9 - 27.3$ eV. The observed enhancement of the 15th harmonic was attributed to its approach toward the range of the $4d^{10}5s^25p$ $^2P_{3/2} \rightarrow 4d^95s^25p^2$ transitions of the Sn II ion generated by the negatively chirped pulse. However, Sn II transitions cannot explain adequately the enhancement of the 17th harmonic when the chirp variations cause a blue shift and its subsequent generation at a wavelength of 45.1 nm ($E = 27.49$ eV). Therefore one can consider the Sn III ion as a potential source of enhancement of this harmonic.

In order to simulate the resonant HHG enhancement in Sn II and Sn III ions one has to calculate the properties of the autoionizing state (AIS) in these ions. In Ref. [23], Duffy et al. recorded 4d–5p photoabsorption spectra of Sn II and Sn III in the 23–33 eV range. The transitions from the $4d^{10}5s^25p$ ground state of Sn II and from the $4d^{10}5s^2$ ground and $4d^{10}5s5p$ excited states of Sn III were observed and successfully identified with the aid of multiconfiguration Hartree–Fock (HF) calculations. These calculations were repeated in the discussed study to replicate these spectra and provide estimates for the autoionizing widths of the dominant transitions.

For the Sn II ($I_p = 14.63$ eV) spectrum configuration interaction calculations were performed in the Russell–Saunders coupling scheme with the HXR mode (Hartree plus exchange plus relativistic corrections) of the Cowan code [24] for transitions from $4d^{10}5s^25p \rightarrow 4d^95s^25p$, np, mf ($5 \leq n \leq 13$, $4 \leq m \leq 13$). The Slater parameters F^k, G^k and the configuration interaction parameter R^k were fixed at 85% of the ab initio values while the spin-orbit parameter was left unchanged. Good agreement was achieved between the calculated transition energies and gf values and the previous work [23], with maximum discrepancies of 0.12 eV and 0.09, respectively, which are attributed to the increased basis set and scaling factors in the present study. Calculations were performed to determine the autoionizing decay widths of the excited $4d^{-1}$ states, which are allowed to decay by the following processes:

$$4d^95s^25p \; np, mf \rightarrow 4d^{10}5s^2 + \varepsilon l \quad (l = 0, 2, 4) \tag{3.1}$$

$$4d^95s^25p \; np, mf \rightarrow 4d^{10}5s5p + \varepsilon' l \quad (l = 1, 3) \tag{3.2}$$

$$4d^95s^25p \; np, mf \rightarrow 4d^{10}5p2 + \varepsilon'' l \quad (l = 2) \tag{3.3}$$

The values of ε, ε', and ε'' are the differences in the configuration average energies of the excited $4d^{-1}$ configuration and each final Sn III ionic configuration. Synthetic spectra were constructed by assuming a Lorentzian line profile, $\sigma(E) = 109.7 f_k \Gamma_k / (2\pi[(E_k - E)^2 + \Gamma_k^2/4])$, where E_k and Γ_k are the energy and autoionization decay width of the transition in electron volts and f_k is the oscillator strength. The synthetic spectrum was then convolved with a Gaussian instrumental function of width 0.030 eV, which is presented in Figure 3.4.

For the Sn III spectra, calculations were performed in jj coupling for transitions from both the $4d^{10}5s^2$ ground and the $4d^{10}5s5p$ excited configuration, which has a configuration average energy of 6.87 eV. In these calculations, the Slater integrals were scaled to 80% except the spin-orbit integral that was left unchanged. The $4d^95s^25p$ configuration (the configuration average energy of 26.91 eV) does not autoionize because it lies below the ionization limit of Sn IV (30.50 eV [25])). The excited configuration $4d^95s5p^2$ has a configuration average energy of 35.22 eV and therefore once populated from the $4d^{10}5s5p$ excited state can autoionize through the following processes:

$$4d^95s5p^2 \rightarrow 4d^{10}5s + \varepsilon l \quad (l = 0, 2, 4) \tag{3.4}$$

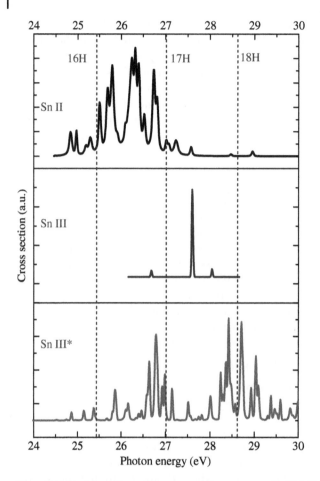

Figure 3.4 The photoabsorption cross-section spectra of Sn II $4d^{10}5s^25p \rightarrow 4d^95s^25p^2$ (top), Sn III $4d^{10}5s^2 \rightarrow 4d^95s^25p$ (middle), and Sn III excited state $4d^{10}5s5p \rightarrow 4d^95s5p^2$ (bottom) convolved with a Gaussian instrumental function of width 30 meV. Dashed lines show 780 nm radiation harmonic frequencies. *Source*: Ganeev et al. 2012 [15]. Reproduced with permission from American Physical Society.

where the values of ε are the differences in the configuration average energies of the excited $4d^{-1}$ configuration and each final Sn IV ionic configuration. The resulting synthetic spectra, which were convolved with a Gaussian instrumental function of width 0.030 eV are also presented in Figure 3.4. The transition properties required for the resonant HHG simulation are listed in Table 3.1. The experimentally observed enhancement of the blue shifted 17th harmonic (at $\lambda = 45.1$ nm, see Figure 3.3, upper graph) can be attributed to the Sn III transitions $4d^{10}5s^2 \rightarrow 4d^95s^25p$ ($E = 27.6$ eV, $\lambda = 44.92$ nm). One can note that

Table 3.1 The autoionizing properties of some transitions of the tin ions.

Ion	Transition	E_{calc} (eV)	gf[a]	Γ (meV)
Sn II	$4d^{10}5s^25p\ ^2P_{3/2} \rightarrow 4d^95s^25p^2\ ^2D_{5/2}$	26.22[b]	1.43	160.0
Sn III	$4d^{10}5s^2\ ^1S_0 \rightarrow 4d^95s^25p\ ^1P_1$	27.6[c]	0.870	—
Sn III*	$4d^{10}5s5p\ (1/2,3/2)_2 \rightarrow 4d^95s5p^2\ (5/2,1/2)_3$[d]	28.48[a]	0.84	47.20

a) The gf value is the product of the oscillator strength f of a transition and the statistical weight g of the lower level.
b) Calculated energies were shifted by 0.46 and −0.65 eV respectively as was the approach in the previous work [23].
c) Calculated energies were shifted by 0.16 eV.
d) The jj coupling is denoted $(J_{core}, J_{nl})_J$ where the subscript J refers to the total angular momentum of the level.
Source: Ganeev et al. 2012 [15]. Reproduced with permission from American Physical Society.

AISs are not involved in this transition, and thus this enhancement cannot be explained via four-step model [13]. The enhancement can be due to better phase-matching near the resonance [26], and/or due to the mechanism of the single-atom response enhancement [27].

Resonant HHG simulation is based on the numerical solution of the 3D time-dependent Schrödinger equation (TDSE) for a model ion in the external laser field. A single-active electron (SAE) approximation, reproducing the interaction with other electrons and with the nucleus with a model potential is used in the discussed case, as was done in Ref. [13]. This method is applicable, in particular, for the description of the resonances in the Sn II and Sn III ions. The following form of the model potential of the parent ion (atomic units are used throughout) was used in that case:

$$V(r) = -\frac{Q+1}{\sqrt{a_0^2 + r^2}} + a_1 \exp\left[-\left(\frac{r - a_2}{a_3}\right)^2\right] \tag{3.5}$$

where Q is the charge state of the generating ion (1 for Sn II and 2 for Sn III), and a_0, a_1, a_2, a_3 are the fitting parameters.

These parameters are chosen to reproduce the properties of certain transitions in Sn II and Sn III* ions. Namely, Sn II has a transition $4d^{10}5s^25p\ ^2P_{3/2} \rightarrow 4d^95s^25p^2\ (^1D)\ ^2D_{5/2}$ with a frequency (26.22 eV) close to the 16th and 17th harmonic frequencies of the 780 nm chirp-free driving radiation and with an oscillator strength essentially exceeding the other transitions in this spectral region. For Sn III* the transition with frequency 28.48 eV has the largest oscillator strength in this spectral region. So, to simulate HHG with Sn III* the model potential parameters were chosen to reproduce properties of this transition. The method of the TDSE solution

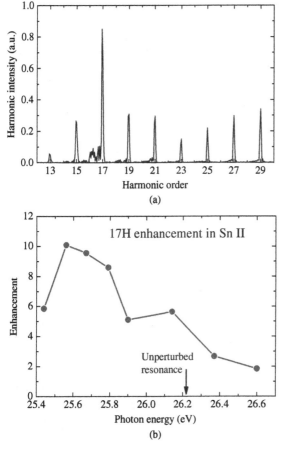

Figure 3.5 (a) Harmonic spectrum calculated for Sn II. The laser intensity is 10^{15} W cm^{-2}. (b) The calculated resonant 17th harmonic enhancement as a function of the harmonic photon energy. The arrow shows the frequency of the transition in the absence of the laser field. *Source:* Ganeev et al. 2012 [15]. Reproduced with permission from American Physical Society.

is described in Ref. [28]. In Figure 3.5a, the spectrum of the microscopic response calculated for the Sn II ion is presented. One can see the pronounced enhancement of the 17th harmonic, in agreement with experimental results presented in Figure 3.2a.

To study the frequency range in which the harmonic is enhanced, the HHG in Sn II was calculated using slightly different fundamental frequencies. The harmonic enhancement as a function of its frequency is presented in Figure 3.5b. One can see that maximal enhancement was achieved for a frequency different to the transition frequency in the absence of the laser, in agreement with experiments [5]. This difference can be attributed to the Stark shift of the autoionizing and ground states in the laser field. The width of the spectral range where the harmonic is enhanced (referenced below as an "enhancement width") is about 0.7 eV (full width at half maximum, FWHM). This is higher than the AIS width calculated in the absence of the laser field (0.160 eV, see Table 3.1) used in the discussed model. This can be explained with the broadening of the AIS due to its photoionization by the laser field.

The tuning of the fundamental frequency was done experimentally in Ref. [12]. This experiment shows the strong dependence of the resonant HHG enhancement on the laser frequency. Unfortunately, the experimentally achievable frequency variation was quite limited. Therefore one can conclude from Ref. [12], only that the "enhancement width" is wider than 0.2–0.3 eV. The upper limit of this width can be roughly estimated from Figure 3.2b: the enhancement of the two harmonics gives the interharmonic distance (1.6 eV) as the approximate upper limit of the "enhancement width." Thus, the calculated value of this width reasonably agrees with the experimental data. Although the exact value of the enhancement width is not measured, the fact of the broadening of the resonance can be nicely seen from comparison of the harmonic enhancement in Figures 3.2 and 3.3 and the photoabsorption cross-section in Figure 3.4. For instance, the cross-section at the 17th harmonic frequency of the chirp-free 780 nm radiation is small, but the harmonic is essentially enhanced. As we mentioned above, the numerical calculations attribute this broadening to the photoionization of the AIS.

Finally, we analyze the comparison of calculated results for Sn II and Sn III and previously reported data on tin harmonics in the case of 800 nm driving radiation. As was mentioned above, one can see from Figure 3.4 that the excited state of Sn III has strong transitions in the frequency range near the 18th harmonic of the 800 nm radiation ($E = 27.90$ eV). Using two-color excitation strong enhancement of this harmonic in Sn III* was found, see Figure 3.6. This figure explains the experimental results reported for the enhanced 18th harmonic of the 800 nm radiation generated in a tin plasma during two-color pump experiments [14]. Under lower intensity the generation takes place in Sn II and thus the 17th harmonic is enhanced. Under higher intensity the generating pulse essentially ionizes the medium, and Sn III* ions contribute to the HHG, providing the enhanced 18th harmonic emission. The isolated 18th harmonic generation is found in the calculation at an intensity of $I = 8 \times 10^{14}$ W cm^{-2}, while experimentally it was found at $I = 2 \times 10^{15}$ W cm^{-2}. At this intensity, in the calculations, the enhancement of the group of harmonics was found (see Figure 3.6c). Again, it can be attributed to the strong photoionization of the AIS. The disagreement could be due to the limited accuracy of used numerical model, as well as the limited precision of the experimental intensity measurement.

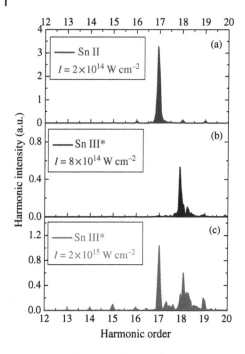

Figure 3.6 Harmonic spectra calculated for Sn II and Sn III* ions in the two-color field (800 + 400 nm). The fundamental intensities are presented in the figures. In graphs (a) and (b) one can see the resonant enhancement of the 17th harmonic in Sn II and the 18th harmonic in Sn III*. The enhancement of the group of harmonics (graph (c)) can be attributed to the ionization broadening of the resonance. *Source*: Ganeev et al. 2012 [15]. Reproduced with permission from American Physical Society.

3.2 Comparative Studies of Resonance Enhancement of Harmonic Radiation in Indium Plasma Using Multicycle and Few-Cycle Pulses

3.2.1 Introduction

The time–frequency analysis of the HHG spectra supported a four-step model [13] where for long-lived resonances, the interference occurs between the populations caused by recollisions in different half cycles. This leads to new possibilities for XUV pulse shaping in the subfemtosecond timescale. Such a shaping of XUV pulse has found confirmation in the studies of HHG in laser-produced manganese plasma using sub-4 fs pulses [8]. The measured HHG spectra exhibited resonant enhancement in the spectral region around the 31st harmonic of a Ti:sapphire laser (~50 eV) where the intensity contrast relative to the adjacent harmonics exceeded 1 order of magnitude. This finding was in sharp contrast with the results of harmonic generation in Mn plasma plumes reported previously for multicycle laser pulses [6] and demonstrated the application of few-cycle pulses, which may significantly change the pattern of resonance harmonics. The reported weak carrier envelope phase (CEP) dependence [8] might reduce the requirements for CEP stabilization of the few-cycle laser pulses. Moreover, without the losses associated with spectral dispersion using gratings thin metal film filtering, it could serve in various applications where the coherent, ultrashort XUV pulses are required.

To address the issue of resonance enhancement of HHG in depth and analyze the validity of the proposed mechanism [8] for subfemtosecond XUV pulse generation, an investigation of the comparative behavior of resonant harmonics generated in an indium plasma at the conditions of multi-(30 fs) and few-cycle (3.5 fs) pulses excitation was reported in Ref. [29]. Those studies have demonstrated the strong influence of the pulse duration on the intensity of the resonantly enhanced harmonics generated from this plasma. Additionally it was shown that, analogously to experiments with multicycle pulses in a tin plasma [15], the enhancement of harmonics in the case of few-cycle pulses does not require the exact coincidence of the frequencies of the plasma ion transitions and the harmonic frequencies of the driving radiation. Finally, theoretical modeling suggested that harmonic emission should depend on the CEP of the driving pulse, while experimental observations show less influence of this parameter on the dynamics of harmonic spectra.

Two femtosecond lasers, which delivered few-cycle and multi-cycle pulses, were used in those experiments. First, a Ti:sapphire laser provided pulses of 25 fs duration and energies per pulse of up to 0.8 mJ at a repetition rate of 1 kHz. These pulses were focused into a 1 m long differentially pumped hollow core fiber filled with neon at a pressure of 3 bar. The spectrally broadened pulses at the output of the fiber system were compressed using chirped mirrors. A pair of fused silica wedges served to fine-tune the pulse compression. High-intensity few-cycle pulses (770 nm central wavelength, 0.2 mJ pulse energy, 3.5 fs duration) were typically obtained from this system. A part of the uncompressed radiation of this Ti:sapphire laser (pulse energy 120 μJ, pulse duration 8 ps, pulse repetition rate 1 kHz) was split from the beam line prior to the first compressor stage and was focused into the vacuum chamber to create a plasma on an indium target placed in a vacuum chamber. The picosecond heating pulses created a plasma plume with a diameter of ~0.5 mm using an intensity on the target surface of $I_{ps} = 8 \times 10^9$ W cm^{-2}. The 3.5 fs pulses were focused, in a direction orthogonal to that of the heating pulse, into the laser plasma using a 400 mm focal length spherical mirror. The delay between plasma initiation and the femtosecond pulse propagation, required for formation and expansion of the plasma plume away from the surface of the target, was set to 33 ns. The intensity of this femtosecond driving pulse in the plasma area was estimated to be $I_{fs} = 3 \times 10^{14}$ W cm^{-2}. The generated harmonics were analyzed by a XUVS consisting of a flat field grating and a microchannel plate coupled to a phosphor screen. The spatially resolved spectra of the generated harmonics were detected by a CCD camera. Longer pulses (30 fs) were also applied for HHG in indium plasma plumes using a 1 kHz repetition rate Ti:sapphire laser and the experimental setup for plasma HHG described in Section 3.1.1 (Figure 3.1). The target (an indium rod of diameter of 10 mm and length of 20 mm) was rotated to minimize overheating and damage of its surface from repeated laser shots (Figure 3.7), thus ensuring more stable ablation conditions and consequently improving harmonic stability [30].

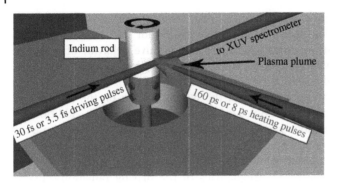

Figure 3.7 Experimental arrangement of rotating target configuration. *Source*: Ganeev et al. 2013 [29]. Reproduced with permission from American Physical Society.

3.2.2 Indium Emission Spectra in the Cases of 40 and 3.5 fs Driving Pulses

The harmonic spectrum of the indium plasma corresponding to harmonic generation of 30 fs pulses showed a strong emission line at ~20.7 eV ($\lambda = 60$ nm) corresponding to the 13th harmonic of the 780 nm driving radiation (photon energy $E_p = 1.59$ eV). This line dominates over the whole spectrum, which also displays weak lines around 17.5 and 23.9 eV, corresponding to the 11th and 15th harmonics, and even weaker higher order harmonics up to the 30th order.

Enhanced 13th harmonic ($\lambda = 61.5$ nm, $E_p \approx 20.14$ eV) from In containing plasmas has been reported earlier [1, 2, 5], using multicycle 10 Hz Ti:sapphire-driving lasers operating at 800 nm, and explained as effect of resonance with the strong In II transition (at 19.92 eV). In the discussed experiments [29], the central wavelength of the 30 fs driving laser pulses (780 nm) was shorter than in those previous studies and the photon energy of the 13th harmonic is detuned from the mentioned In II transition.

The efficiency of the resonant harmonic was maximized by tuning it toward the longer wavelength side, so that the frequency of the 13th harmonic became closer to the In II resonance line. To this aim, a positive chirp was added to the driving pulse by adjusting the separation of the two gratings in the pulse compressor. Figure 3.8 shows both a raw image of the low-order range of the HHG spectrum from the indium plasma, after tuning the 13th harmonic wavelength toward the longer wavelength side, and the line-out of harmonic intensities along the whole harmonic spectrum (up to the 33rd order). The ratio between the intensities of the resonantly enhanced harmonic and the neighboring ones exceeded 40. The dominance of the 13th harmonic in this case becomes even more pronounced as compared with the spectrum obtained before tuning the driving radiation. One can clearly see the excited resonance line (Figure 3.8a, thin vertical line above the left side of the broad

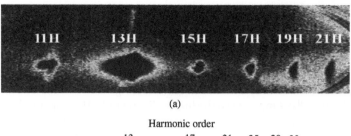

(a)

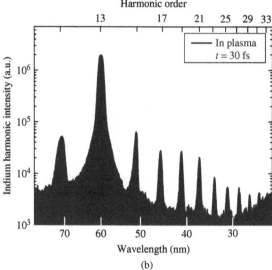

(b)

Figure 3.8 (a) Raw image of the low-order harmonic spectrum generated from an indium plasma in the case of 30 fs (FWHM) driving pulses. The thin line with high divergence above the 13th harmonic is a resonance transition at 19.92 eV. (b) Spectral distribution of indium harmonics, along the whole range of harmonic generation, using multicycle (30 fs) pulses. The highest order obtained is the 33rd harmonic. *Source*: Ganeev et al. 2013 [29]. Reproduced with permission from American Physical Society.

13th harmonic emission) possessing considerably higher divergence than the harmonic emission. The maximum enhancement of 13th harmonic was obtained once the resonance transition was approached from the blue side, at a wavelength that did not exactly match the harmonic frequency with that of the ion transition.

The spectral pattern of the XUV radiation generated from the indium plasma was considerably modified once the driving radiation was changed from multicycle pulses to few-cycle pulses (3.5 fs, $\lambda = 770$ nm, $E_p = 1.61$ eV). At a moderate ablation intensity ($I_{ps} = 5 \times 10^9$ W cm^{-2}), no spontaneous plasma emission (i.e. in absence of excitation with 3.5 fs pulses) was detected in the range of 15–30 eV. At these conditions, focusing the few-cycle pulses in the plasma resulted in the appearance of a strong narrow line at $E_p \approx 20$ eV (Figure 3.9a). No other harmonics were observed at these conditions, contrary to the case of multicycle pulses. Ablation of the target at higher intensity ($I_{ps} = 8 \times 10^9$ W cm^{-2}) also did not result in plasma spontaneous emission. However, in this case, the presence of the few-cycle driving pulses led to additional spectral lines mostly in the shorter wavelength side of the strong ≈ 20 eV line (Figure 3.9b). One can see the highly divergent In II resonance line at 19.92 eV, which displays a maximum intensity in its central part and that is accompanied by some additional lines. The intensity of the ≈ 20 eV emission exceeded that of other emissions in the shorter wavelength range (at 21.9, 23.3, 24.1, and 25.2 eV); however the contrast is lower than in the case shown in Figure 3.8b.

None of these relatively narrow spectral lines, at frequencies above 20 eV, could be assigned to plasma harmonics, which should possess much broader bandwidths than those observed lines. The bandwidths of harmonics are defined by that of the driving radiation, which undergoes considerable broadening after the second stage of chirping and compression from 25 fs (~ 60 nm) to 3.5 fs (~ 215 nm). In the latter case, the positions of the harmonics in the 16–30 eV range were defined using a carbon plasma plume, which allowed the generation of featureless 11th to 15th harmonics (Figure 3.9c). The central wavelength of the driving radiation (770 nm) can be retrieved from this spectrum showing the energies of the 11th ($E_p \approx 17.7$ eV), 13th ($E_p \approx 20.9$ eV), and 15th ($E_p \approx 24.15$ eV) harmonics. The change in the spectral characteristics of the driving radiation leads to corresponding changes of harmonic spectra, which are related with the increase of laser intensity, detuning of harmonic wavelength from the ion transitions, and to some other mechanisms, which will be discussed later. These results show that the strong 20 eV transition observed in the indium plasma spectrum upon interaction with 3.5 fs pulses coincides with neither the 11th harmonic nor the 13th harmonic of 770 nm driving radiation. In the following subsections, we will discuss the origin of this radiation appearing in the case of few-cycle pump and propose a few options for explanation of the features of enhanced narrowband 20 eV emission observed in those studies. Notice that the exact nature of this radiation is still under question.

Commonly, for multicycle pulses, the increase of pulse duration can be accomplished by changing the distance between the gratings in the compressor stage. It is a widely used technique, which enables a tuning of harmonic wavelength from broadband sources (see for example [31, 32]). Particularly, to tune harmonics toward the longer wavelength side, so that the frequency of the 13th harmonic becomes closer to the In II resonance line, a positive chirp was added to the driving pulse by shortening the distance between two gratings in the pulse compressor of 30 fs, 780 nm laser. Only the wavelength at the leading edge of the pulse contributes significantly to the HHG because,

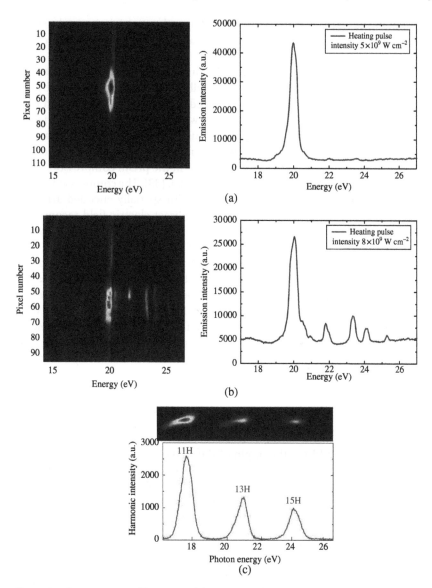

Figure 3.9 Raw images and line-outs of the spectra of the indium plasma emission upon excitation by 3.5 fs (FWHM) pulses using the heating pulse intensities of (a) $I_{ps} = 5 \times 10^9$ W cm^{-2} and (b) $I_{ps} = 8 \times 10^9$ W cm^{-2}. At low heating pulse intensity (a) one can see the highly divergent 19.92 eV resonance emission and its enhanced central part accompanied with some additional lines in (b). In the latter case the intensity of the 19.92 eV emission exceeds that of other emissions (at ~21.9, 23.3, 24.1, and 25.2 eV), though not as strongly as in the case of emission of the harmonics shown in Figures 3.2b and 3.3a. (c) Raw image and line-out of harmonic spectra from a carbon plasma obtained by excitation with 3.5 fs pulses. Broadband harmonics are observed in this case, which could be distinguished from the emission lines shown in (a) and (b). The 11th, 13th, and 15th harmonic energies are centered at ~17.7, ~20.9, and ~24.15 eV, respectively. *Source*: Ganeev et al. 2013 [29]. Reproduced with permission from American Physical Society.

at the intensity used, the strong field generates the multiply ionized plasma that grows increasingly with time, eventually preventing HHG (so-called ionization gating effect).

Another technique for chirping of laser pulses was introduced in the case of few-cycle pulses. The chirp of these broadband pulses was modified by the fine-tuning of the double glass wedges placed after the bunch of chirped mirrors. The change of the path through this wedge pair enabled the variation of spectral distribution along the pulse. Thus the increase of pulse duration in that case was carried out by the movement of the wedges with respect to each other (analogously to the chirp variations in the prism compressors). The description of this technique could be found in Ref. [33]. The compressed pulses were characterized spatially and temporally with spatially encoded arrangement for spectral shearing interferometry for direct electric field reconstruction (SEA-F-SPIDER) [34]. This technique enabled to control the pulse duration during modification of chirp characteristics of a few-cycle pulse.

In the case of few-cycle pulses, the 13th harmonic wavelength cannot be tuned toward the strong In II transition by modifying the laser chirp, as was the case for multicycle pulses. For positively chirped pulses, this was attributed to the largely significant increase of duration for the 3.5 fs pulses, in contrast with the less significant increase (from 30 to 47 fs) for multicycle pulses. Therefore, one can assume that only parts of the whole spectrum of the broadband 11th and 13th harmonics of the driving 770 nm radiation were effectively converted to the $\approx$20 eV emission.

3.2.3 Testing the Indium Emission Spectra Obtained Using 3.5 fs Pulses

In this section, we address the origin of the $\approx$20 eV emission observed when the 3.5 fs pulses propagate through the indium plasma. To define the coherent character of this emission, the influence of the laser polarization on its intensity was tested. The polarization of the driving radiation was changed by rotating a quarter-wave plate inserted in the beam path. With the change of polarization from linear to elliptical, and finally to circular, the intensity of the emission showed a characteristic abrupt decrease, accompanied by the disappearance of shorter wavelength lines. This behavior is a distinctive feature of the harmonic generation mechanism based on the recombination of the returning electron with the parent particle. Notice that in absence of the 3.5 fs excitation pulses, the spontaneous emissions from the In plasma under optimal ablation conditions was not detected.

Another test for the coherent character of this emission was carried out using a Young double-slit interferometer. The double slits were mounted on a translation stage and placed approximately 40 cm from the targets, with the microchannel plate situated at 70 cm from the slits. The two slits, spaced by

Figure 3.10 Spectra of emission generated from the indium plasma using 3.5 fs pulses (a) without and (b) with double slits. *Source*: Ganeev et al. 2013 [29]. Reproduced with permission from American Physical Society.

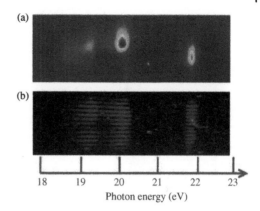

50 μm, were made from tungsten material 6 μm wide and 10 mm long. The images of interference fringes were taken in a subsequent chamber using a flat field grating, a microchannel plate, and a CCD camera.

Figure 3.10 shows the typical XUV emission spectrum and the interference fringes obtained after propagation of the 20 eV emission line through the double-slit interferometer. One can see three enhanced emissions (Figure 3.10a), which do not coincide with the central wavelengths of the 11th, 13th, and 15th harmonics, although as already mentioned, may represent the enhanced contributions of those harmonics. The interference patterns of these three lines are shown in Figure 3.10b. The interferometric visibility of these patterns was measured to be ∼0.66.

The coherence properties of plasma emission from the indium plasma were not studied. However, the comparative measurements of spatial coherence were carried out in the case of zinc plasma, which showed a weak visibility of the interference fringes in the case of spontaneous emission of LPP, while the visibility of harmonics-induced interference pattern (∼0.72) was similar to the case of indium plasma experiments.

The above polarization and interference tests confirmed that the observed enhanced radiation from the indium ablation plume possesses good coherence and was not originated from plasma spontaneous emission.

The change of the gas backing pressure in the hollow fiber allowed to vary the laser pulse duration between 3.5 and 25 fs. In the discussed experiments, the Ne backing pressure in the fiber in was varied order to change the spectrum, pulse duration, and intensity of the ultrashort pulses and subsequently observe the resulting changes in the indium emission spectrum. This is shown in Figure 3.11, which indicates that, by increasing the Ne pressure, the strong ≈20 eV emission is slightly tuned and blue shifted. One can see that there is an "optimal" pressure (2.4 bar) at which generation of the ≈20 eV emission is achieved with maximum efficiency at other equal experimental parameters.

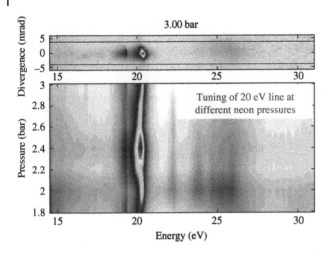

Figure 3.11 Variation of emission from the indium plasma excited by 3.5 fs pulses at different pressures of neon in hollow fiber. Upper panel shows the divergence of excited line at 3 bar of Ne. *Source*: Ganeev et al. 2013 [29]. Reproduced with permission from American Physical Society.

The blue shift of the 20 eV emission can be explained as follows. The increase of Ne backing pressure leads to a considerable broadening of the laser spectrum, which is followed by the shortening of the pulse duration after compression using the chirped mirrors. The corresponding increase of laser intensity induces a change of the free electron density (dN/dt) and a transient variation of the refractive index, a parameter that strongly depends on laser intensity and on the time needed to ionize a considerable fraction of particles. The change in the refractive index N induces a shift $\delta\lambda$ of the central laser wavelength λ that can be defined as [34]:

$$\delta\lambda = -(e^2\lambda^3/2\pi m_e c^3)\, L\, dN/dt \tag{3.6}$$

where L is the length of medium, e and m_e are the charge and mass of electron, and c is the light velocity. The blueshift of the fundamental radiation $\delta\lambda$ leads to a corresponding blueshift $\delta\lambda_q = \delta\lambda/q$ for the qth-order harmonic at $\lambda_q = \lambda/q$. In fact, a blueshift of harmonics has been reported previously in gas HHG experiments [31, 32], as well as in plasma HHG studies [35]. In Ref. [31], the effect of free electrons on the spectral properties of the high-order harmonics was analyzed in a neon gas jet using 30 fs, Ti:sapphire laser pulses. Measurements, carried out at a laser-driving intensity of 4×10^{14} W cm^{-2}, clearly showed the possibility of continuously tuning the harmonic wavelength by taking advantage of the blueshift induced by the transient appearance of free electrons in the medium.

On the other hand, the "optimization" of the 20 eV emission observed at a pressure of 2.4 bar of neon could be related with the effect induced by tuning the harmonic wavelength and the creation of better phase-matching conditions (see the discussion of the possibility of phase matching in the vicinity of resonance presented in the following section). These conditions may appear in some plasma plumes near their ionic or atomic transitions possessing strong oscillation strength. However, the additional influence of the microscopic response cannot be disregarded.

3.2.4 Theoretical Consideration of the Microscopic Response

The 3D TDSE was solved for the model system in the linearly polarized laser field (atomic units are used in this section):

$$i\frac{\partial}{\partial t}\Psi(\mathbf{r}, t) = \left[-\frac{1}{2}\nabla^2 + V(r) - E(t)x\right]\Psi(\mathbf{r}, t) \tag{3.7}$$

where $V(r)$ is the parent ion potential, x is laser field polarization direction. The laser field is given by

$$E(t) = \begin{cases} E_0\left(\sin\left(\frac{\pi t}{2\tau}\right)\right)^2 \cos(\omega_0 t + \varphi), & 0 < t < 2\tau \\ 0, & t > 2\tau \end{cases} \tag{3.8}$$

where ω_0 is the fundamental frequency, and φ is the pulse CEP. Note that under τ equal to integer number of optical half-cycles (more than one), this field satisfies the condition: $\int_{-\infty}^{\infty} E(t)dt = 0$. This condition implies the absence of a static component of the field and thus should be satisfied for any laser field.

Exact description of an AIS assumes multi-electron dynamics study. However, numerical TDSE solution for a multi-electronic system in an intense laser field is an extremely difficult task (practically possible now only for helium). Because of this a SAE approximation was used, and the role of the other electrons was reproduced with a model potential of the parent ion, as it was done in Refs. [13, 15, 36]:

$$V(r) = -\frac{2}{\sqrt{a_0^2 + r^2}} + a_1 \exp\left[-\left(\frac{r - a_2}{a_3}\right)^2\right] \tag{3.9}$$

where the fitting parameters a_0, a_1, a_2, a_3 are chosen to reproduce some properties of In II. Namely, as it was mentioned above, the ion transition $4d^{10}5s^2\,{}^1S_0 \rightarrow 4d^95s^25p\,({}^2D)\,{}^1P_1$ from the ground state to an AIS has a frequency close to that of the 13th harmonic frequency of the 800 nm radiation, and an oscillator strength that substantially exceeds those of the neighboring transitions in the spectral region considered [37]. Thus, it is appropriate to neglect other excited states for this ion and to choose the potential parameters $a_0 = 0.65$, $a_1 = 1.0$, $a_2 = 3.8$, and $a_3 = 1.6$ to reproduce the energies of the ground state and AIS, the AIS width, and the transition oscillator strength.

The numerical method for the TDSE solution is described in Ref. [28]. To characterize the XUV field emission, the second derivative of the dipole moment was calculated. According to the Ehrenfest's theorem, it is equal to the quantum-mechanical expectation value of the force acting on the electron $f(t) = -\langle \Psi(\mathbf{r}, t) | \frac{\partial V(r)}{\partial x} | \Psi(\mathbf{r}, t) \rangle$. Finally, its spectrum $f(\omega) = \int f(t) \exp(i\omega t) dt$ was found.

The spectral intensity $|f(\omega)|^2$ calculated using 800 nm laser pulse with the peak intensity of 10^{15} W cm^{-2} and $\tau = 2^{2\pi}/\omega$ (corresponding to the pulse duration at half-maximum level of intensity $\tau_{\text{FWHM}} = 3.9$ fs) is presented in Figure 3.12. Those results have shown that the enhancement of the XUV generation efficiency near the resonance is even more pronounced, than in the case of a multicycle laser pulse. In Figure 3.13a, the intensity of the XUV emission near the resonance (where the intensity was integrated from $12\omega_0$ to $14\omega_0$) is shown as a function of time. One can see that the resonant XUV is emitted mainly in the trailing edge of the laser pulse. Such behavior is very typical, though the details of this dependence are CEP sensitive. This can be attributed to the following two factors. First, the resonant harmonic intensity is proportional to the AIS population, and this population is accumulated during some time, as it was discussed in Refs. [13, 36]. Thus, the resonant XUV emission is delayed by this time with respect to the laser pulse envelope, which

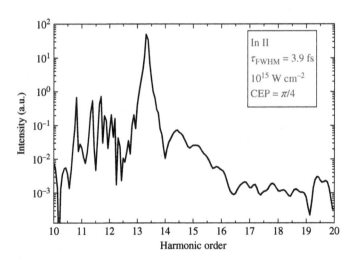

Figure 3.12 Single particle response in the XUV range to the 3.9 fs laser pulse calculated via numerical TDSE solution for a model In II ion. *Source*: Ganeev et al. 2013 [29]. Reproduced with permission from American Physical Society.

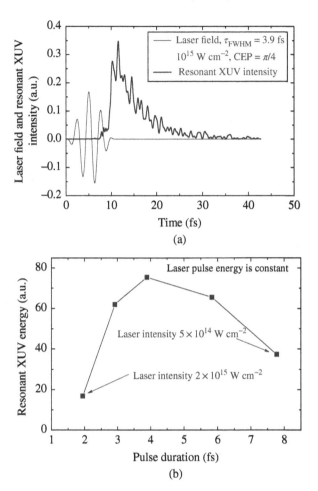

Figure 3.13 (a) Intensity of the XUV emission in the vicinity of the ion transition $4d^{10}5s^2\,{}^1S_0 \rightarrow 4d^9 5s^2 5p(^2D)^1P_1$ from the ground state to an AIS as a function of time. (b) The CEP averaged energy of the resonant XUV emission (from 12ω to 14ω) as a function of the laser pulse duration τ_{FWHM} under constant laser pulse energy. *Source:* Ganeev et al. 2013 [29]. Reproduced with permission from American Physical Society.

corresponds to the AIS lifetime in the field. Although the delay should be less than the field-free resonance lifetime (5.46 fs), due to the photoionization of the AIS, it can be comparable with the laser pulse duration. Secondly, at the trailing edge of the pulse the laser field at the instant of rescattering is less intense, than at the instant of detachment. So the photoionization of the AIS after rescattering is reduced and the XUV emission is more intense. For instance, in conditions of Figure 3.13a almost all XUV radiation is emitted after the laser pulse has turned off.

The resonant XUV energy was estimated as a function of the laser pulse duration assuming that the pulse energy is constant. The result of this simulation is presented in Figure 3.13b, which shows that there is an optimal laser pulse duration of about 4 fs at which maximum yield could be obtained. This result agrees with the experimental findings shown in Figure 3.11. Thus one can conclude that the experimentally found behavior could be attributed (at least partly) to the properties of the microscopic response.

The resonant XUV energy as a function of the laser pulse CEP is shown in Figure 3.14a. This dependence is shown for a given laser intensity, as well the XUV energy averaged over laser intensity (thus taking into account the intensity distribution in the laser beam). One can see that for the given laser intensity, the CEP dependence is significant, but averaging over the laser intensity makes this dependence less pronounced. The resonant XUV energy as a function of the laser intensity is shown in Figure 3.14b. Sharp peaks in this dependence can be attributed to the constructive interference of the contributions to AIS population from different laser half-cycles under specific laser intensities.

3.2.5 Experimental Studies of Harmonic Yield on the CEP of Laser Pulse

Resonance-induced enhancement of a single harmonic of a multicycle pulse in plasma plumes has allowed considerable improvement of harmonic efficiency in some XUV spectral ranges. The goal of the discussed studies was to analyze resonance-induced processes observed in indium ablation plumes using both multi- and few-cycle pulses. As it was already mentioned, the analysis of single harmonic generation in manganese plasma using 3.5 fs pulses has revealed some interesting features of this process [8]. Particularly, the change of the CEP of a few-cycle driving pulse caused only small variations in Mn harmonic spectra. The resonance-enhanced harmonic was generated in the range of 50–51 eV. One can expect that such a pattern of harmonic spectra and weak CEP dependence could be observed as well in the longer wavelength side of XUV spectra. The studies in indium plasma plumes have confirmed this assumption, which can open the doors for development of such coherent, XUV sources and analysis of the spectral properties of emitters through plasma HHG.

Most of the experiments described above were carried out without CEP stabilization (that is, for random CEP values). To address the theoretical assumptions presented in the previous section, HHG experiments using the 3.5 fs pulses with stabilized CEP were carried out. Some differences were found in the emission spectra in the case of two phases ($\varphi = 0$ and $\pi/2$, see the normalized spectra of harmonic and plasma emission in Figure 3.15). In particular, stronger low-order harmonics and a broader resonance emission in the case of $\varphi = \pi/2$ was observed. Overall, the spectral shapes of XUV emissions were approximately similar for these two fixed values of CEP.

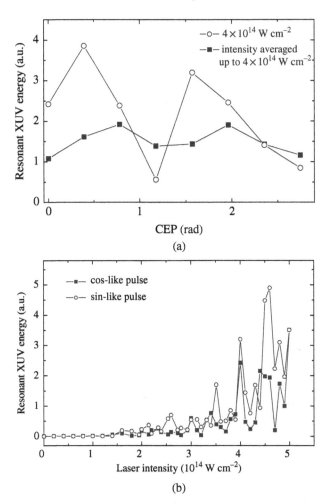

Figure 3.14 (a) Resonant XUV energy for the given laser intensity (empty circles) and averaged over the laser intensity up to 4×10^{14} W cm^{-2} (filled squares) as a function of CEP of the 3.9 fs pulse. (b) The resonant XUV energy as a function of the peak laser intensity for CEP $\varphi = 0$ (cos-like pulse) and CEP $\varphi = \pi/2$ (sin-like pulse). *Source*: Ganeev et al. 2013 [29]. Reproduced with permission from American Physical Society.

The harmonic spectra from the indium plasma plumes showed weak dependence on the CEP of the driving, few-cycle pulses, while the theoretical calculations predicted stronger dependences. In particular, some CEP-induced changes in harmonic spectra were obtained by simulations of the resonant XUV energy as a function of the laser pulse CEP (see Figure 3.14a). The fact that a strong CEP dependence of the harmonic spectra was not observed could be attributed, in the case of 3.5 fs pulses, to the presence of a significant density of

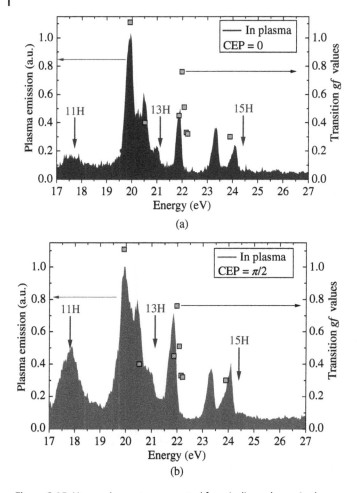

Figure 3.15 Harmonic spectra generated from indium plasma in the case of a fixed CEP ((a) $\varphi = 0$ and (b) $\varphi = \pi/2$) of few-cycle pulses and strong excitation of the indium target. Squares show frequencies and oscillator strengths (*gf* values) of the transitions in In II and In II* taken from Ref. [37] (only transitions with *gf* values exceeding 0.3 are shown). Arrows indicate the frequencies of the harmonics in carbon plasma shown in Figure 3.3c. *Source:* Ganeev et al. 2013 [29]. Reproduced with permission from American Physical Society.

free electrons in the plasma, which might diminish the difference between the HHG spectra recorded for different values of CEP.

The free electrons appearing during propagation of femtosecond pulse through the prepared plasma plume might also influence this process, though the additional concentration of free charges does not considerably increase the initial concentration of electrons. The variation of femtosecond pulse intensity may insignificantly change the already existing free electron concentration, since the conditions were maintained when no significant amount of free electrons appears during HHG at the intensity of femtosecond driving pulse in the plasma area of $\leq 3 \times 10^{14}$ W cm^{-2}. Probably, it is possible to analyze the CEP dependence at variable range of intensities. However, at high intensities, the influence of additionally appeared electrons might decrease the harmonic yield due to the propagation effects induced by the dynamical variation of the dispersion properties of plasma medium. In that case, the CEP dependence might be further blurred compared to other effects.

The phase matching during generation of XUV pulses could also blur the CEP dependence of HHG. Indeed, both phase matching and the microscopic harmonic response depend on the ionization degree. If the HHG efficiency in the experiment is limited by inappropriate phase matching, the lower microscopic response for a given CEP (due to lower ionization degree for this CEP) can be compensated for with better phase matching for this CEP (also due to lower ionization degree). In the plasma plume this effect could be more important than for the HHG in gaseous media due to higher ionization. Such behavior of harmonics was confirmed earlier during calculations performed to study the "phase-matching gating" of the generation of an isolated attosecond pulse [38].

One can see in Figure 3.15 that almost all spectral peaks can be attributed either to broad harmonics with frequencies close to ones found for carbon plasma, or to transitions in In II and In II* described in Ref. [37]. Note that even the spectral peak near 23.3 eV was observed experimentally though not attributed to any transition. This agrees with the presentation of the XUV spectrum in the form:

$$\mu(\omega) = F(\omega)\mu_{nr}(\omega) \tag{3.10}$$

where $\mu_{nr}(\omega)$ is the spectrum generated in absence of resonances and the factor $F(\omega)$ describes the enhancement due to the resonance (equal to unity far from a resonance). In the case of a 30 fs laser pulse, $\mu_{nr}(\omega)$ is a comb of narrow harmonics and proximity to a resonance leads to the enhancement of generation of certain harmonic (or harmonics). In the case of 3.5 fs pulse, the spectrum $\mu_{nr}(\omega)$ is a quasi-continuum, and on top of it the features due to the $F(\omega)$ factor (thus due to resonances) can be seen. Thus the generation of a XUV quasi-continuum with the 3.5 fs pulses allows the observation of spectral features caused by different transitions in the single spectrum.

3.2.6 Discussion

It was already pointed out that the enhanced ≈ 20 eV emission did not coincide with neither the 11th nor the 13th harmonic wavelengths of the 770 nm, 3.5 fs driving laser radiation (as it is evident in Figure 3.9 in the comparison of the emission spectra of indium and carbon plasmas). For few-cycle driving pulses, plasma emissions at around 22, 23.4, and 24 eV are close to the spectral positions of intense In II $4d^{10}5s5p \rightarrow 4d^95s5p^2$ transitions [37]. Since these plasma emission lines do not coincide either with the central wavelength of broad 11th, 13th, and 15th harmonics, one can assume that the spectral wings of the harmonics, in resonance with the approximately above ion transitions, become enhanced due to resonance-induced increase of the nonlinear optical response of the plasma. In that case, propagation effects may play a decisive role for efficient phase matching between the driving and the harmonics fields in the vicinity of abovementioned spectral lines.

Among the factors responsible for the enhancement of individual harmonics, one can note the difference between the phase-matching conditions for different harmonics. The phase mismatch varies as the laser pulse propagates through the plasma plume due to further ionization of the nonlinear medium. For harmonics in the plateau region, the phase mismatch attributed to free electrons is 1–2 orders of magnitude larger than the mismatch due to neutrals and singly charged ions. However, under resonance conditions, when the frequency of a given harmonic becomes close to the frequency of inner-shell atomic transitions, the wavenumber variation for this harmonic caused by atoms may be significant and the free electron effect may be canceled. In these circumstances, it is possible to satisfy the optimal phase condition for a single harmonic, with the consequent increase of conversion efficiency for this only harmonic.

The mechanism for improvement of phase-matching conditions could be as follows. The refractive index of the plasma in the vicinity of a resonant transition (λ_r) can be considerably changed, thus allowing coincidence of the refractive indices of the plasma at the wavelengths of the driving laser and harmonic emission (Figure 3.16). The short-wavelength wing of the resonance showing anomalous dispersion can create the conditions to satisfy this phase-matching condition. One can see that in the area marked by the filled box near λ_{sh}, the condition $n_d \approx n_{sh}$ is fulfilled, where n_d and n_{sh} are the refractive indices of plasma at the wavelength of driving pulse (λ_d) and the short-wavelength side of the harmonic radiation (λ_{sh}), which central wavelength (λ_{ch}) stays far from the resonance line.

To analyze this issue in depth one has to know the bandwidths of these resonances, relative role of the nonlinear enhancement of harmonic emission, and the absorption properties of plasma in the vicinity of resonance, influence of plasma length on the enhancement factor of the part of harmonic, etc. Currently, no information exists on these parameters for indium plasma. Whether this effect affects (or not affects) the phase relations between the interacting waves depends on many factors. Thus propagation effects can play a decisive role in optimizing resonance-induced enhancement of harmonics, especially for very broadband pulses. The joint influence of the processes at the microscale related with mechanisms described in the four-step model and the macroscopic processes related with phase matching of the interacting waves

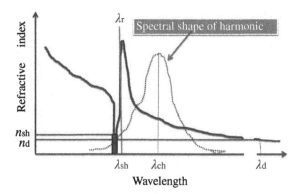

Figure 3.16 Conditions of phase matching between the waves of a broadband driving field and harmonics approaching the short-wavelength side of the resonant transition of the plasma medium. Solid curve shows a dispersion of the refractive index of plasma. Dotted curve represents the spectral shape of a nearby broadband harmonic. The filled area near λ_{sh} shows a possible enhancement range of the part of harmonic spectrum where the phase mismatch became suppressed. Here λ_r denotes the wavelength of the ion resonance, λ_d, λ_{ch}, and λ_{sh} the wavelengths of the driving radiation, the center of the harmonic, and the short-wavelength side of the harmonic respectively and n_d and n_{sh} the refractive indices of plasma at λ_d and λ_{sh}, respectively. *Source:* Ganeev et al. 2013 [29]. Reproduced with permission from American Physical Society.

can create the conditions for the generation of an intense emission possessing the same coherence properties as ordinary harmonics.

Another option to explain the observed In plasma emission lines upon excitation by few-cycle pulses could be related with a lasing effect involving ion transitions. One can assume that, analogously with the X-ray laser operation, excitation of low-ionized plasma by ultrashort pulses, increases the population of the discussed excited ion levels, causing stimulated emission at the corresponding wavelengths once population inversion is established between some ion levels. To support this assumption one has to analyze the $I_e(l \times g)$ dependences, where I_e is the intensity of emitted radiation, l is the length, and g is the gain of the medium, as well as define the saturation condition of the process. The difficulty in explaining the observed spectral peculiarities of the ablation plasma using this approach is related with the unknown values of the lifetimes of the involved excited states and of the ratios between the absorption and the gain of the indium plasma in this spectral range. Moreover, the polarization experiments do not support the lasing assumption, since the gain of the medium should not depend so decisively on the polarization properties of the driving pump laser, as it was observed in the discussed studies.

The important aspect of resonant plasma HHG studies is the temporal characterization of the enhanced harmonic. Measurements of both the femtosecond and attosecond structure of HHG pulses from tin ablation plasma were reported in Ref. [39]. The 17th harmonic of an ~800 nm laser was tuned into resonance with the $4d^{10}5s^25p\ {}^2P_{3/2} \rightarrow 4d^95s^25p^2\ ({}^1D)\ {}^2D_{5/2}$ transition in Sn II. Those studies have shown that the femtosecond envelope of this harmonic does not correspond to a slowly decaying plasma emission, but is indeed slightly shorter than the driving laser pulse. However, the relative phase of the harmonic, which governs the attosecond structure, is significantly perturbed by the resonance. In resonant conditions, the phase locking between the resonant and the neighboring orders is lost, i.e. their relative phase varies significantly within the harmonic spectral width.

At the same time, theoretical modeling of single enhanced harmonic of sub-4 fs pulse from Mn plasma reported in Ref. [8] suggests that this emission could constitute an isolated subfemtosecond pulse. Particularly, the calculations show that the observed 51 eV emission represents subfemtosecond XUV pulses or at least ~1 fs XUV pulses, for different values of CEP. Those studies have demonstrated the observation of a single, broadband (2.5 eV) 31st harmonic. The spectral bandwidth supported assumption about the subfemtosecond pulse duration of this radiation, though no temporal measurements were carried out in those studies.

The discussed measurements of the bandwidths of the enhanced emission from the indium plasma around 20 eV and above, in the case of excitation by 3.5 fs pulses, did not support the subfemtosecond pulse duration of these XUV pulses. Moreover, the divergence of this radiation exceeds the one reported for a resonant harmonic from manganese plasma [8]. The possible explanation of the difference in the spatio-spectral characteristics in these two cases (i.e. resonance emissions from the indium and manganese plasmas excited by few-cycle pulses) could be related with the different above-described mechanisms involved in the generation of enhanced emission from these two plasma samples.

3.3 Indium Plasma in the Single- and Two-Color Near-Infrared Fields: Enhancement of Tunable Harmonics

3.3.1 Description of Problem

As already mentioned, effects of approximately fixed pump laser wavelength in resonant harmonics have been reported long time ago [1] using Ti:sapphire lasers. In those studies, small tuning was accomplished either by using chirping technique within pulse bandwidth or by insignificant variation of master oscillator wavelength. In present studies, the tuning of pump laser wavelength was dramatically improved compared with abovementioned case, which allowed changing of the intensity ratios of various harmonics, particularly near 62 nm in the case of the HHG in the indium plasma, thus forming the conditions for resonance enhancement of different orders of harmonics near the resonances possessing large oscillator strength.

The unavailability of tuning of the wavelength of most frequently used Ti:sapphire lasers significantly restricts the probability of coincidence of the harmonic order and the transition between AIS and ground states possessing large value of oscillator strength. To facilitate the use of plasma harmonic concept for laser-ablation-induced HHG spectroscopy in the XUV range, one should use the tunable sources of laser radiation allowing the fine-tuning of driving pulses and corresponding harmonic wavelengths along the spectral ranges of strong ionic transitions.

In general, the question arises as to why similar resonant harmonics have not been observed with gas media? The tunable femtosecond lasers are now available, and so there should be no problem tuning the laser wavelength to a specific resonance, whether it is of a plasma or gas. The answer on this question is related with the basic principles of the role of resonances in the enhancement of nearby harmonics. In most cases, the harmonic spectra show the absence of enhancement of the harmonics near strong emission lines. In the meantime, previous studies have demonstrated that the resonance enhancement of nearby harmonic close to the resonance can be observed in the case of sufficient amount of excited species (i.e. singly or doubly charged ions). Thus the availability of resonance enhancement depends on the population of the appropriate energetic levels of ions.

The studies of high-order nonlinear processes through exploitation of intermediate resonances show that the proximity of the wavelengths of specific harmonic orders and the strong emission lines of ions does not necessarily lead to the growth of the yield of single harmonic. The nonlinear optical response of the medium during propagation of intense pulses includes, in some particular cases, the resonance-induced enhancement of specific nonlinear optical processes, the absorption of emitted radiation, and the involvement of collective macroprocesses, such as the phase matching between the interacting waves.

In this section, we analyze the studies demonstrating fine-tuning of harmonics in the vicinity of strong In II transition during HHG using the mixture of tunable near-infrared (NIR) source of ultrashort pulses and its second harmonic (H2) in the LPP on the surface of indium target. We also discuss the enhancement of those harmonics, and compare with single-color NIR pump. We also present the theoretical description of observed phenomena [40].

3.3.2 Experimental Arrangements for HHG in Indium Plasma Using Tunable NIR Pulses

Experimental setup consisted on Ti:sapphire laser, traveling-wave optical parametric amplifier (OPA) of white-light continuum, and HHG scheme using propagation of the amplified signal pulses from OPA through the extended LPP (Figure 3.17a). The mode-locked Ti:sapphire laser pumped by diode-pumped, continuous-wave laser was used as the source of 803 nm, 55 fs, 82 MHz, 450 mW pulses for injection in the pulsed Ti:sapphire regenerative amplifier with pulse stretcher and additional double-passed linear amplifier.

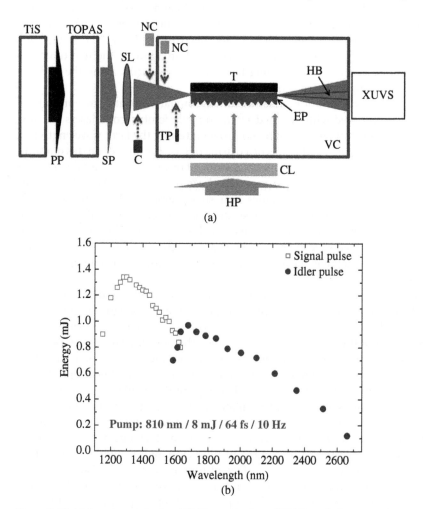

Figure 3.17 (a) Experimental setup. TiS, Ti:sapphire laser; TOPAS, optical parametric amplifier; PP, pump pulse for pumping the optical parametric amplifier; SP, amplified signal pulse from OPA; HP, heating picosecond pulse from Ti:sapphire laser; SL, spherical lens; CL, cylindrical lens; VC, vacuum chamber; T, target; EP, extended indium plasma; NC, nonlinear crystal (BBO); C, calcite; TP, thin glass plates; HB, harmonic beam; XUVS, extreme ultraviolet spectrometer. (b) Spectral tuning of mid-infrared signal and idler pulses. *Source*: Ganeev et al. 2016 [40]. Reproduced with permission from American Physical Society.

The output characteristics from this laser were as follows: central wavelength 810 nm, pulse duration 350 ps, pulse energy 5 mJ, 10 Hz pulse repetition rate. This radiation was further amplified up to 22 mJ in the home-made three-passed Ti:sapphire linear amplifier. Part of amplified radiation with pulse energy of 6 mJ was separated from a whole beam and used as a heating pulse for homogeneous extended plasma formation using the 200 mm focal length cylindrical lens installed in front of the extended indium target placed in the vacuum chamber. The indium plasma was analyzed as the medium for harmonic generation using the tunable source of ultrashort pulses. The 5 mm long sample of bulk indium was installed in the vacuum chamber for laser ablation. The intensity of the heating pulse on the target surface was varied up to 4×10^9 W cm^{-2}. The ablation sizes were 5×0.08 mm^2.

The remaining part of amplified radiation was delayed with regard to the heating pulse. After compression and pump of OPA, the signal or idler pulse from parametric amplifier propagated through the LPP 35 ns from the beginning of target irradiation by heating pulses. After propagation of compressor stage the output characteristics of Ti:sapphire laser were as follows: pulse energy 8 mJ, pulse duration 64 fs, 10 Hz pulse repetition rate, pulse bandwidth 17 nm, central wavelength 810 nm. This radiation pumped the OPA. Signal and idler pulses from OPA allowed tuning along the 1200–1600 nm and 1600–2600 nm ranges, respectively (Figure 3.17b).

In most of the discussed HHG experiments, the signal pulses from OPA were used, which were 1.5 times stronger than the idler pulses. These experiments were carried out using the 1 mJ, 70 fs pulses tunable in the range of 1250–1400 nm. This variation of driving pulse wavelength was sufficient for tuning the harmonics along various resonances of ionic species. The spectral bandwidth of tunable pulses was 45 nm. The intensity of the 1310 nm pulses focused by 400 mm focal length lens through 3 mm thick MgF$_2$ window of vacuum chamber into extended plasma was 2×10^{14} W cm^{-2}. These driving pulses were focused at a distance of ~150 μm above the target surface. The plasma and harmonic emissions were analyzed using an XUVS containing a cylindrical mirror and a 1200 grooves/mm flat field grating with variable line spacing. The spectrum was recorded on a microchannel plate detector with the phosphor screen, which was imaged onto a CCD camera.

The experiments were carried out using both the single-color and two-color pumps of LPP. The reasons for using the double beam configuration to pump the extended plasma are related to small energy of the driving NIR signal pulse (~1 mJ). The $I_H \propto \lambda^{-5}$ rule (I_H is the harmonic intensity and λ is the driving field wavelength) led to a significant decrease of harmonic yield in the case of longer wavelength sources compared with the 810 nm pump and did not allow the observation of strong harmonics from the single-color NIR (1310 nm) pulses. Because of this the second-harmonic (H2) generation of signal pulse was used to apply the two-color pump scheme (NIR + H2) for plasma HHG. The 0.5 mm thick BBO crystal was installed inside the vacuum chamber on the path of focused signal pulse (Figure 3.17a). The BBO was cut at $\theta = 21°$ and adjusted to the phase matching for 1310 nm wavelength. At these conditions, it was not necessary to tune the crystal for each OPA wavelength due to small dispersion in the 1240–1400 nm range. The conversion efficiency of 655 nm pulses was ~27%. The spectral bandwidth of H2 pulse was 22 nm. The two orthogonally polarized pump pulses were sufficiently overlapped both temporally and spatially in the extended plasma, which led to a significant enhancement of odd harmonics, as well as generation of the even harmonics of similar intensity as the odd ones.

The variation of the relative phase between fundamental and second harmonic waves was analyzed by insertion of the 0.15 mm thick glass plates on the path of these beams. The plates were introduced between the BBO crystal and LPP, as shown in Figure 3.17a. The group velocity dispersion for the driving and second harmonic waves propagating through such plates makes possible the variation of both the relative phase of two pumps and the delay between the envelopes of these pulses.

The two-color pump of plasma using different polarizations of interacting waves was analyzed at the conditions when the BBO crystal was installed outside the vacuum chamber. The 2 mm thick calcite plate was inserted in front of the BBO crystal at the conditions when the NIR pulses generate H2 in the 0.7 mm thick BBO placed between the focusing lens and input window of the vacuum chamber. The rotation of calcite affected only driving pulse by changing the polarization from linear to elliptical and to circular. In that case, the two-color pump consisted on the linearly polarized H2 wave and circularly (or elliptically) polarized fundamental NIR pulses.

3.3.3 Experimental Studies of the Resonance Enhancement of NIR-Induced Harmonics in the Indium Plasma

The harmonics generated in indium plasma using NIR pulses (1 mJ, 1330 nm) were significantly weaker compared with the case of 8 mJ, 810 nm pump due to the abovementioned wavelength-dependent yield of harmonics. The comparison of these two pumps at similar energies of pulses (1 mJ) showed the sevenfold decrease of harmonic yield in the plateau region using 1330 nm pulses compared with the 810 nm pulses (Figure 3.18a), while the theoretical prediction of this ratio was $I_{1330nm}/I_{810nm} = (810 \text{ nm}/1330 \text{ nm})^{-5} \approx 12$. The observed harmonic cutoff in the case of NIR pulses was approximately similar to the one from the 810 nm pump (27 (H29) and 26 eV (H17), respectively), contrary to the theoretically expected extension of the cutoff energy for longer wavelength pump ($E_{cutoff} \propto \lambda^2$), due to very small conversion efficiency in the case of NIR pulses, which did not allow the observation of harmonics below the 40 nm spectral region. The weak idler pulses from OPA, with the energy varying in

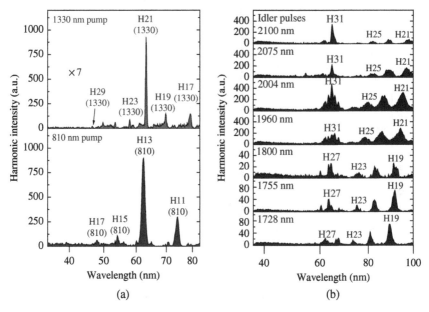

Figure 3.18 (a) Harmonic spectra using 1330 (upper panel) and 810 nm (bottom panel) pulses. Upper panel magnified with a factor of 7 for better comparison of the harmonic spectra generated using different pumps. (b) Spectra of tunable harmonics in the case of single-color pump of indium plasma using idler (1728 – 2100 nm) pulses of OPA. *Source:* Ganeev et al. 2016 [40]. Reproduced with permission from American Physical Society.

the range of 0.3–0.5 mJ, were also used for harmonic generation in plasma. Even those small energies of driving longer wavelength pulses were sufficient to observe the resonance enhancement of different harmonic orders near the AIS of indium in the case of the idler pulses tunable along the 1730–2100 nm range (Figure 3.18b).

These studies showed that the use of tunable NIR pulses caused generation of mostly single enhanced harmonic (H21) close to the AIS and a few weak odd harmonics in the longer wavelength range of XUV (Figure 3.19a). The insertion of the BBO crystal into the path of the focused driving beam drastically modified the harmonic spectra. Extension of the observed harmonic cutoff, significant growth of the yield of odd harmonics, comparable harmonic intensities for the odd and even orders along the whole range of generation, tuning of harmonics allowing the optimization of resonance-induced single harmonic generation, as well as a few neighboring orders close to AIS of indium, were among the advanced features of these two-color experiments (Figure 3.19b). These studies showed that the advantage of the resonance-induced enhancement of harmonics using NIR + H2 pulses is the opportunity of fine-tuning of this high-order nonlinear optical process for spectral enhancement of the

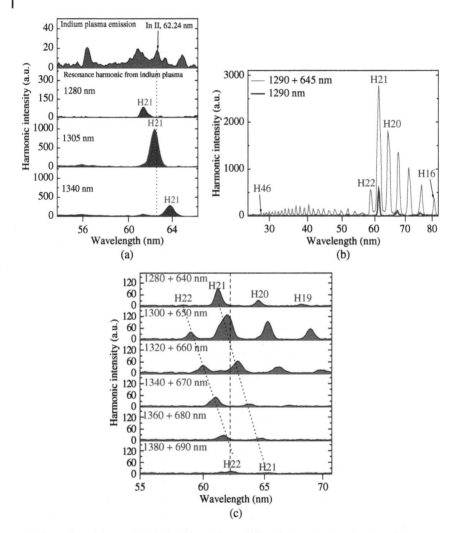

Figure 3.19 (a) Tuning of H21 along the In II resonance. Upper panel shows the indium plasma emission spectrum. Three bottom panels show resonantly enhanced harmonic using single-color pump (1280, 1305, and 1340 nm, respectively). Dotted line shows the position of the resonance transition of In II. (b) Comparative harmonic spectra from indium plasma using single-color (1290 nm; thick curve) and two-color (1290 and 645 nm; thin curve) pumps. (c) Relative variations of H21 and H22 yields using different wavelengths of NIR and H2 pump radiation. Dotted lines show the tuning of H21 and H22. Dashed line shows the position of resonance transition. *Source*: Ganeev et al. 2016 [40]. Reproduced with permission from American Physical Society.

harmonic yield. Another advantage is closely related with the analysis of the oscillator strengths of some ionic transitions using the HHG approach.

Various schemes of the two-color pump for HHG were introduced in Refs. [41–48]. As underlined in Ref. [48], a strong harmonic generation in the case of two-color pump is possible due to formation of a quasi-linear field, selection of a short quantum path component, which has a denser electron wave packet, and higher ionization rate compared with the single-color pump. The orthogonally polarized second field also participates in the modification of the trajectory of accelerated electron from being two-dimensional to three-dimensional that may lead to removal of the medium symmetry. With suitable control of the relative phase between the fundamental and second-harmonic pumps, the latter field enhances the short path contribution while diminishing other electron paths, resulting in a clean spectrum of harmonics.

The use of tunable broadband NIR radiation and its second harmonic allowed the analysis of the enhancement of the groups of harmonics close to the resonance possessing strong oscillator strength (Figure 3.19c). One can see the growth of harmonic yield in the 62.3 nm region of AIS for different groups of harmonics during tuning of the wavelength of driving pulses. These studies using tunable NIR pulses and their second harmonics showed a fine-tuning of the resonance-enhanced harmonic and change of the order of this harmonic. The tunability is practically unlimited since the tuning of fundamental wavelength (Figure 3.17b) allowed shifting the wavelength of high-order harmonic over the wavelength of neighboring harmonic, which means the overlap of a full octave.

Two upper panels of Figure 3.20a show the harmonic spectra in the case of using two different schemes when the BBO crystal was installed either inside or outside the vacuum chamber. In the latter case, the linearly polarized 1330 nm pulses generated H2 in the 0.7 mm thick BBO placed outside the vacuum chamber. In the case shown in the bottom panel of Figure 3.20a, the polarization of driving 1330 nm pulses was changed from linear to circular by inserting the calcite plate in front of BBO crystal. In that case, the pump radiation after propagation of BBO crystal consisted on the circularly polarized 1330 nm pump and linearly polarized 665 nm pump. This apparently led to generation of high-order harmonics only from the 665 nm pump, which is seen in the bottom graph of this figure, showing the odd harmonics of H2 radiation.

In the two-color pump HHG experiments in gases, the variation of relative phase between fundamental and second harmonic waves of 30 fs pulses allowed observing the threefold beatings between the long- and short-trajectory induced harmonics [49]. The use of the focusing optics inside the XUV spectrometer in discussed experiments did not allow distinguishing the influence of the short and long trajectories of accelerated electrons on the divergence of harmonics. However, the difference in the

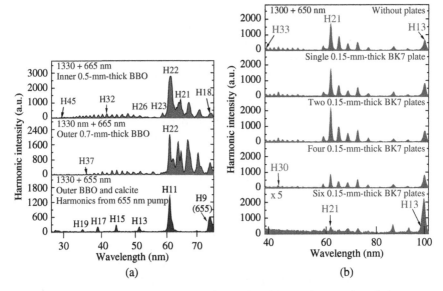

Figure 3.20 (a) Harmonic spectra generated using the BBO crystal inserted inside (upper panel) and outside (middle panel) the vacuum chamber. Bottom panel shows generation of the odd harmonics of 655 nm radiation once the calcite plate was installed in front of BBO crystal, which led to variation of the polarization of NIR pulse from linear to circular. (b) Harmonic spectra for variable phase difference between NIR and H2 pulses. One can see a gradual change of the relative intensities of H13 and resonance-enhanced H21 using different number of inserted 0.15 mm thick BK7 plates. *Source*: Ganeev et al. 2016 [40]. Reproduced with permission from American Physical Society.

plasma harmonic spectra was observed when the thin (0.15 mm) glass plates were introduced between BBO crystal and plasma (Figure 3.17a). The examples of both decrease of the two-color field in the plasma area and decrease of the resonance enhancement of H21 are presented in Figure 3.20b. The thin glass plates were inserted after BBO crystal to analyze the variation of the relative phase between two pumps and to compare the change of relative intensities of "resonant" (H21) and "nonresonant" (H13) harmonic yields. One can see a significant departure from the large ratio H21/H13 in the case of the absence of the glass plates, which are actually the relative phase modulators (upper panel of Figure 3.20b), toward the low ratio of these harmonics in the case of propagation through six 0.15 mm thick plates (bottom panel). Thus the variation of relative phase between pumps may diminish the role of AIS in the single harmonic enhancement. The decrease of pump intensity (due to the growth of Fresnel losses during reflection from a few plates) may also cause a decrease of the overall nonlinear response.

The relation between the addition of single 0.15 mm thick BK7 plate and relative phase shift between fundamental and second harmonic waves depends on the group velocity dispersion in the glass. The group velocity dispersion leads to a change of the relative phase between waves, and, in the case of thick samples, may lead to the temporal walk-off of two pulses. Particularly, due to this effect in the BK7, the 650 nm pulse (2ω) was delayed in the 0.15 mm thick glass by $\Delta_{BK7} = d[(n^o_\omega)_{group}/c - (n^e_{2\omega})_{group}/c] \approx 1.3$ fs with respect to the 1300 nm pulse (ω) due to $n^o_\omega < n^e_{2\omega}$ in this positive optical element. Here Δ_{BK7} is the delay between two pulses after leaving the thin glass, d is the glass thickness, $c/(n^o_\omega)_{group}$ and $c/(n^e_{2\omega})_{group}$ are the group velocities of the ω and 2ω waves in the BK7, c is the light velocity, and n^o_ω and $n^e_{2\omega}$ are the refractive indices of BK7 at the wavelengths of the ω and 2ω pumps. In the case of NIR pulses, the group velocity dispersion between fundamental and H2 waves was notably smaller compared with the case of Ti:sapphire laser radiation. One can also mention the delay between two pulses in the output of BBO crystal. In the case of 0.5 mm thick BBO, the calculated walk-off between 1300 and 650 nm waves was 24 fs. Note that $t_{2\omega}$ at the output of the 0.5 mm long BBO crystal was estimated to be 76 fs, while the 1300 nm pulse duration was 70 fs. The additional plates of BK7 led to the change of the relative phase and some decrease of the delay between the envelopes of fundamental and second harmonic waves. The relative phase, ϕ, is proportional to the delay between envelopes, ΔT, and the carrier frequency, v_0 ($\phi \approx 2\pi \Delta T v_0$) [50]. Thus the single BK7 plate provides ~0.1π variation of the relative phase.

In discussed experiments, the HHG occurred in quite a long medium (5 mm). In the following, we discuss the role of the propagation effect in this kind of experiment. To achieve higher HHG conversion efficiency, the length of the medium might be increased, provided that the phase mismatch between the laser field and the harmonic radiation maintains low [51]. For the medium lengths where the reabsorption can be neglected and for the optimum phase-matching conditions, the harmonic intensity increases as the square of medium length. However, once the medium length exceeds the coherence length, the harmonic intensity shows the oscillations due to phase mismatch [52]. To analyze this propagation process in the LPP, one has to carefully define the best conditions of plasma HHG in the extended medium, while taking into account the peculiar properties of the indium used for laser ablation.

Initially, the dependence of the harmonic yield on the length of indium plasma was analyzed. It was found that the slope of this curve was close to 2 at the appropriately ablated plasma until maximum length is used in these experiments (5 mm). The following studies were carried out at the conditions of this "optimal" LPP. Note that application of stronger ablation of indium target led to decrease of the harmonic yield due to phase mismatch caused by large amount of free electrons. In that case, the coherent length of high-order harmonics became shorter than the length of the homogeneous extended plasma. The abovementioned dependence was deviated from the slope of 2. To overcome this propagation effect, one can use the quasi-phase-matching concept earlier demonstrated during HHG studies in the LPP [53] based on the formation of the group of short jets instead of the imperforated extended plasma plume.

3.3.4 Theory of the Process

In the following, we discuss the results of the numerical simulations of two-color NIR-driven HHG in indium ion. The three-dimensional TDSE was used to describe the interaction between the two-color NIR field and the indium ion (atomic units are used throughout, unless otherwise stated):

$$i\frac{\partial \Psi(\mathbf{r},t)}{\partial t} = \left[-\frac{1}{2}\nabla^2 + V(\mathbf{r}) - E(\mathbf{r},t)\cdot\mathbf{r}\right]\Psi(\mathbf{r},t) \tag{3.11}$$

where $V(\mathbf{r})$ is the potential of the plasma system. Here the model potential introduced in Ref. [13] was adopted to reproduce the properties of the indium ion:

$$V(\mathbf{r}) = -\frac{2}{\sqrt{a^2 + r^2}} + b\exp\left[-\left(\frac{\mathbf{r}-c}{d}\right)^2\right] \tag{3.12}$$

This potential can support the metastable state by a potential barrier, which corresponds to the AIS of the ion. The parameters a, b, c, and d are chosen to be 0.65, 1.0, 4.0, and 1.6, respectively. There is one dominant transition from AIS to ground state ($4d^{10}5s^2\,^1S_0 \rightarrow 4d^95s^25p^1\,P_1$) in indium ion using this potential [29]. The two-color driving field is synthesized by an x-polarized NIR fundamental field and a y-polarized second harmonic assistant field. The driving field is given by:

$$E(r,t) = \begin{cases} E_0\sin^2(\pi t/T)\cos(\omega_0 t)x \\ \quad +\sqrt{0.25}E_0\sin^2(\pi t/T)\cos(2\omega_0 t + \phi)y, & 0 < t < T \\ 0, & t > T \end{cases}$$

$$\tag{3.13}$$

where E_0, ω_0 are the amplitude and central frequency of the 1300 nm laser field. The CEPs of the 1300 nm field and the 650 nm field are set as 0, ϕ is the relative phase between the two fields, T is the pulse duration and is given by $T = 10T_0$, where T_0 is the optical cycle of the 1300 nm pulse. The peak intensities of the 1300 nm pulse and the 650 nm pulse were assumed to be 2×10^{14} and 5×10^{13} W cm^{-2}, respectively. Eq. (3.11) can be numerically solved by using the split-operator method [54–56]. Once the evolution of the electron wave function $\Psi(\mathbf{r},t)$ is found, the time-dependent dipole acceleration $a(t)$ can be calculated with the Ehrenfest's theorem [57]

$$a(t) = -\left\langle \Psi(\mathbf{r},t)\left|\frac{\partial[V(\mathbf{r}) - E(\mathbf{r},t)\cdot\mathbf{r}]}{\partial r}\right|\Psi(\mathbf{r},t)\right\rangle \tag{3.14}$$

The harmonic spectrum $a(\omega)$ is then obtained from the Fourier transform of the dipole acceleration, which is given by

$$a(\omega) = \int_0^T a(t) \exp(-i\omega t) \mathrm{d}t \qquad (3.15)$$

The harmonic spectrum with the two-color NIR field is presented by the thick curve in Figure 3.21. For comparison, the harmonic spectrum with the fundamental field alone is also presented by the thin curve. As shown in Figure 3.21, the resonant harmonic around 62.2 nm is much more intense than other harmonics. This corresponds to the transition between the ground state and AIS in indium ion. Moreover, adding a second harmonic as the assistant field significantly modifies the spectrum. One can clearly see the enhancement of resonant harmonic yield (the 21st harmonic), the extension of harmonic cutoff, and the generation of even harmonics (the 20th and 22nd harmonics) in the two-color field compared to the single-color case. Figure 3.22 presents the two-color harmonic spectra with different wavelengths of the NIR fields. Other laser parameters are the same with that in Figure 3.21. It is shown that when tuning the wavelength of the driving pulses, different groups of harmonics around the region of AIS can be selectively enhanced. As the wavelength of the driving pulse varies from 1280 to 1380 nm, the tuning of the wavelength of the enhanced radiation is clearly observed, and the resonant harmonic changes from 21st to 22nd order. The theoretical predictions here agree well with the features of two-color resonant HHG observed in experiment as shown in Figure 3.19b,c.

In Figure 3.23a, the variation of intensity of the resonant XUV emission (from $20\omega_0$ to $22\omega_0$) was presented as a function of time in the case of single-color (thick solid curve) and two-color (thin solid curve) pumps. Here ω_0 is the frequency of 1300 nm fundamental pulse. One can see that the XUV radiation is delayed with respect to the laser pulse envelope. Almost all of the resonant

Figure 3.21 Simulated harmonic spectra from indium ion with single-color (thin line) and two-color (thick line) fields. The relative phase in the two-color field is 0. *Source*: Ganeev et al. 2016 [40]. Reproduced with permission from American Physical Society.

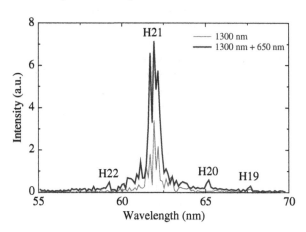

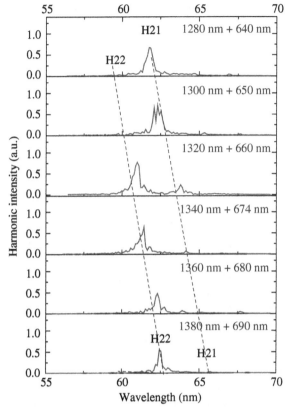

Figure 3.22 Simulated variations of high-order harmonic yield using different wavelengths of NIR laser fields. Dashed lines show the tuning of the 21st and 22nd orders of harmonics. *Source*: Ganeev et al. 2016 [40]. Reproduced with permission from American Physical Society.

harmonics are emitted after five optical cycles. Besides, there are still XUV emissions after the laser pulse has turned off (from 10 to 20 optical cycles). Such behavior is related to the accumulation of the AIS population in ion system. Compared to the single-color case, the intensity of the resonant harmonic is significantly increased from 12 to 16 optical cycles in the two-color field. In addition, the harmonic yield in the two-color field shows a more stable distribution around the maximum. In the two-color field, the ionization rate of the ground state is increased by the assistant field. Then more ionized and laser-accelerated electrons can be captured into the AIS. Therefore the transition probability between the AIS and ground state is increased and the intensity of the resonant harmonic is enhanced.

It is worth mentioning that the delay time of the two-color resonant harmonics depends on the parameters of two-color field. For example, in Figure 3.23a one can see the XUV emission in the two-color field is more delayed than that in the single-color case with the relative phase $\phi = 0$. But if ϕ became varied to 0.5π, the maximum emission of the two-color resonant harmonics occurs at 7.5 optical cycle, while the maximum emission in the single-color field is at

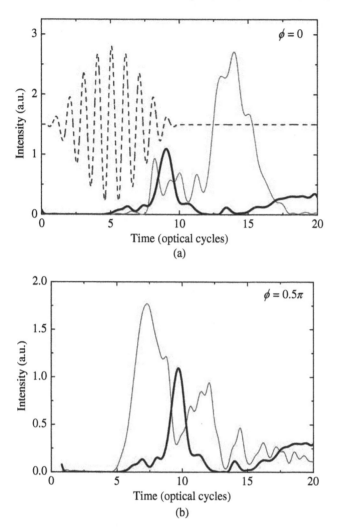

Figure 3.23 Temporal profiles of resonant emission (from H20 to H22) in the case of single-color (thick solid curve) and two-color (thin solid curve) pumps. The relative phases in Figure 3.23a,b are 0 and 0.5π, respectively. The dashed line shows the electric field of 1300 nm pulse. *Source*: Ganeev et al. 2016 [40]. Reproduced with permission from American Physical Society.

9 optical cycle (see Figure 3.23b). This indicates that the delay effect of the resonant emission in the two-color field is less obvious than that in the single-color field with $\phi = 0.5\pi$. The variation of the temporal profiles of resonant harmonic emission in the two-color field is related to the modulation of AIS dynamics by the control field.

The influence of the relative phase on the harmonic yield was also analyzed. The variations of harmonic yield for the resonant 21st and nonresonant 19th harmonics are presented by the solid and dashed curves in Figure 3.24a. The yield has been normalized by the case of $\phi = 0$ for each curve in order to make a clear comparison. Figure 3.24b shows variation of the relative intensities of 21st and 19th harmonics (H21/H19). In Figure 3.24c the comparative harmonic

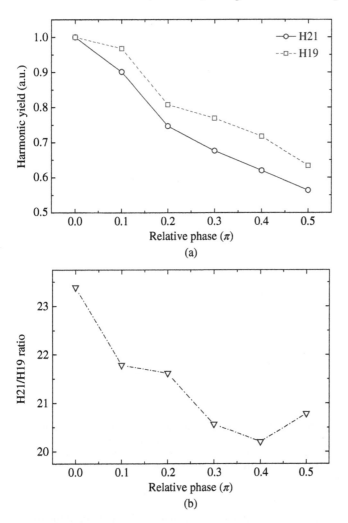

Figure 3.24 (a) Dependences of H19 and H21 yields on the relative phase between the 1300 nm and its second harmonic pulses. (b) Variation of the relative intensities of the 21st and 19th harmonic with relative phase. (c) Simulated harmonic spectra with different relative phases of 0 and 0.5π. *Source*: Ganeev et al. 2016 [40]. Reproduced with permission from American Physical Society.

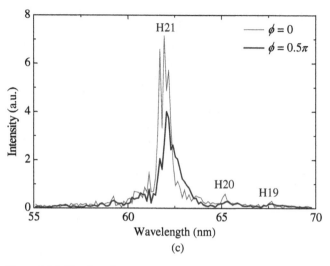

Figure 3.24 *(Continued)*

spectra with different relative phases of 0 and 0.5π are presented. One can see that by varying the relative phase of the two-color field one can influence the yield of the resonant and nonresonant harmonics. It was found that the change of the harmonic yield in the resonance region is more significant than that in the nonresonance region. When ϕ varies from 0 to 0.5π, the yield of the 21st harmonic decreases faster than that of the 19th harmonic. Correspondingly, the ratio of relative intensities for H21/H19 changes from large values to smaller ones. This indicates that the relative phase of the two-color field may play a role in single harmonic enhancement during resonance-affected HHG. Besides, it was found that the variation of the harmonic spectrum at different ϕ is related to the pulse envelope used in the calculations. When using the trapezoid envelope, the influence of ϕ on the harmonic spectrum is smaller than that using "$\sin^2$" envelope.

3.3.5 Discussion and Comparison of Theory and Experiment

The main message of this work is the first demonstration of the use of tunable sources for analysis of the influence of strong resonance on the harmonic efficiency for different orders, contrary to previous applications of the sources with fixed wavelengths, which do not allow the nonlinear spectroscopy of ionic media. The effect of the resonance on the HHG, though not new, is still debated in literature (see, for example [58]). Note that no resonance enhancement of harmonics was reported using gaseous media, while harmonic generation in the plasmas is a proven road for the analysis of the spectroscopic features of various ionic species using the method of high-order nonlinear spectroscopy, needless to say that this method allows achieving high fluencies of XUV photons. In the following, we analyze the difference with previous studies of the influence of resonances on the HHG. The application of multicycle pulses from OPA for gas harmonics has shown the perspectives for generation of attosecond pulses. Thus, the proposed approach could be considered as a road for overcoming the restriction in generation of ultrashort pulses. Plasma harmonics, being proven as an alternative to gas ones, already showed the advantages, which did not achieve by the latter methods. The use of OPA allows achieving further steps in understanding of the peculiarities of resonance enhancement.

The application of two-color pump makes available a significant enhancement of harmonic yield compared with single-color pump. This claim is not a new one, since it has been proven by many researchers. The novelty is the use of this feature for the analysis of resonant HHG in the NIR field, which cannot be realized using single-color scheme due to λ^{-5} rule. The use of only fundamental pump is almost impossible and unpractical for the tasks of this research, since no strong harmonics were achieved, or they were extremely weak. The use of second wave is not a trick but rather a necessity to study the resonant high-order harmonic generation using the mid-infrared (MIR) pump.

Moreover, different changes were introduced in the properties of these two pumps, which allowed observing the variable response of the medium. One of these changes was the modification of the relative phase between two pumps. This modification resulted in the significant change in the relative intensities of resonance-enhanced harmonic and other ordinary harmonics (compare upper and bottom panels of Figure 3.20b). This finding clearly indicates the stronger influence of the relative phase between pumps on the resonance-enhanced harmonics rather than "non-resonant" harmonics. The calculations of the role of relative phase showed that it was less strong than the experiment shows. The importance of these data underlined by the fact that they show the relative influence of the phase on the "resonant" and "non-resonant" harmonics, since the additional impeding factors (optical losses, imperfect overlap of pulses, etc.) equally influence both harmonics.

The variation of the temporal profiles of resonant emission in the two-color field is related to the modulation of resonant HHG process by the control field. However, it is very difficult to fully analysis the behavior of AIS in the exterior laser field (variation of energy, width, decay time, etc.), because this information is entangled in the evolution of the wave function in the discussed TDSE model. The used theoretical model (i.e. numerical solving of the 3D TDSE) allowed to reproduce the experimental observations and characterize the new spectral and temporal characteristics of the two-color resonant HHG compared to the single-color case. Those results have indicated that the dynamics of the AIS can be turned by the two-color field, therefore the temporal profile of the resonant emission in the two-color field depends on the relative phase.

Since the role of micro- and macroprocesses in resonance enhancement is still debated, this result points out the stronger influence of the former processes. The interpretation of these results probably requires additional studies. Notice that the study [58] also points out the same peculiarity, while analyzing the variations of the coherence length of these harmonics. Thus the analysis of resonance processes by various means may offer different options in definition of the relative influence of single particle response and collective response of the medium on the harmonic yield.

The HHG efficiency in the indium LPP was optimized with respect to various parameters. In particular, small delay between 350 ps heating pulses and 70 fs driving pulses (<5 ns) did not allow the observation of harmonics at the used distance between the target and the axis of driving beam propagation. The harmonic generation efficiency abruptly increased once the delay exceeded 5 ns. In the case of indium plasma, the maximal harmonic yield was observed at 35 ns delay. At longer delays, for a fixed distance between the target and the laser beam, the harmonic yield started gradually decreasing until the entire disappearance at ~150 ns. However, it was possible optimizing the conversion efficiency at the delays larger than 35 ns by increasing the distance between the target and the laser beam. In discussed studies [40], the maximum efficiency of HHG for a 35 ns delay was achieved when the distance between the focal region of the driving beam and the target surface was ~150 μm.

The optimal delay depends on the target material, particularly on its atomic number (Z). Different targets were analyzed to reveal the optimal plasma medium at a certain delay between the pulses, corresponding to the maximum conversion of femtosecond radiation into the harmonics. It was shown that, at relatively short delays, targets with smaller Z values provide higher conversion efficiency in comparison with heavy targets due to larger velocities of the former particles. These measurements were performed at 20 ns delay between the ablated and converted pulses. One might expect this value to be optimal for light ions and atoms because, for heavier ions such as indium ($Z = 49$, atomic weight 115), the time of flight from the target surface to the region of femtosecond radiation propagation exceeds the delay between the pulses because of the lower velocities of In ions in comparison with lighter species.

In the following, we briefly discuss the plasma formation above the indium target surface. This process cannot be explained by simple heating of the target surface, its successive melting and evaporation, and the runaway of particles with thermodynamic velocities. These relations are valid for the fairly slow processes induced by long pulses, where the atomic velocity at a target heating temperature of 1000 K is about 7×10^2 m s^{-1}. At these conditions, atoms and ions move by only 15 μm from the target surface during 20 ns. On the assumption that laser plasma formation is determined in this case by such a slow process, the generation of harmonics would be impossible because the radiation to be converted passes at a distance of 150 μm above the target surface. In this case, the particles would reach the interaction region only 200 ns later. At the same time, efficient generation of harmonics (e.g. in indium plasma) started to be observable even at a delay of 20 ns between the pulses and barely occurs at the delays above 150 ns. This also holds true for other plasma formations.

This inconsistency between the thermal model of propagation of evaporated material and the observed efficient HHG in LPP at small delays between the ablating and driving pulses is explained within another model of plasma formation, specifically, the plasma explosion during target ablation by short pulses. The dynamics of plasma front propagation during laser ablation by short pulses has been analyzed in a number of studies (see, e.g. [59, 60] and references therein). A numerical analysis of plasma formation from a target irradiated with a single laser pulse was reported in monograph [59]. The velocities of plasma front in accordance with plasma explosion model were in the range of 1×10^4–1×10^5 m s^{-1}.

In experiment, the dynamics of plasma origin and propagation can be analyzed using the shadowgraph technique. The spatial characteristics of the laser plasma formed by short pulses under similar conditions on the surfaces of lighter (boron, $Z = 5$) and heavier (manganese, $Z = 25$) targets were reported in Ref. [61]. In the case of heavier target (Mn), the plasma front propagated at a velocity of 6×10^4 m s^{-1}. The plasma front passed a distance of 130 μm after a few nanoseconds, rather than several hundreds of nanoseconds, in accordance with the above-described thermal model of plasma expansion. One can assume that, in the case of indium ions possessing twice heavier mass than manganese, the slower movement causes later appearance of the plasma cloud in the area of laser beam propagation. Obviously, the formation of optimal plasma is not limited by the occurrence of plasma front in the propagation region of radiation to be converted. To this end, a specific time is necessary for the concentration of particles responsible for harmonic generation to reach a certain value.

The application of fundamental waves together with assistant field dramatically changes the efficiency of HHG and allows the observation of the resonantly enhanced harmonic alongside with the enhanced nearby harmonics. The influence of assistant (H2) field on HHG did not crucially depend on the relative intensities of two pumps. In discussed studies [40], the maximal efficiency of H2 generation using the 1310 nm pulses (27%) means the ratio of the H2 pulse energy and a whole 1310 nm pulse energy (measured before the propagation of BBO). This efficiency shows that the ratio of assistant (655 nm) and driving (1310 nm) pulse energies inside the plasma was approximately 1 : 3. The theoretical calculations took into account different ratios of two interacting pulses. The variation of this ratio between 1 : 5 and 1 : 3 did not change definitely the conclusions defined from the calculations of the used theoretical model. In discussed studies, we analyzed the calculations of harmonic spectra using the 1 : 4 ratio. The role of second field in the used ratio of pulse energies was similar to the influence of this assistant field at rather smaller ratios. Particularly, in the case of using the 1 : 5 and 1 : 20 ratios, similar strong enhancement of the odd harmonics, as well as equal odd and even harmonic intensities were achieved in the previous studies using the 800 nm class lasers [62, 63].

3.4 Resonance Enhancement of Harmonics in Laser-Produced Zn II and Zn III Containing Plasmas Using Tunable Near-Infrared Pulses

The change of excitation of matter may cause the appearance of some additional excited states, which can enhance the neighboring harmonics as well. This variation of plasma properties makes it difficult to analyze the complex behavior of harmonic generation in LPP. Among such species, the zinc plasma can be considered as an interesting subject of studies, since it allows the variation of conditions of excitation of various AIS depending on the conditions of ablation of the bulk target. The variation of target excitation allows the analysis of the nonlinear response of such plasma by comparing the enhancement of different harmonics, provided they coincide or stay close to those resonances. In the following, we discuss the fine-tuning of harmonics in the vicinity of neutral and ionic transitions of zinc during HHG using NIR source of ultrashort pulses and its second harmonic propagating through the LPP. We also analyze the enhancement of those harmonics depending on the plasma-formation conditions [64].

Experimental setup was similar to the one described in Section 3.3.2. Zinc plasma was analyzed as the medium for harmonic generation. The 5 mm long zinc plate was installed in the vacuum chamber for laser ablation. Previously, this plasma has shown some attractive features, particularly the single harmonic enhancement using few-cycle pulses [65] and a strong low-order harmonic yield in the case of multicycle pulses [66]. In both cases, the fixed wavelength of Ti:sapphire laser was used for the excitation of harmonics in the Zn plasma.

3.4.1 Single- and Two-Color Pumps of Zinc Plasma

The harmonic yield using NIR pulses (1 mJ, 1300 nm) was significantly weaker compared with the 8 mJ, 806 nm pump due to the $I_H \propto \lambda^{-5}$ rule. The comparison of these two pumps at similar energies of pulses (1 mJ) showed the eightfold decrease of harmonic yield from zinc plasma in the case of 1320 nm pulses compared with the 806 nm pulses (Figure 3.25a), close to the theoretical prediction of this ratio (11.8). The harmonic cutoff in the case of NIR pulses was also lower compared with the 806 nm pulses, contrary to the expectations in the extension of cutoff for the longer wavelength pump ($E_{cut-off} \propto \lambda^2$), due to very small

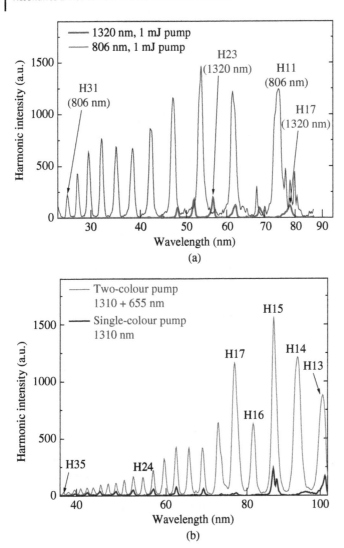

Figure 3.25 (a) Harmonic spectra from zinc plasma using 806 nm (thin curve) and 1320 nm (thick curve) pulses measured at similar conditions. (b) Comparative spectra from Zn plasma in the case of single-color (thick curve) and two-color (thin curve) pumps at similar conditions of experiment. *Source*: Ganeev et al. 2016 [64]. Reproduced with permission from © IOP Publishing.

conversion efficiency, which did not allow the observation of harmonics below the 40 nm spectral region.

Thus, the use of 1 mJ NIR pulses for plasma harmonic studies caused generation of weak odd harmonics in the longer wavelength range of XUV. Because of this the two-color pump of plasma was used. The insertion of the BBO crystal on the path of focused driving beam drastically modified the harmonic spectrum. The extension of harmonic cutoff compared with single-color pump 6–10 times growth of the yield of odd harmonics, comparable harmonic intensities for the odd and even orders along the whole range of generation, and tuning of harmonics allowing the optimization of resonance-induced single harmonic generation were among the advanced features of those experiments (Figure 3.25b). Most important feature of discussed studies was the availability in the analysis of the influence of excited states on the harmonic distribution at different excitation conditions of ablating target. In the following, we analyze the results of harmonic spectra modification using tunable driving two-color (NIR + H2) orthogonally polarized pulses leading to the observation of resonance enhancement of some harmonic orders caused by different species of plasma.

As mentioned in previous subsections, the advantage of the studies of resonance-induced enhancement of harmonics using tunable NIR + H2 pulses is the opportunity of fine-tuning of this high-order nonlinear optical process for spectral enhancement of the harmonic yield. Another advantage is closely related with the analysis of the oscillator strengths of some ionic transitions of metals and semiconductors using the HHG approach. Thus this approach could be considered as a new method of nonlinear spectroscopy. Here we discuss some examples of resonance enhancement of the harmonics of NIR radiation using the Zn target ablated at different heating pulse fluencies. The use of tunable broadband NIR radiation allowed the analysis of the enhancement of the groups of harmonics close to the resonances possessing strong oscillator strengths.

3.4.2 Modification of Harmonic Spectra at Excitation of Neutrals and Doubly Charged Ions of Zn

Different fluencies of heating pulses on the Zn target allowed formation of the LPP contained various species (i.e. excited neutrals, singly and doubly charged ions, and electrons). The goal of discussed studies was to analyze the dynamics of harmonic spectra at weak, medium, and strong ablation of zinc by 350 ps pulses. These terms mean the conditions of plasma, when harmonics are not canceled by a large amount of free electrons leading to phase mismatch but rather influenced by the presence of different species in the plasma. Plasma excitation was controlled by analyzing the emission spectra in the visible and XUV ranges. At weak ablation of target (i.e. at a fluency of $0.6 \, \mathrm{J \, cm^{-2}}$), no XUV emission lines were seen in those spectra. At these conditions, a set of harmonic spectra was measured using the tuning of NIR pulses with a step of 30 nm. The HHG spectra demonstrated the featureless pattern when each next (odd and even) order gradually decreased starting from longest observable harmonic (H13) up to the H24 (Figure 3.26a), while tuning the NIR pulses between 1280 and 1460 nm. Thus weak excitation of Zn did not lead to the formation of the conditions for the resonance-induced enhancement of single harmonic.

The growth of fluency (0.8 J cm^{-2}) led to the appearance of the emission of excited singly charged zinc ions. This plasma demonstrated emission of the group of Zn II lines at the wavelengths of 75.5, 76.7, and 77.9 nm, as well as 88.1 and 89.3 nm. At these conditions, the plasma harmonic spectra started to modify and show the enhancement of the harmonics tuned along the 75.5–77.9 nm range (Figure 3.26b). The third panel from the top of this figure shows the enhanced H17 compared with H16 and H18. The enhancement factor of this harmonic as varied between one and eight times depending on the wavelength of NIR pulses, while other harmonics showed the featureless plateau-like spectrum. The variation of the wavelength of NIR pulses with the step of 10 nm allowed the variation of the wavelength of H17 with the step of 0.6 nm. With this tuning, the maximally enhanced harmonic occurred at the conditions of coincidence of its wavelength with the strongest emission line (third panel from the top, 76.7 nm).

Further growth of fluency (1.3 J cm^{-2}) led to the appearance of strong emission lines (Figure 3.26c, upper panel). At these conditions, the harmonic generation was deteriorated due to large amount of free electrons. A decrease of excitation of target (down to ~1 J cm^{-2}) allowed generation of harmonics in the presence of plasma emission lines. The observed HHG spectra at these conditions showed the resonance-induced enhancement at different wavelengths corresponding to different transitions. Particularly, the 1300 + 650 nm pump of plasma led to the generation of enhanced H15 and H17. The enhancement was originated from the influence of Zn II resonances (76.7 and 88.1 nm). Moreover, the appearance of a strong line of Zn III (67.8 nm, see upper panel of Figure 3.26c) led to enhancement of harmonics in the vicinity of this spectral region. Particularly, it was clearly seen in the case of 1360 + 680 nm pump, when H20 ($\lambda = 68$ nm) was almost coincided with the above ionic line. One can see the fivefold enhancement of H20 with regard to the neighboring harmonics. At the same time, this spectrum shows the absence of enhancement of the harmonics near Zn II lines (76.5 and 88.9 nm), since they were detuned from these regions. Note that the enhancement of the part of broadband H11 of few-cycle (4.5 fs) pulses in the vicinity of 67.8 nm (18.3 eV, 3d^{10} → 3d^{9} (^{2}D)4p) was reported previously in Ref. [65].

The discussed experiments have demonstrated that the resonance enhancement of nearby harmonic close to the resonance can be observed in the case of sufficient amount of excited species (i.e. singly or doubly changed ions of

Figure 3.26 Tunable harmonic spectra in the cases of (a) weak, (b) moderate, and (c) strong excitation of Zn target (see text for further details). Dashed lines show the tuning of harmonics. Solid lines show the ionic transitions responsible for enhancement of harmonics. *Source*: Ganeev et al. 2016 [64]. Reproduced with permission from © IOP Publishing.

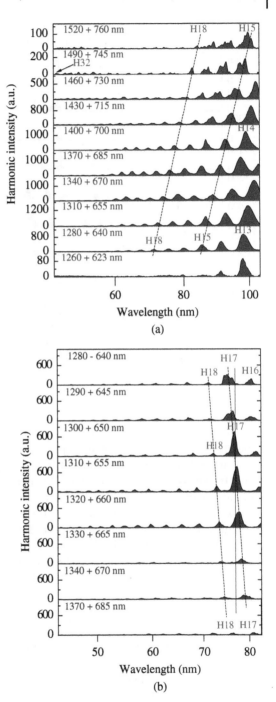

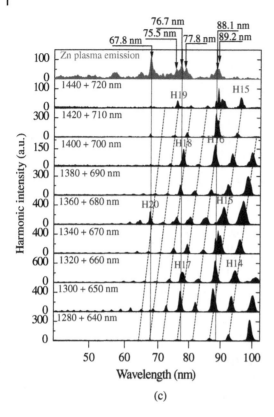

Figure 3.26 (*Continued*)

zinc in this particular case). Thus the availability of resonance enhancement depends on the population of appropriate energetic levels of ions. Note that only part of harmonic was enhanced in the vicinity of 67.8 nm. Figure 3.27 shows the emission spectrum of H18–H22 in the case of 1350 + 675 nm pump. One can see that the bandwidth of the enhanced 20th harmonic (~0.4 nm) was approximately four times narrower than the one of the neighboring harmonics (~1.6 nm). This enhancement of the part of harmonic shows a similarity with the recently reported partial enhancement of single harmonic in Ar gas, which was attributed to the influence of Fano resonance [67]. There are also a few other assumptions, which will be discussed in the following section.

3.4.3 Peculiarities of HHG in Zinc Plasma Using Tunable Pulses

Previous studies have shown the advanced properties of extended Zn plasma for generation of the strongest lower order harmonics of Ti:sapphire laser among other plasma species. The range of energy fluencies of ~0.3–1 J cm^{-2} seems satisfied with the requirement for the formation of the optimal plasma,

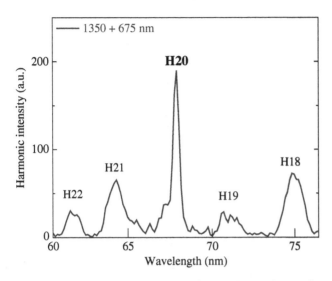

Figure 3.27 Enhancement of the part of H20 at the conditions of coincidence of this harmonic with the 67.8 nm transition of Zn III. *Source*: Ganeev et al. 2016 [64]. Reproduced with permission from © IOP Publishing.

which has previously allowed the generation of highest harmonic orders. Those studies have also shown a strong emission of 9th harmonic (89.1 nm) of Ti:sapphire laser, which considerably exceeded the neighboring orders. The ratio of the intensities of this and neighboring higher order harmonics (~8×) analyzed at unsaturated conditions of registration was considerably larger compared with other plasmas (C, Au, Cu). This observation points out that the mechanism of the enhancement of 89.1 nm radiation was other than the prevalence of the lower orders over the higher orders at the beginning of the plateau-like range of harmonic distribution. The analysis of plasma emission during excitation of the Zn target, without the propagation of the driving pulse through the plasma, showed the presence of some ionic lines attributed to the Zn II and Zn III transitions. Note the observed closeness of two $3d^{10}4s-3d^94s4p$ transitions of Zn II (88.1 and 89.3 nm) with the wavelength of 9th harmonic of 802 nm driving radiation.

Further, previously reported observation of the strong emission of 18.3 eV transition ($3d^{10}-3d^9$ (^{2}D)4p) of Zn III at the conditions of plasma excitation by few-cycle broadband pulses centered at 770 nm [65] was attributed to the enhancement of the part of 11th harmonic of this radiation, though the wavelength of this harmonic (70 nm) did not exactly match with the wavelength of $3d^{10}-3d^9$ (^{2}D)4p transition (67.8 nm). In those studies, a narrowband-enhanced emission was similar to the one observed in the discussed experiments [64] and could also be attributed to the influence of resonances and propagation processes. As we already mentioned, there are some other explanations of the

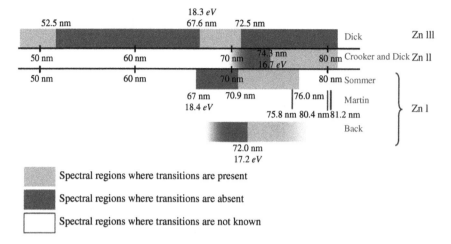

Figure 3.28 Spectral regions of lines corresponding to the transitions to excited states (in particular, to AIS) in Zn I, Zn II, Zn III according to the published data (from top to bottom [68, 69, 72, 73], and [70] correspondingly). *Source*: Ganeev et al. 2016 [64]. Reproduced with permission from © IOP Publishing.

resonance enhancement of harmonics, which are mostly based on the analysis of the micro- and macroprocesses. In the following, we address various options for explanation of resonance enhancement of harmonics.

The autoionizing levels of Zn have been reported in a few studies [68–74]. Figure 3.28 summarizes published data on the lines in Zn atoms, singly and doubly charged Zn ions in 50–100 nm spectral region. To make the comparison more obvious, the spectral regions are shown where the lines are present or absent, but not the positions of all the numerous lines. One can see that the HHG enhancement in the spectral region 67.6–72.5 nm can take place only in Zn III. So the enhanced coherent XUV pulses at 67.8 nm in Figure 3.26c show the presence of essential ratio of these ions in plasma under stronger excitation. The absence of the enhancement at this wavelength in Figure 3.26a,b shows that the number of such ions is negligible under weak and moderate plasma excitation. The enhancement in the spectral region longer than 74.3 nm can be due to Zn II ions but not Zn III; this enhancement is observed in both cases of moderate and strong excitation, showing that in both cases essential number of Zn II ions is present in the plasma. Disappearing of the resonant enhancement for the case of the weak excitation in Figure 3.26a shows that (i) Zn II ions are almost absent in this case and (ii) the Zn neutrals (which thus dominate in the plume) do not provide the resonant HHG enhancement. Note that Zn has relatively high first ionization potential (9.4 eV), so domination of neutrals in the laser plume under the weak excitation is very natural; this is not the case for other metals used in plasma HHG experiments.

It is shown in Figure 3.28 that transitions to excited states (in particular, to AIS) corresponding to the wavelengths longer than 71–72 nm are present in neutral Zn. However, these transitions do not provide HHG enhancement in Figure 3.26a. This can be attributed to the low oscillator strengths of these transitions for the wavelength shorter than 104 nm [75].

In Refs. [13, 36, 76] the numerical approach to study resonant HHG was developed. It is based on the numerical solution of the TDSE for an electron in a model potential and the external laser field. The model potential has an excited quasi-stable state, which models the AIS in the real atom or ion. This approach was developed to study numerically the HHG in Zn ions. Note, that the resonances due to transitions to the highly excited bound states, not to the AIS, were observed. However, these bound states are broadened due to photoionization in the intense laser field. So in this sense there is no fundamental difference between transitions to AIS (enhancing above ionization threshold harmonics) and transitions to highly excited bound states (enhancing below-threshold harmonics). Moreover, the generation of the below-threshold harmonics also includes the quasi-free electronic motion as one of the steps of the process [77, 78], so the resonant enhancement of this generation can be described within the similar four-step model.

The model potential suggested in Ref. [13] was used for these studies. Its parameters were chosen to reproduce the resonant HHG enhancement by Zn II ions near 76.7 nm. The TDSE is solved numerically for the electron in this potential and the external laser field with intensity of 2×10^{14} W cm^{-2} and different wavelengths. The calculated spectra are shown in Figure 3.29. One can see that the enhancement is maximal when the harmonic frequency coincides with the transition one, and substantially decreases when the fundamental wavelength is detuned by only 10 nm. So the calculated results reasonably well reproduce the experiment presented in Figure 3.26b. The calculated enhancement is higher than the observed one.

The width of the resonantly enhanced H17 is close to the ones of the other harmonics, both in calculated and experimental result. This is not the case for the experimentally observed enhancement of H20 in Zn III (see Figure 3.27.) This difference can be explained using an analytical theory based on the four-step model [76]. In particular, this theory shows that the shape of the resonantly enhanced harmonic line is a product of the harmonic line, which would be emitted in the absence of the resonance, and the enhancement factor, which essentially exceeds one time within the spectral width of the excited state. The enhancement of H17 in Zn II is due to the state, which is very close to the ionization threshold (the difference is approximately 2 eV), so it was broadened due to photoionization. Thus the enhancement takes place in the spectral region comparable with the harmonic width, and the latter remains approximately the same as for the nonresonant harmonics. The state in Zn III enhancing H20 is 20 eV lower than the ionization threshold, so it remains

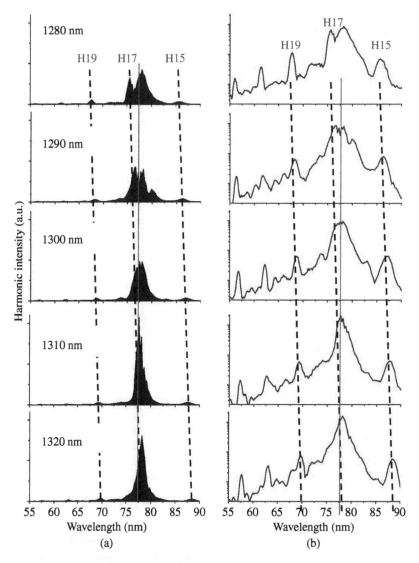

Figure 3.29 The harmonic spectra calculated via TDSE numerical solution in the linear (a) and logarithmic (b) scales. The fundamental wavelengths are presented in the graphs. The dashed lines show the tuning of harmonics, the solid line shows the ionic transition. *Source*: Ganeev et al. 2016 [64]. Reproduced with permission from © IOP Publishing.

narrow. The enhancement region is narrower than the harmonic line width, so the H20 line shape is close to one of the surrounding harmonics but with a narrow maximum due to the resonant enhancement.

The narrow plasma emission line of Zn II (67.8 nm) coincides with the central wavelength of the broadband 20th harmonic originated from 1350 + 675 nm two-color pump. One can assume that the part of this harmonic, being in resonance with the above ion transition, became enhanced due to resonance-induced increase of the nonlinear optical response of the plasma.

The demonstrated method of optimization of the driving pulses wavelength for the resonance-enhanced HHG opens new opportunities of harmonic generation of ultrashort pulses in the LPP plumes and can be considered as a tool for various studies of different species (particularly large molecules and clusters) in the ablated conditions. The use of tunable two-color pump of LPP allows the analysis of the ionic states in plasmas, and broadens the subjects of studies compared with presently used HHG in gases. With these studies, the application of NIR radiation for the analysis of the dynamics of nonlinear optical response of ablated solids compared with the 800 nm class lasers commonly used for plasma HHG studies was demonstrated. The method allows a search for new opportunities in improvement of HHG conversion efficiency using the NIR laser sources, particularly using the multicolor pump of plasmas and optimization of single harmonic yield using tunable radiation.

3.5 Application of Tunable NIR Radiation for Resonance Enhancement of Harmonics in Tin, Antimony, and Chromium Plasmas

In this section, we discuss the role of resonances in the Sn, Sb, and Sn plasmas on the HHG efficiency at the conditions of tuning the driving pulses in the NIR range [11].

3.5.1 Experimental Results

Experimental setup was described in Section 3.3.2. Sn, Cr, and Sb plasmas were used as the media for harmonic generation by applying the tunable source of ultrashort pulses. Previously, those plasmas have shown the single harmonic enhancement using the fixed wavelength of pump (Ti:sapphire) lasers. The 5 mm long samples of above elements were installed in the vacuum chamber for laser ablation.

The advantage of the studies of resonance-induced enhancement of harmonics using OPA is the opportunity of fine-tuning of this high-order nonlinear optical process for spectral enhancement of the harmonic yield. In the following, we show some examples of the resonance enhancement of the odd and even harmonics using Sn, Sb, and Cr ablations and the adjustment of different harmonics with regard to the same ionic transition.

Prior to resonance-enhancement studies of harmonics using NIR + H2 pulses the above targets were analyzed using the pulses from the conventional Ti:sapphire laser ($\lambda = 806$ nm, $E = 3$ mJ, $I = 6 \times 10^{14}$ W cm^{-2}). Figure 3.30a shows the enhancement of 17th harmonic (H17) in tin plasma, H21 in antimony plasma, and H29 in chromium plasma marked by parallelepipeds. Those targets were chosen due to known ionic transitions responsible for abovementioned enhancements. The studies of the oscillator strengths of those transitions were reported in Refs. [23, 79–82].

This figure shows the comparative studies of the HHG in three above plasmas. One can conclude the relative efficiencies obtained from these plasmas by comparing the Y-axes. The highest conversion efficiency was observed in the case of Sb plasma.

The use of tunable broadband NIR radiation allowed the analysis of the enhancement of the groups of harmonics close to the resonances possessing strong oscillator strengths, which caused the harmonic enhancement in the case of relatively narrowband 806 nm pulses. Chromium plasma showed the enhancement of harmonics close to the 27 nm region where strong transitions of Cr II significantly modify the featureless decay of plateau-like harmonic spectrum (see the raw images of harmonics obtained using CCD camera, Figure 3.30b). One can see the growth of harmonic yield in this spectral region for different groups of harmonics during tuning of the wavelength of driving pulses. Particularly, harmonics starting from H46 and higher were stronger compared with lower order ones in the case of 1280 nm + H2 pump (upper panel). Similar feature was observed for other groups of harmonics in the case of 1310 nm + H2, 1380 nm + H2, 1420 nm + H2, and 1460 nm + H2 pumps (other panels). Dashed lines show the tuning ranges of two harmonics (H20 and H25). In the case of 1420 nm + H2 pump the harmonics were extended above the 60s orders of NIR radiation.

The purpose in showing raw images is to acquaint the reader with real collection data and visually demonstrate the appearance of the separated group of enhanced harmonics in the case of chromium plasma. The saturated images were intentionally chosen to present the spectra for better viewing. Note that for the line-outs of the HHG spectra shown in other figures the unsaturated images were used. The x-axis is shown in the figure on the basis of the calibration of XUV spectrometer for better viewing of the distribution of harmonics along the short-wavelength region. The HHG spectrometer was calibrated using the plasma emission from the used ablated species, as well as other ablating targets.

The enhanced H29 from the 800 nm class lasers in the case of chromium plasma was earlier reported in Ref. [83]. In Figure 3.30c, the comparative line-outs of harmonic spectra in the case of 1300 nm + H2 and 806 nm

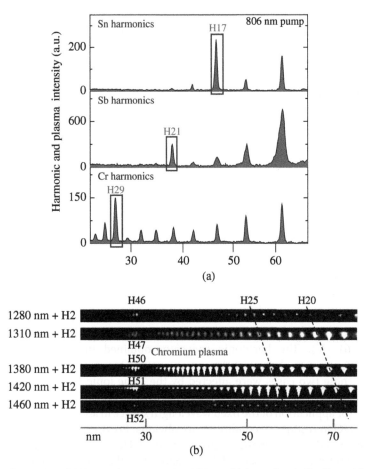

Figure 3.30 (a) Harmonic spectra using 806 nm driving pulses in Sn, Sb, and Cr plasmas showing the resonance enhancement of the single harmonics (H17, H21, and H29, respectively). (b) Raw images of the tunable harmonic spectra using the pump of chromium plasma by NIR + H2 pulses. NIR pulses were tuned in the range of 1280–1460 nm. Dashed lines in this and other figures show the tuning of specific harmonics. One can see the notable enhancement of harmonics in the vicinity of 27 nm and significant decrease of harmonic yield in the range of 29.5–31 nm. (c) Comparative spectra of the harmonics generated in Cr plasma using the 1300 + 650 nm and 806 nm pumps. The NIR-induced curve was shifted along the *y*-axis for better visibility and comparison with the 806 nm induced harmonic spectrum. *Source*: Ganeev et al. 2016 [11]. Reproduced with permission from © IOP Publishing.

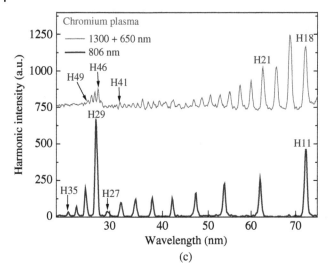

Figure 3.30 *(Continued)*

pumps are shown. In the latter case (bottom curve), one case see a significant suppression of the 27th harmonic ($hv = 41.53$ eV) followed with the enhanced H29 ($hv = 44.61$ eV). The harmonic spectrum obtained in the case of 1300 + 650 nm pump is shown in the upper curve. One can see that the harmonics in these two cases were affected by the same ionic transitions of chromium ions.

The reasons for application of the fixed wavelength source in above-described case for the HHG in resonance conditions are obvious. Previous experiments, which demonstrated the resonance enhancement of H29 in the chromium plasma using 800 nm class lasers, did not allow the thorough analysis of this process from the point of view of the role of the resonances on the growth of conversion efficiency in the case of single harmonic. The discussed studies, which used the tunable pump waves, allowed defining the role of these resonances. One can see in Figure 3.30c that only H29 (and partially H31) was enhanced, while in the case of 1300 + 650 nm pump the maximum enhancement (H46) did not coincide with the wavelength of H29 of 806 nm pump. The wavelength of H29 (27.97 nm) of 806 nm pump rather positioned between the wavelength of H47 (27.66 nm) of 1300 nm pump (upper panel of Figure 3.30c) and the wavelength of H46 (28.26 nm) of the same pump. Thus it became obvious that there is some transition corresponding to the group of resonances of Cr II spectra, which lies out of the wavelength of H29 of 806 nm pump and affects the nonlinear optical response of the plasma. This is a clear example of the nonlinear spectroscopy of plasma using high-order processes, which could not be realized in the case of the fixed wavelength sources. That is why HHG from 806 nm and 1390 nm + H2 pumps should be compared for better understanding of the peculiarities of resonance processes.

Previous studies of photoabsorption and photoionization spectra of Cr plasma in the range of 41–42 eV [79] have demonstrated the presence of strong transitions, which could be responsible for a suppressed pattern of harmonic spectrum in the wavelength region of 29.5–31 nm. The region of "giant" 3p → 3d resonances (44–45 eV, $gf = 0.63$) of Cr II spectra was analyzed in Refs. [79, 82] and the strong transitions that could enhance the nonlinear optical response of the plume were revealed.

Similar feature was observed in the case of harmonic generation in the tin LPP. The studies of the resonant enhancement of HHG in a tin plasma using the 806 nm driving pulses (Figure 3.30a, upper panel) showed a strong 17th harmonic analogously to those reported in previous studies of this plasma medium [12, 15]. In the discussed experiments a stronger 17th harmonic was observed, with an enhancement factor of 8× compared with neighboring harmonic orders. The following studies using the tunable NIR pulses and their second harmonics have shown a fine-tuning of the resonance-enhanced harmonic and change of the order of this harmonic (Figure 3.31a). The maximally enhanced harmonics, for which both microprocesses and macroprocesses were optimized to generate highest photon yield, were changed from H27 in the case of 1290 nm + H2 pump (bottom panel) to H31 in the case of 1450 nm + H2 pump (upper panel). In all these cases, the preceding harmonics were suppressed compared with resonance-enhanced ones (for example, compare H29 and H27 in the case of 1370 nm + H2 pump), though not as strongly as in the case of chromium plasma.

One can find from the published data of Sn II transitions in the studied spectral region that the 17th harmonic of 806 nm radiation ($hv = 26.15$ eV, $\lambda = 47.41$ nm) is close to some transitions between autoionizing and ground states of tin. The frequencies of these transitions, some of which possess reasonably large oscillator strengths, lie in the photon energy range of 24.9–27.3 eV. The reported enhancement of this harmonic, as well as the H27–H31 of the NIR pulses, were attributed to their closeness with the $4d^{10}5s^25p \; {}^2P_{3/2} \rightarrow 4d^95s^25p^2 \; ({}^1D)^2D_{5/2}$ transition of the Sn II ion ($gf = 1.52$ [23]).

Antimony LPP has previously been analyzed as the nonlinear medium in Refs. [4, 83] where single harmonic of Ti:sapphire laser at the wavelength of 37.7 nm was approximately 1 order of magnitude stronger with regard to the neighboring harmonic orders. Those studies allowed exploring a small tuning of the used Ti:sapphire laser (783–797 nm [4]), which led to definition of the maximally enhanced H21 in the case of 791 nm pump. In discussed present studies using 806 nm pulses of Ti:sapphire laser, the sixfold enhancement of this harmonic ($hv = 32.30$ eV) with regard to the neighboring ones was observed (Figure 3.30a, second panel). The application of tunable

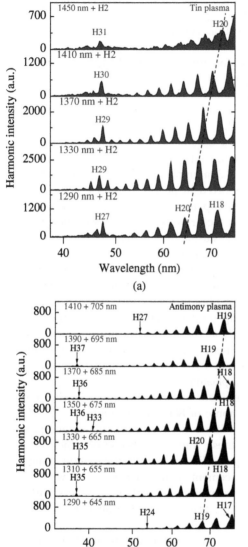

(a)

(b)

Figure 3.31 (a) Tuning of resonance-enhanced harmonics generated in the 47 nm region using the two-color pulses propagated through the tin plasma. (b) Harmonic spectra from Sb plasma at different NIR + H2 pumps. (c) Tuning of harmonics in the region of strong ionic transition of antimony (37.5 nm). *Source*: Ganeev et al. 2016 [11]. Reproduced with permission from © IOP Publishing.

Figure 3.31 (*Continued*)

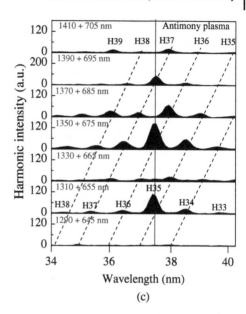

(c)

NIR + H2 pulses allowed precisely determining the optimal wavelength of pump radiation at which the maximum enhancement of harmonics in the same spectral region can be achieved ($\lambda = 37.5$ nm, Figure 3.31b). The tuning of harmonics along this region led to gradual growth and decrease of H35 to H37 (Figure 3.31c). This figure shows the details of the harmonics variations in the narrow spectral range in the vicinity of the resonance transition (37.5 nm) responsible for the enhancement of tuned harmonics.

The resonance enhancement of these harmonics in antimony plasma in the 37–38 nm region is attributed to the strong Sb II transitions $(4d^{10}5s^25p^2\,^3P_2 \rightarrow 4d^95s^25p^3\,(^2D)\,^3D_3$ and $4d^{10}5s^25p^2\,^1D_2 \rightarrow 4d^95s^25p^3\,(^2D)\,^3F_3)$ at the wavelengths of 37.82 nm (33.78 eV) and 37.55 nm (33.02 eV). The XUV spectra of antimony plasma have been analyzed in Ref. [80]. The oscillator strengths of above transitions have been calculated to be 1.36 and 1.63, respectively, which were a few times larger than those of the neighboring transitions. The smaller enhancement factor ($\sim$3×) of resonance-related harmonics (H35–H37) observed in the present study compared with previous reports (10× [4], 20× [83]) is attributed to broader pump and harmonic bandwidths, as well as longer wavelength of driving pulses.

Although the above studies were carried out using orthogonally polarized pumps, one can carry out the HHG using parallel polarized pumps. For these purposes the idler pulses and 806 nm pump were used. Those studies showed that both orthogonal and parallel polarizations of the two pumps led to enhancement of the whole harmonic spectrum. In other words, independently on the variation of the spatial trajectory of accelerated electron (i.e. two- and three-dimensional movement) the appearance of even, sum, and difference harmonics, together with odd ones, point out the role of second field as a driving force for broadening the number of coherent frequency components in the XUV spectrum.

Independently on the polarization state of assistant field, the resonance-enhanced mechanism of harmonic amendment remains unchanged. Adding H2 wave, which is orthogonal to the fundamental one, changes the laser-field-driven dynamics of the ionized electron on the microscopic level. In this case, an enhancement of the HHG was observed for atomic harmonics [42], i.e. not only for plasma harmonics.

In the two-color pump HHG experiments in gases, the variation of relative phase between fundamental and second harmonic waves of 30 fs pulses allowed observing the threefold beatings between the long- and short-trajectory induced harmonics [49]. The use of the focusing optics inside the XUV spectrometer in discussed experiments did not allow distinguishing the influence of the short and long trajectories of accelerated electrons on the divergence of harmonics. However, the difference in the plasma harmonic spectra was observed when the thin (0.15 mm) silica glass plates, which are actually the relative phase modulators, were introduced between BBO crystal and plasma to analyze the variation of the relative phase between two pumps and to compare the change of the relative intensities of "resonant" and "nonresonant" harmonic yields. A significant departure from the large ratio of resonant and nonresonant harmonics in the case of the absence of the glass plates toward the low ratios of these harmonics in the case of propagation through six 0.15 mm thick plates was observed. Thus the variation of relative phase between pumps may diminish the role of AISs in the single harmonic enhancement.

Notice that, in the case of 806 nm pump, the second harmonic conversion efficiencies and the temporal overlaps of the driving and second harmonic waves, as well as the pulse durations of 403 nm radiation in the cases of different BBO crystals, were analyzed. In the case of abovementioned lengths of crystals allowing the 0.4%, 5%, 9%, 11%, and 13% second harmonic conversion efficiencies, the ratios of overlapped pulses inside the plasma were 0.004, 0.03, 0.04, 0.01, and 0.007, respectively. One can see the prevalence of using the 0.3 and 0.5 mm long BBO, since other crystals allowed the observation of less efficient odd and even harmonics generation due to insignificant overlap inside the plasma plume. Moreover, the duration of second harmonic pulse increases in the case of longer crystals (65, 72, 105, 140, and 195 fs correspondingly; compare with the 64 fs driving pulse). The decrease of the intensity of second harmonic also diminishes the role of this radiation in the variation of HHG spectra. In the case of NIR pulses, lesser dispersion of BBO crystal has diminished the role of walk-off.

3.5.2 Theoretical Analysis of Resonance-Enhanced Harmonic Spectra from Sn, Sb, and Cr Plasmas

In the case of single-color laser field, a time-periodic (period $T = 2\pi/\omega$) linearly polarized laser field, with the electric field vector (in dipole approximation) given by

$$E_L(t) = E_0 \sin(\omega t + \phi)\hat{e}_x \qquad (3.16)$$

where E_0 is the electric field amplitude, and ϕ is an arbitrary phase, was used. For orthogonally polarized bichromatic laser field the electric field vector lies in the xy plane and is defined by

$$E_L(t) = E_{L1} \sin(r\omega t + \phi_r)\hat{e}_x + E_{L2} \sin(s\omega t + \phi_s)\hat{e}_y \qquad (3.17)$$

where E_{Lj} ($j = 1, 2$), is the jth electric field vector amplitude, $\hat{e}_x$ and $\hat{e}_y$ are the unit polarization vectors along the x- and y-axes, respectively, the component frequencies $r\omega$ and $s\omega$ are integer multiples of the same fundamental frequency ω, and ϕ_r and ϕ_s are arbitrary phases. The analysis was restricted to the case of linearly polarized bichromatic field components. A more general case of elliptically polarized field components has been analyzed in Ref. [84]. In the discussed studies, the case of perpendicular polarization of the components of two-color field was analyzed. One can expect that for an arbitrary angle between the polarizations of the field components the HHG spectra will be modified. In this case, the electron trajectory between the ionization and recombination time would be different so that the harmonic yield and the cutoff position may change.

One can assume that the target material is such that there is a high radiative transition probability between the ground state with the energy E_1 and a low-lying state having the energy E_2. If the laser frequency is such that the condition $\Delta\omega = E_2 - E_1 = n_R\omega$, n_R – integer, is fulfilled, then, during the single-state HHG process, a coherent superposition of the ground state and this excited state will be formed. In this case, the strength for emission of a harmonic having the frequency Ω takes the form [85, 86]:

$$D(\Omega) = a_1^2 D_{11}(\Omega) + a_2^2 D_{22}(\Omega) + a_1 a_2 [D_{21}(\Omega) + D_{12}(\Omega)] \qquad (3.18)$$

where a_1 and a_2 are the initial amplitudes of the bound states in the superposition of states 1 and 2, and $D_{jj'}(\Omega)$ is the Fourier transform of the time-dependent dipole $d_{jj'}(t)$, which is given in Ref. [87]. This simple model, which will be further discussed in the following section (see Ref. [84] for more detail and Ref. [88] for application), is able to explain qualitatively the abovementioned experimental data.

Note that due to the definition of the electric field vector, both x and y components of $D(\Omega)$ contribute to the harmonic intensity, which is defined by the relation $\Omega^4|D(\Omega)|^2$. The hydrogen-like atom model was used in these studies, while taking into account only the 1s $(j = 1)$ and 2p $(j = 2)$ states. Some numerical examples of HHG from a coherent superposition of Sn^+, Sb^+, and Cr^+ states for $a_1 = a_2 = 1/2^{0.5}$ are presented in the following. In the case of single-color pump, the laser-field intensity was assumed to be 2×10^{14} W cm^{-2} and $\phi = 0$, while for the case of two-color pump $(r = 1, s = 2)$ the laser-field intensities were $I_1 = 1.5 \times 10^{14}$ W cm^{-2} and $I_2 = 0.5 \times 10^{14}$ W cm^{-2}.

In Figure 3.32a, the harmonic yield is presented as a function of the harmonic order for Sn^+ whose ionization potential is 14.63 eV. For single-color laser field the results are presented by solid line with filled circles while the numerical results for two-color laser field are presented by solid line with filled squares. In the case of two-color field, both phases are equal to zero. In order to achieve a better visibility, numerical spectra for single-color field are shifted down by 2 orders of magnitude. In the top (bottom) panel of Figure 3.32a, the numerical results for fundamental wavelength 806 nm (1370 nm) are presented. In the case of 806 nm, the resonant harmonic is H17, while in the case of fundamental wavelength of 1370 nm the resonant harmonic is H29, in accordance with the experimental results.

Figure 3.32b shows the harmonic yield as a function of the harmonic order for Sb^+. The ionization potential is 16.63 eV, while the other laser parameters are the same as in Figure 3.32a. Here the fundamental wavelength is 806 nm and the corresponding resonant harmonic is H21. In the case of two-color field and for the fundamental wavelength of 1350 nm, the resonant harmonic is H36 and this was also in accordance with the experimental results. In the case of single-color field with fundamental wavelength of 1350 nm, emission of even harmonics is forbidden due to the inversion symmetry. In this case $\Delta\omega$ was slightly changed in order to obtain the adjacent resonant harmonic H35. The resonant harmonic is denoted by corresponding solid lines in the upper right corner of the bottom panel in Figure 3.32b.

In Figure 3.32c, analogous numerical results are presented for the Cr^+ whose ionization potential is 16.48 eV. The other laser parameters are the same as in Figure 3.32a. For the fundamental wavelength of 806 nm the corresponding resonant harmonic is H29. In the case of two-color pump and for the fundamental wavelength of 1280 nm the resonant harmonic is H46, which was similar to the experimental results. From the same abovementioned reasons as in the case of Sb^+ ions, in the case of single-color field it is not possible to obtain this resonant harmonic. Again, $\Delta\omega$ was slightly changed in order to obtain the adjacent resonant harmonic H47. The resonant harmonic is denoted by the corresponding solid lines in the upper right corner of the bottom panel in Figure 3.32c.

In order to show the influence the relative phase between two orthogonal components of the field given by Eq. (3.17), in Figure 3.33 the numerical results for Sn II are presented. The fundamental wavelength is 1370 nm, $\phi_s = 0$ and

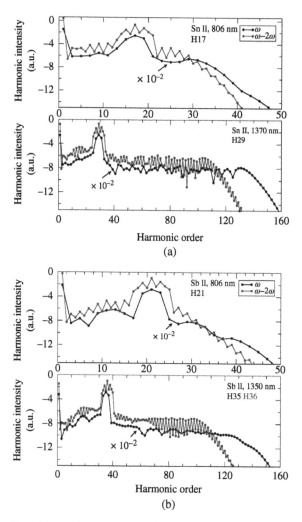

Figure 3.32 (a) Harmonic intensities as functions of the harmonic order for HHG from Sn II. The laser field intensity is 2×10^{14} W cm^{-2} and $\phi = 0$ for single-color field, and $I_1 = 1.5 \times 10^{14}$ W cm^{-2} and $I_2 = 0.5 \times 10^{14}$ W cm^{-2} with $r = 1$, $s = 2$, $\phi_r = \phi_s = 0$ for two-color field. The fundamental wavelengths are indicated in the upper right corner of both panels as well as corresponding resonant harmonics. Numerical results for single-color laser field are presented by solid line with filled circles while the numerical results for two-color laser field are presented by solid line with filled squares. (b) Same as in (a), but for Sb II. The resonant harmonic for 806 nm is H21, while the resonant harmonic for single-color (two-color) field with 1350 nm is H35 (H36). (c) Same as in (a), but for Cr II. The resonant harmonic for 806 nm is H29, while the resonant harmonic for single-color (two-color) field with 1280 nm is H47 (H46). *Source:* Ganeev et al. 2016 [11]. Reproduced with permission from © IOP Publishing.

Figure 3.32 (*Continued*)

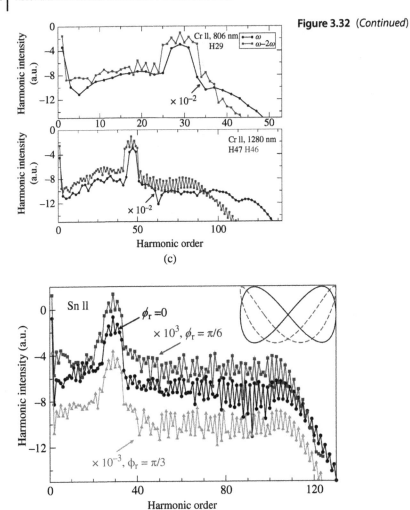

Figure 3.33 Harmonic intensities as functions of the harmonic order for HHG from Sn II by two-color field whose components have the same intensities and wavelengths as in the lower panel of Figure 3.32a, for $\phi_s = 0$ and for three different phases $\phi_r = 0$, $\pi/6$, and $\pi/3$ as denoted. The inset in the upper right corner shows the corresponding electric field vectors. *Source*: Ganeev et al. 2016 [11]. Reproduced with permission from © IOP Publishing.

the three different phases $\phi_r = 0$, $\pi/6$, and $\pi/3$ are indicated. In order to achieve better visibility, harmonic spectrum for $\phi_r = \pi/6$ ($\phi_r = \pi/3$) is shifted up (down) by 3 orders of magnitude. As it can be seen in Figure 3.33, the cutoff position of the harmonic yield only slightly depends on the relative phases, while the impact on the group of resonant harmonics is negligible. As an inset, in the upper right corner of Figure 3.33, the electric field vector is shown for phases

$\phi_r = 0$ (solid line) and $\phi_r = \pi/6$ (dashed line; the field for $\phi_r = \pi/3$ looks the same as the field for $\phi_r = \pi/6$, while the corresponding vector potential $A_L(t)$, $E_L(t) = -dA_L(t)/dt$, changes the sign). From these polar plots of the field one can see that there are long time intervals during which the field is close to linear. If, in the first step of the HHG process, the electron appears in the continuum at the beginning of such intervals it is able to return to the parent ion at the end of these intervals and recombines emitting a high harmonics. The situation is similar to the case of HHG by bicircular field [89].

In Figure 3.34, the harmonic yield is presented as a function of the harmonic wavelength for Sb$^+$ exposed to the field given by Eq. (3.17). The fundamental wavelength changes from 1290 (bottom panel) to 1410 nm with steps of 20 nm. In all presented panels, $\Delta\omega = 33.02$ eV was assumed. The corresponding wavelength is denoted in the upper left corner of each panel. As it can be seen in Figure 3.34, the resonant harmonics H35, H36, and H37 correspond to the fundamental wavelengths 1310, 1350, and 1390 nm, respectively. The intensities of

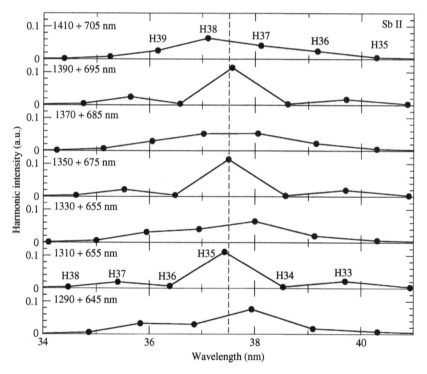

Figure 3.34 Harmonic intensities as functions of the harmonic order for HHG from Sb II for different fundamental wavelengths starting from 1290 nm (bottom panel) to 1410 nm (top panel). The other laser parameters are as in the lower panel of Figure 3.32b. *Source*: Ganeev et al. 2016 [11]. Reproduced with permission from © IOP Publishing.

these resonant harmonics decrease when the fundamental wavelength changes in steps 20 nm (see the first, third, fifth, and seventh panels of Figure 3.34 (from bottom to the top)). One can find a similarity in the behavior of harmonic yield in this calculation and the experiment shown in Figure 3.31c.

3.5.3 Discussion

The discussed studies have demonstrated new opportunities in the analysis of the strength of some ionic transitions responsible for enhancement of single harmonic in the plateau region. The example of such analysis is shown in Figure 3.31a, where the intensity ratio of the various harmonics near 47 nm vary as one tunes the pump laser wavelength. It is obvious that the enhancement of single nearby harmonic strongly depends on the detuning out of ionic transition. Correspondingly, the enhancement factor, or intensity ratio between the "resonance" harmonic and neighboring ones, will be obviously changed once the wavelength of enhanced harmonic tunes toward or outward the transition. The fact that this process is reproduced for various harmonic orders just confirms the consideration of the fundamental role of the ionic transitions possessing large oscillator strength in the enhancement of harmonics. This phenomenon is qualitatively reproduced in the calculations showing the enhancement of different harmonic orders near the same ionic transition.

Next, in Figure 3.31c, the 37.5 nm harmonic is in resonance with the pump laser for three cases, 1390 + 695 nm, 1350 + 675 nm, and 1310 + 655 nm. However, the intensity of the resonant 37.5 nm harmonic varies considerably. The reason in the variation of enhancement factor is also related with different energies of NIR radiation. The largest energy of NIR pulses in these experiments was observed in the case of 1350 nm pulses. The OPA was just tuned, which allowed observation of the above phenomenon of intensity ratio variations. Meanwhile, Figure 3.34 presents the simulations for the enhancement of different harmonic orders assuming similar energies of pump radiation.

The 806 nm and NIR + H2 induced enhancements of resonance harmonics in chromium plasma are shown in Figure 3.30c. However, it is difficult to compare the enhancement factors in these two cases. In the case of former pump, 5 mJ pulses were used, while in the case of latter pump it was approximately 0.7 + 0.2 mJ energy of those pulses. Even at these unfavorable conditions (significantly less energy of main (NIR) pulse and abovementioned wavelength-dependent rule) only fourfold decrease of the enhancement of resonance harmonic using these pulses was observed. Once 1 mJ pulses of 806 nm pump were used, no harmonics were generated below 40 nm. So from this point of view one can admit the dramatic comparative enhancement of resonant harmonic in the case of used scheme at similar conditions of pump energies.

One can speculate on the availability of using the high energy transitions related to deeper core resonance. It is not clear whether deeper core resonances can decisively influence the HHG. Deeper core resonances may appear for multiple ionized plasma ions (see Ref. [90] for inert gas ions). However, the formation of multiple ionized ions would lead to the growth of free electron concentration, which may strongly affect the phase-matching condition. In other words, the prevalence of the phase mismatch over the single-atom related growth of harmonics due to deeper core resonances may fully cancel the advantages of using the resonance concept.

Although the approaches considering the role of resonances as the main reason of experimentally observed growth of the yield of some single harmonics presently prevail, one has to point out another approach developed in Ref. [91]. As it has been shown in those studies using the nonperturbative theory of light–matter interaction the energy shifts of atomic states approach to the free atom energy level difference in the case of laser fields of near-atomic strength. Such behavior can be considered from the point of view of atomic field strength definition. It was suggested that the interpretation based on the presence of resonances with ionic transitions is doubtful, because in the laser fields of near-atomic strength the motion of atomic electron obeys the action of two equal forces that are due to the intra-atomic and external laser fields. The above approach is probably the reason why one cannot take into account the role of resonances (with laser dressing all spectrum is shifted into resonance). But this is just only theoretical approach developed in abovementioned work, contrary to many other approaches and publications, and thus it cannot draw the general conclusions. This and other approaches are just a matter of extensive discussion of the topics of resonance studies.

Figure 3.30c contains two dependences of the harmonics as the functions of the wavelength in the XUV region using the fixed wavelengths of the driving fields (1300 + 650 nm, and 806 nm, respectively). One can clearly see that they do nothing with the tuning of the driving field. Meanwhile, the dependences of the harmonic yield on the wavelength of tuned radiation are shown in Figures 3.30b and 3.31a–c.

Depending on the wavelength of driving field the order of maximally enhanced harmonic varies (see H46–H51 in the case of chromium plasma, Figure 3.30b; H27–H31 in the case of tin plasma, Figure 3.31a; H31–H35 in the case of antimony plasma, Figure 3.31b). In the meantime, the wavelengths of those enhanced harmonics did not exactly match with the wavelengths in corresponding ionic transitions of abovementioned species. They just correspond to the odd and even harmonics of the NIR pulses used in each concrete case.

The maximum enhancement in the case of antimony plasma (for other species this optimal wavelength correspondingly varies depending on the position of ionic transition) of the harmonic in the vicinity of resonance (37.5 nm) was achieved in the case of 1350 nm NIR pulses and their second harmonics (Figure 3.31b, middle panel). At this pump wavelength, the enhanced harmonic (H36) was strongest among other enhanced harmonics using different pump wavelengths (compare the intensity of this harmonic shown in the middle panel with the intensities of other enhanced harmonics in this spectral region shown in other panels).

3.6   Model of Resonant High Harmonic Generation in Multi-Electron Systems

The four-step model [13] gives reasonably good qualitative estimates of the efficiencies of resonant HHG for different experimental conditions by substituting experimentally measured decay widths of the resonant levels into SAE-based analytical formula to account for many-electron effects. There are however some uncertainties within this model, because SAE approximation does not consider real structure and dynamics of intra-atomic electronic structure. Namely, the electrons responsible for the resonant transitions with high oscillator strength definitely belong to inner shells of the systems in which resonant HHG was experimentally demonstrated. These electrons are in general very unlikely to be ionized directly by tunneling. So the corresponding excited states are not necessarily the AIS of the atomic systems, even if the energy of resonant transition is close to the ionization energies of the atoms. One should also take into consideration the dressing of higher Rydberg states, which may greatly modify the real AIS of the system.

Another common drawback of all models of resonant HHG, which assume the population of certain resonant level, is the unclear physical nature of transition of electron to the ground state with corresponding emission of coherent radiation at the frequency of a harmonic. If we consider these transitions spontaneous due to finite lifetime of the resonant level or another laser-dressed state, then the coherence and duration of the resonant radiation will differ greatly compared to nonresonant harmonics. If the process is considered a stimulated emission then the absence of population inversion will lead to decrease of harmonics near the resonance due to higher probabilities of stimulated absorption even when the degeneracy of the resonantly populated state is high, although the coherence problem of resonant harmonics will be solved. Experimental confirmation of the existence of spontaneous part of resonant harmonics is an important tool for verification of the proposed approaches. It is still more favorable to consider depletion of resonantly populated level as a lasing without inversion (LWI) process, which cancels the stimulated absorption cross-section due to interference of two absorption channels in the presence of two fields coupled to the corresponding transitions [92].

In this section, we discuss the model of resonant HHG in multi-electron systems [93]. We analyze the theory of resonant elastic scattering in the presence of energetically forbidden inelastic scattering channel in the approximation of strong coupling of two scattering channels. It will be shown that inelastic scattering channel cannot be completely closed and is extremely important for LWI processes within resonant HHG. However, the nature of the approach is similar to the model of channel closing, which has been successfully applied to studies of anomalous above-threshold ionization (ATI) spectra near ponderomotively upshifted Rydberg states [94, 95]. Note that the structure of energy levels considered for ATI is quite different that prevents direct generalization to resonant HHG. We also discuss experimentally demonstrated enhancement of different harmonic orders in the In plasma using the 806, 1431, and 1521 nm pumps, which support the proposed model.

3.6.1 Theory

Considering two-particle scattering problem it is useful for particle x (electron for HHG processes) and target A (atomic or ionic core) to denote elastic scattering as

$$x + A \rightarrow x + A \tag{3.19}$$

and inelastic scattering where the target can be either excited to a bound excited state or a continuum excited state or ionized as

$$x + A \rightarrow x' + A^* \tag{3.20}$$

In the approximation of infinitely heavy target A and neglecting identity of electrons for simplicity full Hamiltonian of the system $(x \cup A)$ can be written as:

$$H = \hat{H}_A + \hat{K} + \hat{V} = \hat{H}_0 + \hat{V} \tag{3.21}$$

where $\hat{K} = -(\hbar^2/2\mu)\nabla^2$ is the kinetic energy operator of the particle, $\hat{V} = \hat{V}(\xi, \mathbf{r})$ is operator of interaction of particle with the target, $\hat{H}_A$ is the Hamiltonian of noninteracting target satisfying stationary Schrödinger equation:

$$\hat{H}_A \Phi_n(\xi) = E_n \Phi_n(\xi), \quad n = 0, 1, 2, \dots \tag{3.22}$$

$\Phi(\xi) = \sum \Phi_n(\xi)$ is the expansion of target's wavefunction in the basis of eigenfunctions of noninteracting target, which form a full set:

$$\langle \Phi_n | \Phi_{n'} \rangle = \delta_{nn'}$$
$$\sum \Phi_n(\xi) \Phi_n^*(\xi') = \delta(\xi - \xi') \tag{3.23}$$

here ξ is a set of inner variables of the target. Note that although HHG process is nonstationary and highly nonperturbative, the problem of scattering of electron with given energy on target can be solved perturbatively as it contains no expression for pump field. Note that eigenfunctions actually describe both bound and continuous states of the noninteracting target and these eigenfunctions in general do not describe any bound states for interacting target. The sum sign denotes summation over bound state and integration over continuous states. The cross-section of inelastic scattering that excites the target into state n is then given by:

$$\left(\frac{d\sigma}{d\Omega}\right)_{|0\rangle \to |n\rangle} = \frac{p_f}{p_i}\left(\frac{\mu}{2\pi\hbar^2}\right)^2 |\langle \exp(ik_i r)\Phi_0(\xi)|\hat{V}|\exp(ik_f r)\Phi_n(\xi)\rangle|^2$$

(3.24)

In case of $n=0$ Eq. (3.24) gives us elastic scattering cross-section. Here indices i and f denote initial and final parameters of the scattered electron, $k_i = p_i/\hbar$, $k_f = p_f/\hbar$ and according to energy conservation law:

$$\frac{p_f^2}{2\mu} + E_n = \frac{p_i^2}{2\mu}$$

(3.25)

When E_n is the energy of resonant interatomic transition, Eq. (3.24) is the correspondence between inelastic collision cross-section and the kinetic energy of the electron accelerated by the laser pump field. The perturbation operator $\hat{V}$ is the operator of interaction of the incident electron with the remaining electrons of the atom:

$$\hat{V} = \hat{V}(r; r_1, \dots r_z) = -\frac{Ze^2}{r} + \sum_{j=1}^{Z} \frac{e^2}{|r - r_j|}$$

(3.26)

here **r** is the radius-vector from the center of the atom to the incident electron. The second term in this operator assumes significant influence of electronic correlation on the process of resonant HHG as it has been shown by analysis of influence of precision of representation of electron–electron correlation on resonant HHG in Ref. [87]. During HHG process one should consider also time-dependent part $\hat{V}(t) = rE(t)$, this part influences the scattering process only via AC-Stark shift and does not describe target–particle interaction at all. Using Eq. (3.26), one can outline the total scattering amplitude in Born approximation as [96]:

$$F_{n0}^{(B)}(k_f, k_i) = -\frac{\mu}{2\pi\hbar^2}\int d^3 r e^{iqr}\int \Phi_n^*(r_1, \dots, r_Z)\left(-\frac{Ze^2}{r} + \sum_{j=1}^{Z}\frac{e^2}{|r - r_j|}\right)$$

$$\times \Phi_0(r_1, \dots, r_Z) d^3 r_1 \dots d^3 r_Z$$

(3.27)

Denoting matrix element by atomic wavefunctions as

$$\left\langle n \left| \sum_{j=1}^{Z} e^{iqr_j} \right| 0 \right\rangle \equiv \int \Phi_n^*(r_1, \ldots, r_Z) \left(\sum_{j=1}^{Z} e^{iqr_j} \right)$$
$$\times \Phi_0(r_1, \ldots, r_Z) d^3r_1 \ldots d^3r_Z \tag{3.28}$$

and using the relation

$$\int e^{iqr} \frac{1}{|r - r_j|} d^3r = e^{iqr_j} \int \frac{e^{iq(r-r_j)}}{|r - r_j|} d^3r = \frac{4\pi}{q^2} e^{iqr_j} \tag{3.29}$$

one can get the final cross-section of elastic scattering in the presence of inelastic scattering and inelastic scattering correspondingly as

$$\frac{d\sigma_{elas}}{d\Omega} = Z^2 \left(\frac{d\sigma}{d\Omega} \right)_R |-1 + F_e(q)|^2 \tag{3.30}$$

$$\frac{d\sigma_n}{d\Omega} = \frac{p_f}{p_i} \left(\frac{d\sigma}{d\Omega} \right)_R |F_{n0}(q)|^2 \tag{3.31}$$

Here $\left(\frac{d\sigma}{d\Omega} \right)_R$ is the Rutherford's formula for differential cross-section of electrons on point unit charge, $F_e(q)$ is the electronic density form-factor, $F_{n0}(q)$ is the inelastic electronic form-factor:

$$F_e(q) = \frac{1}{Z} \int \left\langle 0 \left| \sum_{j=1}^{Z} \delta(r - r_j) \right| 0 \right\rangle e^{iqr} d^3r \tag{3.32}$$

$$F_{n0}(q) = \left\langle n \left| \sum_{j=1}^{Z} e^{iqr_j} \right| 0 \right\rangle, \, |n\rangle \neq |0\rangle \tag{3.33}$$

For purposes of resonant HHG the inelastic electronic form-factor can be simplified to:

$$F_{n0}(q) = iq \left\langle n \left| \sum_{j=1}^{Z} r_j \right| 0 \right\rangle \tag{3.34}$$

In case of linearly polarized field, zero initial electron velocity and when magnetic part is not considered, the accelerated electron is moving one-dimensionally in the polarization direction of the field. So change of r to z in this case is acceptable as very good approximation, if z-axis is the polarization direction of electric field. In case of one-dimensional movement of the scattered electron near the return to origin, which is a good approximation for single-color HHG, $r_j \to z_j$ and taking into account the expression for the oscillator strength:

$$f_{n0} = \frac{2m_e \varepsilon_n}{\hbar^2} \left| \left\langle n \left| \sum_j z_j \right| 0 \right\rangle \right|^2, \quad \sum_n f_{n0} = Z \tag{3.35}$$

one can write the inelastic to elastic cross-section ratio in the conditions of allowed inelastic cross-section:

$$\frac{d\sigma_n}{d\sigma_{elas}} = \frac{\frac{p_f}{p_i}q^2\left(\frac{\hbar^2 f_{n0}}{2m_e E_n}\right)}{Z^2\left(-1+\frac{1}{Z}\int \rho_e(r)e^{iqr}d^3r\right)}, \quad \left(\frac{p_i^2}{2\mu}>E_n\right) \tag{3.36}$$

Thus, there is a proportionality of excitation of inner-shell electron to a higher state by inelastic scattering to oscillator strength of the considered transition. It can be also shown that optically allowed transitions are excited most strongly in this process:

$$\begin{cases} L_n = L_0 \pm 1, L_0; & \Delta L = 0, 1; \\ J_n = J_0 \pm 1, J_0; & \Delta J = 0, 1 \\ \pi_n = -\pi_0 \end{cases} \tag{3.37}$$

This restriction is much weaker than the restriction of single-electron-only transitions $\Delta\ell = \pm 1$, $\Delta m = 0, \pm 1$ and actually allows collective electronic excitations to a metastable resonant state, which may be a collective excitation of inner-shell electrons as well and is most likely responsible for resonant HHG enhancement in C_{60} fullerene laser plasma. However, the state $|n\rangle$ should not be able to experience Auger decay. All the transitions used in resonant HHG experiments satisfy these conditions. As it was pointed out in Ref. [86] the excited state can in fact be the origin of emission of other HHG electrons, which then contribute to the harmonics that are close to the resonant one. However, this effect has not been confirmed by experiments, so HHG where origin state of the accelerated electron is an excited state is not considered important for resonant HHG. Relation of probability of spontaneous radiative decay of excited state to coherent HHG photon generation probability is determined by Einstein coefficients A_{21} and B_{21} and intensity of nonresonant harmonics as a stimulating emission. Following Ref. [13], the excited state is populated by inelastic scattering and depopulated via photoionization, where photoionization's cross-section is close to ω (in atomic units (au)). As a result, the resonance is in general sufficiently populated.

In case proper conditions for resonant HHG enhancement by stimulated emission are achieved, either by maintaining population inversion between the excited and some other lower state or by phasing out stimulated absorption in LWI process, the total emitted radiation is given by the equation:

$$W = \int_v q(v)\rho_v B_{21}dv \tag{3.38}$$

$$q(v) = \frac{1}{2\pi}\frac{\Delta v_{Lorentz}}{(v-v_0)^2+\Delta v_{Lorentz}^2/4} \tag{3.39}$$

In case of HHG in laser plasmas with relatively low density the collision of particles is not so frequent to consider a variety of relaxation processes. Low-density laser plasma has concentration in the range of 10^{16}–10^{18} cm^{-3}, while concentration of air at normal conditions is close to 10^{19} cm^{-3}. One can safely neglect recombination of accelerated electrons on nonparent ions at such conditions. In addition, laser–plasma interaction time is too short for the Doppler effects to manifest. So $\Delta v_{\text{Lorentz}} \approx \Delta v_0 = A_{21}/2\pi$. Relation (3.38) determines also the applicability of Relation (3.24) in Ref. [13] due to proportionality to the width of the resonant level. In addition, if resonant width Γ in Ref. [13] is assumed to be $\Delta v_{\text{Lorentz}}$, both models would show a cubic dependence. However, the difference is only visual as the resonant level presented in Ref. [44] has a different meaning of resonance in elastic cross-section only, because the resonant growth of elastic cross-section is possible only in conditions, where there is no inelastic scattering. In addition, Relation (3.38) has a very strong dependence on the correlation between the spectral width of the corresponding nonresonant harmonic and the transition. Because of this, one should consider the elastic cross-section in the presence of the inelastic cross-section channel closing:

$$\left(\frac{p_i^2}{2\mu} < E_n \right) \tag{3.40}$$

Separating contribution of particle and target into the wavefunction of the scattering system to a finite set of field-free eigenfunctions gives

$$\Psi(\xi, r) = \sum_n u_n(r)\Phi_n(\xi) \tag{3.41}$$

where so-called channel functions satisfy

$$(\hat{h}_n - E)u_n(r) = -\sum_{m \neq n} V_{nm}(r)u_m(r), \tag{3.42}$$

and the operator of target–particle interaction in the nth channel is

$$\hat{h}_n = \frac{\hat{p}^2}{2\mu} + E_n + V_{nn}(r). \tag{3.43}$$

In the approximation of two bound channels ($n = 1, 2$) under the condition leading to Eq. (3.38)

$$u_1(r)|_{r \to \infty} = e^{ikr} + f(k', k)\frac{e^{ikr}}{r} \tag{3.44}$$

u_1 is the radial-dependent wavefunction of free particle, which interacts with the target that is in ground state, $|1\rangle = \Phi_1(\xi)$ and u_2 interacts with the target that is in bound excited state, $|2\rangle = \Phi_2(\xi)$. So, Eq. (3.44) describes the process in which the target in its ground state is excited by flow of particles from infinity, which have momentum k_1. In the case when the energy of the scattering system is close to energy of resonant transition, that is, to eigenenergy of the discrete spectrum of the Hamiltonian $\hat{h}_2$ (3.43), that is, $\hat{h}_2\varphi_\lambda = E_\lambda\varphi_\lambda$ and $E \approx E_m$, equation:

$$u_1 = \psi_{1,k_1}^{(+)}(r) + [\hat{G}_1(E,r,r')V_{12}(r)\varphi_0]\frac{1}{E - E_m}\langle\varphi_0|V_{21}|u_1\rangle \tag{3.45}$$

can be solved with respect to u_1, yielding amplitude of elastic scattering near resonance as

$$f(k',k) = f_{pot}(k,k) - \frac{\mu}{2\pi\hbar^2}\frac{\langle\psi_{1,k'}^{(-)}|V_{12}|\varphi_0\rangle\langle\varphi_0|V_{21}|\psi_{1,k}^{(+)}\rangle}{E - (E_m + \Delta) + i\frac{\Gamma}{2}} \tag{3.46}$$

where $f_{pot}(k,k)$ is the amplitude of potential scattering, μ is electron's mass, and Δ is the decay rate:

$$\Delta = \sum_v \frac{|\langle\varphi_0|\hat{V}_{21}|\psi_v\rangle|^2}{E - E_v} + \int\frac{|\langle\varphi_0|\hat{V}_{21}|\psi_{1,k'}\rangle|^2}{E - E'}\rho(k')d\Omega\,dE'$$

$$\Gamma = 2\pi\int|\langle\psi_{1,k'}|\hat{V}_{21}|\varphi_0\rangle|\rho(k')d\Omega \tag{3.47}$$

$\rho(k')$ is the density of final states, $d\Omega$ is elementary solid angle, and $\psi_{1,k_1}^{(+)}(r)$ is the exact solution of the problem of elastic scattering, in case influence of all inelastic scattering channels is neglected, $\hat{G}_1(E,r,r')$ is Green's function that describes movement of particle x in the channel 1 neglecting its connection to other channels, $V_{12}(r) = V_{21}^*(r) = \langle 1|\hat{V}|2\rangle = \int\Phi_1^*(\xi)\hat{V}(\xi,r)\Phi_2(\xi)d\xi$, $\varphi_0 = e^{ikr}|_{k=0}$ is the bound state of the particle, $E - E_m$ is the difference between the full energy E of the system and the energy E_l of one of eigenstates of discrete spectrum of Hamiltonian $\hat{h}_1$. $u_2(r)|_{r\to\infty}$ is the exponentially decaying wave, which simply means that for such kinetic energies of incident particle the inelastic scattering can take place only virtually. ψ_v is determined by the discrete part of spectral representation of operator:

$$\frac{1}{E^{(+)} - \hat{h}_1} = \sum_v\frac{|\psi_v\rangle\langle\psi_v|}{E - E_v} + \int\frac{|\psi_{1,k'}\rangle\langle\psi_{1,k'}|}{E^{(+)} - E'}\rho(k')d\Omega\,dE' \tag{3.48}$$

It should be noted that processes of excitation to elastic and inelastic states do not interfere if states E_n and E_m are not the same states of target's Hamiltonian, because of the channel-closing condition (Eq. (3.40)). Indeed, the electron can be trapped in the elastic scattering state due to channel closing (Eq. (3.40)) only if its kinetic energy is lower than the energy difference between the ground state and inelastic scattering state. The kinetic energy of the electron at the instant of return to its origin is determined only by instant of ionization and time-dependence of pump field. But according to the uncertainty principle $\Delta t\Delta E \geq \hbar$, so assuming condition $E \approx E_m$ for (Eq. (3.45))

Figure 3.35 Level splitting in two-channel approximation in the presence of resonance near channel-closing. *Source*: Redkin and Ganeev 2017 [93]. Reproduced with permission from © IOP Publishing.

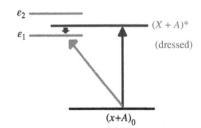

one can never tell if the electron participates in elastic or inelastic scattering when there is no state E_m with energy lower than E_n. Fortunately, this level E_m can easily be produced by AC-Stark splitting of the considered resonant level of the unperturbed target. So the elastic scattering population can be achieved when there is AC-Stark shift in the system. In case the population of elastic scattering state Λ is nonzero (e.g. due to laser dressing) LWI is possible if transition $\Lambda \Leftrightarrow |n\rangle$ is permitted in the presence of the pump field, which is resonant to energy difference between them. Fulfillment of these conditions can greatly increase the yield of resonant HHG for higher frequency harmonics. The most important consequence of Eq. (3.46) is that resonant state can be split in the presence of laser field (Figure 3.35) where the efficient cross-section of the splitting is given by Fano equation. It has been stated in Ref. [92] that in laser-dressed states the coherence may be nonzero and thus LWI can be possible even in conditions of no optical coherences. So the observation of resonant HHG can be considered as the experimental evidence of three-dimensional laser dressing of states. However, the key feature of the presented model is possibility of population of bound states in the process of HHG and this feature had to be demonstrated by computer simulation. The existence of such excitation in computer simulation is necessary, but not sufficient to prove the presented model.

3.6.2 Calculations

Quantum mechanical simulation was performed to ensure that the excited state can, in fact, be populated in HHG experiments by inelastic scattering. The main part of any HHG simulation is the solution of TDSE. Complexity of wavefunction for systems with more than 2 degrees of freedom was the reason to restrict most of HHG simulations to SAE problems. There are only two ways to make a TDSE solution feasible for multi-electron systems, which are based on time-dependent density functional theory (TDDFT) and multiconfigurational time-dependent Hartree–Fock (MCTDHF) approaches, respectively. Although TDDFT calculations are much faster for large problems, they are not very convenient in investigation of the nature of HHG, because nothing can be said about which electrons actually participate in the process.

Additionally, available TDDFT solution packages are restricted to plane-wave basis functions, which is reliable for most solid-state physics problems, but not guaranteed to give proper picture of ionization. Below the MCTDHF approach will be used, which not only gives a good combination of exactness of direct TDSE solution and speed of HF approximation, but also has a support of system's splitting, not included in HF approach.

The MCTDHF approach treats the wavefunction of multi-electronic system as

$$\Psi(Q_1, \dots, Q_f, t) = \sum_{j_1=1}^{n_1} \dots \sum_{j_f=1}^{n_f} A_{j_1 \dots j_f}(t) \prod_{\kappa=1}^{f} \phi_{j_\kappa}^{(\kappa)}(Q_\kappa, t) \qquad (3.49)$$

where $Q_1, \dots, Q_f$ are the coordinates of electrons, $A_{j_1 \dots j_f}$ is the antisymmetrized A-vector for all n_κ time-dependent expansion functions $\varphi_{j_\kappa}^{(\kappa)}$ for every degree of freedom κ. Setting $n_\kappa = n_1$ describes direct solution of TDSE and $n_\kappa = 1$ simplifies the wavefunction to ordinary time-dependent HF approximation. For direct observations of collisional excitation of core electrons, one should also reduce this representation to that of original MCTDHF approach, thus considering A-vector nonsymmetrized and the particles-distinguishable, which modifies electron–electron scattering cross-section only quantitatively (due to exchange interaction). This modification will not destroy the physical meaning of the observed processes.

The equations of motion in the MCTDHF approach are derived from modified variational principle:

$$\left\langle \delta \Psi_{\text{MCTDHF}}(t) \left| i\frac{d}{dt} - H(t) \right| \Psi_{\text{MCTDHF}}(t) \right\rangle = 0 \forall t \qquad (3.50)$$

The choice of basic functions was checked to ensure that the system actually supports bound-level structure and free particle movement. For the solution of MCTDHF equations of motion Heidelberg MCTDHF package was used, which has a good support of various basis functions in discrete variable representation (DVR). For description of angular motion of electron and dependence of pseudopotential on the orbital momentum of electron the Legendre type DVR (spherical harmonics in laser polarization plane) was used:

$$\chi_{l-m+1}(\theta) = \sqrt{\frac{2l+1}{2} \frac{(l-m)!}{(l+m)!}} P_l^m \cos(\theta) \qquad (3.51)$$

For longitude movement of electron a careful choice of basis is required, because ground-state basis functions, which can be obtained from molecular geometry programs, have a good representation of ground-level structure and can give ionization dynamics and cross-sections very good, but they do not include support for a particle in a free space. It has been shown in Ref. [97] that pseudopotentials not only reduce the computational effort, but also make plane-wave basis sets more appropriate for the real systems, so the first evident choice is the exponential DVR:

$$\chi_j(x) = (|x_{max1} - x_{01}|)^{-1/2} \exp(2i\pi j(x - x_{01})/(|x_{max1} - x_{01}|)) \qquad (3.52)$$

It was supposed that discrete level structure can be better represented by combined these exponential DVR and Laguerre (hydrogen-like) basis functions:

$$\chi_n(x) = x^{1/2}(|x_{max2} - x_{02}|)\sqrt{(n-1)!/n!}\,\exp(-x/2) \qquad (3.53)$$

The pseudopotentials for indium ionic backgrounds have been generated by means of open-source pseudopotential interface and unification module (OPIUM) pseudopotential generator [34] for neutral, singly-, and doubly ionized indium. Resonant HHG in indium is attributed to $4d^{10}5s^2{}_1S^0 \rightarrow 4d^9 5s^2 5p\ {}^1P_1$ transition in In$^+$ ion, so only three-electron system is sufficient for this simple case, if correct pseudopotentials are chosen. However, when there are multiple excitations, the potential changes nonadiabatically, so for example, in case of two-particle impact excitation the Hamiltonian will have the form:

$$\hat{H} = \begin{pmatrix} H_{11} & H_{12} \\ H_{21} & H_{22} \end{pmatrix} \qquad (3.54)$$

where H_{11} is the neutral potential, H_{21} and H_{12} are the corresponding potentials for the 1st and the 2nd core electrons removed and H_{22} is the potential for both core electrons being excited. The states are electronically coupled to each other. The resulting HHG spectrum was obtained by Fourier transformation of the induced dipole.

As a driving pulse for all calculations the Gaussian-shaped linearly polarized beam with FWHM of 35 fs and carrier frequency of 0.057 au ($\lambda = 800$ nm), and peak intensity 5×10^{14} W cm^{-2} was used. In Figure 3.36 (dashed line), the absorption spectrum of a given system (indium ion, external electron removed, that is, H11 pseudopotential) is presented, which has been obtained by adding impulse to every electron in it and letting the system propagate freely for a long time (35 fs) and taking spontaneous emission spectrum afterward. It is seen that there are some prominent absorption peaks in the area of 13th harmonic, which can be predicted even from time-independent simulations. Then propagation of the system in a laser field was performed and the results for spontaneous-emission-based estimate of HHG were obtained (Figure 3.36, solid line). The cutoff was observed for the 39th harmonic and nonresonant harmonics had approximately the same intensity in the plateau region. It is actually not surprising that resonant HHG has been observed, although with the intensity only 10–20 times higher than nonresonant harmonics, which is less than those obtained from experimental studies.

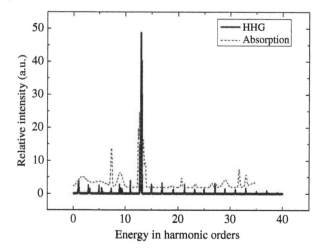

Figure 3.36 Absorption spectrum of singly charged indium ion (dashed line, shifted by 2 units of relative intensity for better visibility) and resonant HHG spectrum (solid line) of the full system. *Source*: Redkin and Ganeev 2017 [93]. Reproduced with permission from © IOP Publishing.

To investigate the process deeper, the movement of all particles in the system was monitored. Notice that even in case they were not treated as antisymmetric, but as distinguishable, even the particles not influenced by laser have in general been transferred to excited potential surfaces. This evidently means that for some accelerated electron energy the system would have less energy when this energy is exchanged during inelastic collision, that is, collision excitation is possible, and this excitation also happens coherently due to short pump pulses.

To check if the population transfer indeed occurs due to inelastic collision, and not some direct process or radiationless electron capture, the possibility for the second particle to occupy the excited state was compared by analyzing the indices of product functions from $l = 2$ d-ground state to $l = 1$ p-state, with the maximum kinetic energy of the accelerated particle at the instant of recombination derived from semiclassical assumptions. It is clearly seen that the population of excited state has time-dependent peaks (Figure 3.37) with maximums close to instants when kinetic energy at return time is in general very close to that of resonant 13th harmonic. On the contrary, if resonant HHG would originate from the accelerated particle itself (direct recombination or electron capture), these peaks for indistinguishable particles would correspond to instants when kinetic energy at instant of recombination is equal to the energy of resonant harmonic minus ionization energy and for distinguishable particles there should have been no clear dependence at all. It is seen that only the kinetic energy of the accelerated particle is transferred for the excitation of another one, so collision excitation of inner electrons during

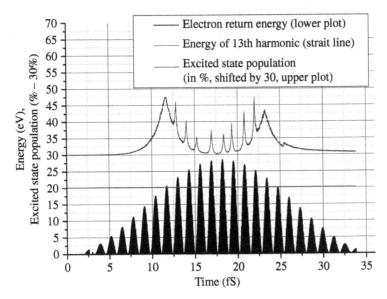

Figure 3.37 Time dependence of the electron return energy (black filled area) and the excited state population (upper line) for 35 fs pulses with Gaussian envelope; peak intensity: 5×10^{14} W cm^{-2}. The maximum excited state population reached 17.5% (upper line, which shifted upward by 30 units for better visibility). *Source*: Redkin and Ganeev 2017 [93]. Reproduced with permission from © IOP Publishing.

inelastic scattering is possibly the base of resonant HHG. The population of the excited state also experiences some Rabi-like oscillations, which can, in principle, show the possibility of LWI processes.

The population in Figure 3.37 is normalized to 13 for better visibility, and 13 corresponds to the maximum observed occupation of excited state of approximately 17.5%, which means that there is no population inversion during HHG process, so stimulated-emission-based enhancement is not possible. From that one can deduce that resonant HHG is also in general a spontaneous emission.

3.6.3 Experiment

To compare the modeling of resonance enhancement with experiment the following studies were carried out, which demonstrated the enhancement of different harmonics in the vicinity of the $4d^{10}5s^2$ $_1S^0 \rightarrow 4d^95s^25p$ $^1P_1$ transition in In II ion (19.92 eV, 62.24 nm). This transition is energetically close to the 13th harmonic ($h\nu_{H13} = 19.99$ eV or $\lambda = 62$ nm) of 806 nm radiation, the 23rd harmonic ($h\nu_{H23} = 19.93$ eV or $\lambda = 62.21$ nm) of 1431 nm radiation, and the 25th harmonic ($h\nu_{H25} = 20.38$ eV or $\lambda = 60.84$ nm) of 1521 nm radiation, thereby allowing the resonance-induced enhancement of their intensity provided the plasma conditions allow maximally efficient harmonic generation.

The experimental setup consisted of a Ti:sapphire laser, a traveling-wave OPA of the white-light continuum, and a HHG scheme using propagation of the amplified signal pulses from the OPA through the extended LPP. The details of experimental setup were described in Section 3.3.2.

Upper panel of Figure 3.38 shows the plasma emission spectrum of indium plasma at the conditions of ablation, which do not allow efficient HHG. The harmonics generated in the indium plasma at the conditions when plasma

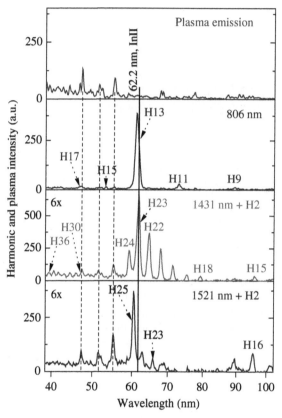

Figure 3.38 Plasma and harmonic emission spectra generating in the indium plasma. Upper panel: plasma emission at overexcitation of ablating target. Second panel: single-color pump (806 nm) of optimally formed indium LPP. Third panel: two-color pump (1431 + 715.5 nm) of optimally formed indium LPP. Fourth panel: two-color pump (1521 + 760.5 nm) of optimally formed indium LPP. Two bottom panels were magnified with the factor of 6× compared with the second panel due to notably smaller conversion efficiency in the case of longer wavelength pumps. Solid line corresponds to the In II transition responsible for the enhancement of harmonics. Dotted lines show the emission lines of In plasma observed in all these cases. *Source:* Redkin and Ganeev 2017 [93]. Reproduced with permission from © IOP Publishing.

emission was suppressed due to lower fluence of heating pulses. Three bottom panels correspond to the latter conditions. In the case of 806 nm pump (second panel from the top), extremely strong 13th harmonic (H13) dominated the whole spectrum of emission. This harmonic was close to the 62.24 nm transition of In II shown as the solid line. H13 was 15–40 times stronger than the neighboring harmonics depending on the plasma formation conditions.

The use of the two-color pump of NIR pulses (1431 + 715.5 nm, third panel from the top) also demonstrated the enhancement of harmonics in the vicinity of 62.24 nm transition of indium ion. One can see the enhancement of odd and even harmonics close to this wavelength. H23 exactly matched with the above resonance transition and correspondingly showed highest yield. In the case of another two-color pump (1521 + 760.5 nm, fourth panel from the top) the optimal conditions of most efficient conversion were maintained for the H25.

Thus the transition of In II significantly affected the harmonic spectrum generating from the indium LPP. Notice that application of most other targets did not lead to enhancement of specific harmonic but rather to the featureless homogeneous harmonic distribution gradually decreasing up to the cutoff region. These studies confirmed the earlier reported studies on the significant influence of some ionic transitions on the harmonic distribution, which were also observed in the case of Cr, Mn, Sn, and some other plasmas. The enhancement factors of specific harmonics noticeably depended on the oscillator strengths of the involved ionic transitions. Note that in some cases the conditions for resonant recombination were fulfilled for more than one harmonic provided two-color fields were used. So there should not be a single resonance populated during HHG. Any multiphoton-based resonant excitation cannot explain such multiple resonant enhancement, while the discussed approach for resonant HHG excitation by inelastic scattering remains valid as inelastic scattering channels of different spectral terms (here $4d^{10}5s^x5p^y \rightarrow 4d^95s^x5p^{y+1}$) can be populated by accelerated electron with probability proportional to square of oscillator strength. The radiation of harmonics is also the stimulated emission as resonant HHG greatly decreased at conditions of spectral detuning of NIR pulses from resonances. The lifetime of resonant state is also much greater than duration of both main pulse and resulting harmonics to consider a significant spontaneous emission at harmonic frequency an important process. It has been shown both theoretically and experimentally for In, Sn, Sb, Cr that electronic transitions from inner d-subshell to outer p-subshell (that corresponds to the subshell where the first detached electron originally resided) of species with different ionization degrees lie close to each other. Control of degree of ionization of laser plasma can thus be used to fine-tune the frequencies of harmonics, which can be resonantly generated. Model of multi-electron resonant recombination restricts the choice of spectral transitions, which are promising to resonant HHG to permitted dipole transitions from the inner subshells to the lowest unoccupied subshell.

Similar results that show resonant enhancement can be obtained using some fixed model potential, e.g. described in Ref. [13]. However, potential [13] has two major drawbacks, one resulting from another (i) this is a fitting potential that has no obvious physical explanation or direct relation to certain spectral transition lines; (ii) it cannot be calculated ab initio or derived from any theoretical assumption or approximation for a given system, all four parameters can be fitted only using experimental data. So, in fact its precision is a result of fitting to an experiment. Theory presented in Ref. [13] is quite precise and describes many experimentally observed features of resonant HHG in given systems, which makes it really valuable. But it has little relation to the described potential barrier. On the contrary, the matrix transition element is a core part of the described model.

One should be very careful when speaking of precise coincidence of theoretical and experimental quantitative results. There is a problem of reproducibility of probe beam – experimental results are mostly averaged over shots, while in theory and calculations the laser beams tend to be perfect Gaussian beams. There is also a problem of phase matching that relates to concentration of laser-ablated plasma particles. The concentration of laser plasma is in fact very difficult to be calculated. In addition, in most calculations the ablation beam itself is also not exactly reproducible. Every theory inflicts additional approximation errors. All this should be considered when respecting precision of calculation as the best proof for validity of theory. But the model potential in Ref. [16] has immense advantage of reducing a multi-electron problem to one-electron problem. So better binding of its parameters to spectral properties of given systems can be considered promising as well.

As for derivation of Eq. (3.24) in Ref. [13], one should note that no relaxation-related process is actually included in it. On the contrary, the analytical model of this equation considers only some way of population of a reaction channel, and even in Ref. [13] this channel is viewed as an inelastic scattering channel. To sum it up, the discussed model is in fact supported by calculations given in Ref. [13], and some possible difference in modeling is not too important.

References

1 Ganeev, R.A., Suzuki, M., Baba, M. et al. (2006). *Opt. Lett.* 31: 1699.
2 Ganeev, R.A., Singhal, H., Naik, P.A. et al. (2006). *J. Opt. Soc. Am. B* 23: 2535.
3 Ganeev, R.A., Naik, P.A., Singhal, H. et al. (2007). *Opt. Lett.* 32: 65.
4 Suzuki, M., Baba, M., Kuroda, H. et al. (2007). *Opt. Express* 15: 1161.

5 Ganeev, R.A., ElougaBom, L.B., Kieffer, J.-C., and Ozaki, T. (2007). *Phys. Rev. A* 75: 063806.

6 Ganeev, R.A., ElougaBom, L.B., Kieffer, J.-C., and Ozaki, T. (2007). *Phys. Rev. A* 76: 023831.

7 Suzuki, M., Baba, M., Ganeev, R.A. et al. (2007). *J. Opt. Soc. Am. B* 24: 2686.

8 Ganeev, R.A., Witting, T., Hutchison, C. et al. (2012). *Opt. Express* 20: 25239.

9 Boltaev, G.S., Ganeev, R.A., Kulagin, I.A., and Usmanov, T. (2014). *J. Opt. Soc. Am. B* 31: 436.

10 Ganeev, R.A., Zheng, J., Wöstmann, M. et al. (2014). *Eur. Phys. J. D* 68: 325.

11 Ganeev, R.A., Odžak, S., Milošević, D.B. et al. (2016). *Laser Phys.* 26: 075401.

12 Suzuki, M., Baba, M., Ganeev, R. et al. (2006). *Opt. Lett.* 31: 3306.

13 Strelkov, V. (2010). *Phys. Rev. Lett.* 104: 123901.

14 Ganeev, R.A., Chakera, J.A., Naik, P.A. et al. (2011). *J. Opt. Soc. Am. B* 28: 1055.

15 Ganeev, R.A., Strelkov, V.V., Hutchison, C. et al. (2012). *Phys. Rev. A* 85: 023832.

16 Brugnera, L., Frank, F., Hoffmann, D.J. et al. (2010). *Opt. Lett.* 35: 3994.

17 Iaconis, C. and Walmsley, I.A. (1998). *Opt. Lett.* 23: 792.

18 Chang, Z., Rundquist, A., Wang, H. et al. (1998). *Phys. Rev. A* 58: R30.

19 Kim, H.T., Kim, J.H., Lee, D.G. et al. (2004). *Phys. Rev. A* 69: 031805.

20 Tosa, V., Kim, H.T., Kim, I.J., and Nam, C.H. (2005). *Phys. Rev. A* 71: 063808.

21 Kim, C.M. and Nam, C.H. (2006). *J. Phys. B* 39: 3199.

22 Ganeev, R.A., Singhal, H., Naik, P.A. et al. (2009). *Phys. Rev. A* 80: 033845.

23 Duffy, G., van Kampen, P., and Dunne, P. (2001). *J. Phys. B* 34: 3171.

24 Cowan, R.D. (1981). *The Theory of Atomic Structure and Spectra*. Berkeley, CA: University of California Press.

25 D. R. Lide (ed) *CRC Handbook of Chemistry and Physics* (73rd edn) (1992).

26 Ganeev, R.A. (2009). *Open Spectrosc. J.* 3: 1.

27 Oleinikov, P.A., Platonenko, V.T., and Ferrante, G. (1994). *J. Exp. Theor. Phys. Lett.* 60: 246.

28 Strelkov, V., Sterjantov, A., Shubin, N., and Platonenko, V. (2006). *J. Phys. B* 39: 577.

29 Ganeev, R.A., Witting, T., Hutchison, C. et al. (2013). *Phys. Rev. A* 88: 033838.

30 Hutchison, C., Ganeev, R.A., Witting, T. et al. (2012). *Opt. Lett.* 37: 2064.

31 Altucci, C., Bruzzese, R., de Lisio, C. et al. (2000). *Phys. Rev. A* 61: 021801.

32 Froud, C.A., Rogers, E.T.F., Hanna, D.C. et al. (2006). *Opt. Lett.* 31: 374.

33 Robinson, J.S., Haworth, C.A., Teng, H. et al. (2006). *Appl. Phys. B* 85: 525.

34 Witting, T., Frank, F., Arrell, C.A. et al. (2011). *Opt. Lett.* 36: 1680.

35 Ganeev, R.A., Hutchison, C., Witting, T. et al. (2012). *J. Phys. B* 45: 165402.

36 Tudorovskaya, M. and Lein, M. (2011). *Phys. Rev. A* 84: 013430.

37 Duffy, D. and Dunne, P. (2001). *J. Phys.* B34: L173.

38 Strelkov, V.V., Mével, E., and Constant, E. (2008). *New J. Phys.* 10: 083040.

39 Haessler, S., Strelkov, V., ElougaBom, L.B. et al. (2013). *New J. Phys.* 15: 013051.

40 Ganeev, R.A., Wang, Z., Lan, P. et al. (2016). *Phys. Rev. A* 93: 043848.

41 Cormier, E. and Lewenstein, M. (2000). *Eur. Phys. J. D* 12: 227.

42 Kim, I.J., Kim, C.M., Kim, H.T. et al. (2005). *Phys. Rev. Lett.* 94: 243901.

43 Mauritsson, J., Johnsson, P., Gustafsson, E. et al. (2006). *Phys. Rev. Lett.* 97: 013001.

44 Pfeifer, T., Gallmann, L., Abel, M.J. et al. (2006). *Opt. Lett.* 31: 975.

45 Yu, Y., Song, X., Fu, Y. et al. (2008). *Opt. Express* 16: 686.

46 Liu, X.-S. and Li, N.-N. (2008). *J. Phys. B* 41: 015602.

47 Charalambidis, D., Tzallas, P., Benis, E.P. et al. (2008). *New J. Phys.* 10: 025018.

48 Kim, I.J., Lee, G.H., Park, S.B. et al. (2008). *Appl. Phys. Lett.* 92: 021125.

49 Brugnera, L., Hoffmann, D.J., Siegel, T. et al. (2011). *Phys. Rev. Lett.* 107: 153902.

50 Telle, H.R., Steinmeyer, G., Dunlop, A.E. et al. (1999). *Appl. Phys. B* 69: 327.

51 Tosa, V., Takahashi, E., Nabekawa, Y., and Midorikawa, K. (2003). *Phys. Rev. A* 67: 063817.

52 Lange, H.R., Chiron, A., Ripoche, J.F. et al. (1998). *Phys. Rev. Lett.* 81: 1611.

53 Ganeev, R.A., Suzuki, M., and Kuroda, H. (2014). *Phys. Rev. A* 89: 033821.

54 Krause, J.L., Schafer, K.J., and Kulander, K.C. (1992). *Phys. Rev. A* 45: 4998.

55 Protopapas, M., Keitel, C.H., and Knight, P.L. (1997). *Rep. Prog. Phys.* 60: 389.

56 Castro, A., Appel, H., Oliveira, M. et al. (2006). *Phys. Status Solidi B* 243: 2465.

57 Burnett, K., Reed, V.C., Cooper, J., and Knight, P.L. (1992). *Phys. Rev. A* 45: 3347.

58 Rosenthal, N. and Marcus, G. (2015). *Phys. Rev. Lett.* 115: 133901.

59 Hora, H. (1991). *Plasmas at High Temperature and Density*. Heidelberg: Springer.

60 Rus, B., Zeitoun, P., Mosek, T. et al. (1997). *Phys. Rev. A* 56: 4229.

61 Ganeev, R.A., Suzuki, M., Baba, M., and Kuroda, H. (2005). *Opt. Spectrosc.* 99: 1000.

62 Ganeev, R.A., Singhal, H., Naik, P.A. et al. (2010). *Phys. Rev. A* 82: 053831.

63 Ganeev, R.A., Hutchison, C., Zaïr, A. et al. (2012). *Opt. Express* 20: 90.

64 Ganeev, R.A., Suzuki, M., Yoneya, S. et al. (2016). *J. Phys. B* 49: 055402.

65 Ganeev, R.A., Abdelrahman, Z., Frank, F. et al. (2014). *Appl. Phys. Lett.* 104: 021122.

66 Ganeev, R.A., Baba, M., Suzuki, M. et al. (2014). *J. Appl. Phys.* 116: 243102.

67 Rothhardt, J., Hädrich, S., Demmler, S. et al. (2014). *Phys. Rev. Lett.* 112: 233002.
68 Crooker, A.M. and Dick, K.A. (1968). *Can. J. Phys.* 46: 1241.
69 Dick, K.A. (1968). *Can. J. Phys.* 46: 1291.
70 Back, C.G., White, M.D., Pejčev, V., and Ross, K.J. (1981). *J. Phys. B* 14: 1497.
71 Mansfield, M.W.D. (1981). *J. Phys. B* 14: 2781.
72 Martin, N.L.S. (1984). *J. Phys. B* 17: 1797.
73 Sommer, K., Baig, M.A., and Hormes, J. (1987). *Z. Phys. D* 4: 313.
74 Predojević, B., Šević, D., Pejčev, V. et al. (2003). *J. Phys. B* 36: 2371.
75 Marr, G.V. and Austin, J.M. (1969). *J. Phys. B* 2: 107.
76 Strelkov, V.V., Khokhlova, M.A., and Shubin, N.Y. (2014). *Phys. Rev. A* 89: 053833.
77 Yost, D.C., Schibli, T.R., Ye, J. et al. (2009). *Nat. Phys.* 5: 815.
78 Xiong, W.-H., Geng, J.-W., Tang, J.-Y. et al. (2014). *Phys. Rev. Lett.* 112: 233001.
79 McGuinness, C., Martins, M., Wernet, P. et al. (1999). *J. Phys. B* 32: L583.
80 D'Arcy, R., Costello, J.T., McGuinnes, C., and O, Sullivan, G. (1999). *J. Phys. B* 32: 4859.
81 McGuinness, C., Martins, M., van Kampen, P. et al. (2000). *J. Phys. B* 33: 5077.
82 West, J.B., Hansen, J.E., Kristensen, B. et al. (2003). *J. Phys. B* 36: L327.
83 Ganeev, R.A., Suzuki, M., Baba, M., and Kuroda, H. (2005). *Appl. Phys. Lett.* 86: 131116.
84 Milošević, D.B. (2015). *J. Phys. B* 48: 171001.
85 Milošević, D.B. (2010). *Phys. Rev. A* 81: 023802.
86 Milošević, D.B. (2007). *J. Phys. B* 40: 3367.
87 Redkin, P.V. and Ganeev, R.A. (2010). *Phys. Rev. A* 81: 063825.
88 Ganeev, R.A. and Milošević, D.B. (2008). *J. Opt. Soc. Am. B* 25: 1127.
89 Milošević, D.B., Becker, W., and Kopold, R. (2000). *Phys. Rev. A* 61: 063403.
90 Popmintchev, D., Hernández-García, C., Dollar, F. et al. (2015). *Science* 350: 1225.
91 Andreev, A.V., Ganeev, R.A., Kuroda, H. et al. (2013). *Eur. Phys. J. D* 67: 22.
92 Agarwal, G.S. (1991). *Phys. Rev. A* 44: R28.
93 Redkin, P.V. and Ganeev, R.A. (2017). *J. Phys. B* 50: 185602.
94 Kopold, R., Becker, W., Kleber, M., and Paulus, G.G. (2002). *J. Phys. B* 35: 217.
95 Paulus, G.G., Grasbon, F., and Walther, H. (2001). *Phys. Rev. A* 64: 021401.
96 Balashov, V.V. (2000). *Quantum Scattering Theory*. Moscow: MAKS Press.
97 Grinberg, I., Ramer, N.J., and Rappe, A.M. (2000). *Phys. Rev. B* 62: 2311.

4

Resonance Enhancement of Harmonics in Metal-Ablated Plasmas: Early Studies

4.1 Indium Plasma: Ideal Source for Strong Single Enhanced Harmonic

4.1.1 Strong Resonance Enhancement of Single Harmonic Generated in Extreme Ultraviolet Range

In this subsection, the first observation of an exceptionally strong resonance enhancement of a single harmonic in a plateau region is described. This phenomenon was observed during investigations of harmonic generation from femtosecond pulse interacting with indium plasma. A very strong 13th harmonic generation ($\lambda = 61.2$ nm), almost 2 orders of magnitude, exceeding the intensities of neighboring harmonics in the plateau range was observed in those studies [1].

The pump laser used in these studies was a chirped-pulse amplification Ti:sapphire laser system (Spectra-Physics, Tsunami/TSA10F), operating at a 10 Hz pulse repetition rate, whose output was further amplified using a three-pass amplifier. A portion of the uncompressed radiation (pulse energy $E = 15$ mJ, pulse duration $t = 210$ ps, central wavelength $\lambda = 796$ nm) was split from the main beam by a beam splitter (BS) and used as the heating pulse. This pulse was focused normally by a spherical lens on the indium target located in a vacuum chamber, which produced an ablation plume predominantly consisting of neutrals and low-charged ions. The intensity of picosecond pulse on the target surface was varied between 5×10^9 and 9×10^{10} W cm^{-2}. After some delay, a femtosecond main pulse ($E = 15$ mJ, $t = 150$ fs, $\lambda = 796$ nm) was focused on the indium plasma from a direction parallel to the target surface, using an off-axis parabolic gold mirror with 350 mm focal length. The maximum intensity of the main femtosecond pulse was 1×10^{15} W cm^{-2}.

The high-order harmonics were spectrally resolved using a flat-field grazing-incidence extreme ultraviolet (XUV) spectrometer with a Hitachi 1200 grooves/mm grating. An additional gold-coated grazing-incidence cylindrical mirror was placed in front of the grating, to spatially image

Resonance Enhancement in Laser-Produced Plasmas: Concepts and Applications,
First Edition. Rashid A. Ganeev.

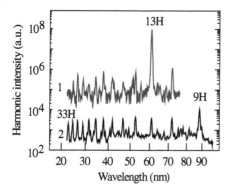

Figure 4.1 High-order harmonic spectra from (1) indium and (2) silver plumes. *Source*: Ganeev et al. 2006 [1]. Reproduced with permission from Optical Society of America.

emission from the plasma onto the detector. The XUV spectrum was detected by a microchannel plate (MCP) with phosphor screen and recorded by a charge-coupled device (CCD) camera.

In Figure 4.1, a typical spectrum obtained using the above experimental set-up is shown. High-order harmonics up to the 33rd order were observed in these experiments with indium plasma and showed a plateau pattern (Figure 4.1, curve 1). The harmonic spectrum generating from silver plasma under the same experimental conditions showed analogous characteristic plateau region for the harmonics higher than the 11th order (Figure 4.1, curve 2) and extending up to the 57th order. The conversion efficiency at the plateau region in the case of the In plasma was measured to be 8×10^{-7}. The most intriguing feature observed in these studies was a very strong 13th harmonic, whose intensity was almost 2 orders of magnitude higher compared with those of its neighbors. The conversion efficiency to the 13th harmonic was 8×10^{-5}, and for pump energy of 15 mJ, this corresponded to 1.2 μJ in the single 13th harmonic.

After the observation of such an unusual harmonic distribution, the question arises as to whether the strong emission associating with the 13th harmonic ($\lambda = 61.2$ nm) originates from amplified spontaneous emission, re-excitation of plasma by femtosecond beam, or nonlinear optical process related with enhancement of separate harmonic due to its spectral proximity to a resonance transitions. To clarify this question, the polarization and spectrum of the main pump laser were varied, which allowed investigating their influence on the 13th harmonic output.

First, to confirm that the strong line emission near 61 nm in the harmonic spectrum originated from a nonlinear process, rather than from plasma ionic emission, the effects of pump laser polarization on high-order harmonic generation (HHG) was studied. A quarter-wave plate was installed in front of the off-axis focusing mirror to vary the polarization of the main beam from linear to elliptical, circular, and finally again linear. Figure 4.2 shows the output of the 61.2 nm emission for different angles of rotation of the quarter-wave plate.

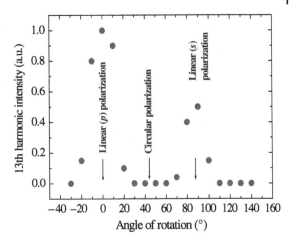

Figure 4.2 The 13th harmonic intensity as a function of the angle of rotation of quarter-wavelength plate. *Source*: Ganeev et al. 2006 [1]. Reproduced with permission from Optical Society of America.

Small deviation from linear polarization led to considerable decrease in the 61.2 nm emission intensity, which is a typical behavior for high-order harmonics. The 22.5-degree rotation of the quarter-wave plate led to the complete disappearance of 61.2 nm emission, as it should be assuming the origin of HHG. One can note that the excited lines of plasma spectrum observed at different polarizations of main beam remained unchanged, which clearly shows that the strong emission at 61.2 nm is the 13th harmonic. The difference between harmonic output for pump beams with linear p- and s-polarizations could be explained by different reflection efficiency of the grating for these polarizations.

Next, the wavelength of the main pump beam was varied to analyze whether the excited transitions from indium plasma has an influence on the plateau pattern of harmonic distribution. The central wavelength of the output radiation of laser was changed between 770 and 810 nm. The 13th harmonic output was considerably decreased with deviation of fundamental wavelength from 796 nm. At the same time a strong enhancement of the 15th harmonic at 775 and 782 nm was observed, while other harmonics remained relatively unchanged. All these observations show the influence of atomic and/or ionic transitions on the nonlinear susceptibilities of individual harmonics. The harmonic spectra observed at three different wavelengths of fundamental radiation (796, 782, and 775 nm) are presented in Figure 4.3. In Figure 4.3d, the spectrum of indium plasma is shown when the heating pulse energy is increased, but without irradiation of the main femtosecond pulse. These studies have demonstrated that the 21 nm shift in the central wavelength of the main beam (which corresponded to a 1.5 nm shift in the wavelength of the 13th harmonic) considerably changed the overall pattern of harmonic distribution at the plateau region.

An insight into the mechanism of this enhancement can be obtained by comparing the wavelength of the intensity-enhanced harmonics with those

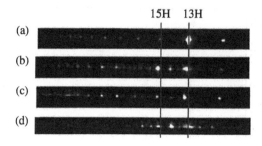

Figure 4.3 Harmonic spectra from indium plasma using (a) 796, (b) 782, and (c) 775 nm main pulse. (d) Spectrum of In plasma generated at high intensity of heating pulse radiation. *Source*: Ganeev et al. 2006 [1]. Reproduced with permission from Optical Society of America.

of past works. Figure 4.3d shows the spectrum from an In ablation plume (without irradiation by the main high-intensity femtosecond laser pulse) generated by a heating pulse with an energy higher than that used to generate optimum harmonic output. The ionization level of this ablation plume was too high to produce efficient harmonics. Comparison with past work on In plasma emission [2] shows that most of the emission observed in Figure 4.3d is due to radiative transition to the ground state $(4d^{10}5s^2\,^1S_0)$ and lowly lying state $(4d^{10}5s5p)$ of In II. This transition can be driven into resonance with the 13th harmonics (61.2 nm, 20.26 eV) by the AC-Stark shift, thereby resonantly enhancing its intensity.

One can also observe in Figure 4.3 a less pronounced but similar resonance effect for the 15th harmonic. For pump laser with central wavelength of 796 nm (Figure 4.3a) the 15th harmonic is weak, even compared with the 17th harmonic. However, for 782 nm the 15th harmonic shows enhancement, with its intensity comparable to the lower 13th harmonics. When the pump wavelength is further shifted to 770 nm, the enhancement of the 15th harmonic disappears. This behavior can be attributed to the tuning and detuning of the 15th harmonic to the $4d^{10}5s5p^3P_2 \rightarrow 4d^95s5p^2\,(^3P)\,^3F_3$ transition of In II at 23.85 eV (51.99 nm), which has a *gf* of 0.30. As the pump wavelength is changed from 796 nm to 782 nm and 775 nm, the photon energy of the 15th harmonic changes from 23.3 eV to 23.8 eV and 24.0 eV. Combined with AC-Stark shifts of the transition, the detuning is minimal for the 782 nm pump laser. The lower enhancement factor for the 15th harmonic as compared with that for the 13th harmonic can be explained by the slight detuning and lower *gf* of the 23.85 eV In II transition.

Past simulations have revealed that the resonance at a specific nth harmonic order should result in intensity enhancement effects for multiple harmonics with nth order and higher [3, 4]. This is in contrast to the present result, where only a single harmonic that is in resonance is enhanced. This discrepancy between simulation and experimental results can be explained by comparing two works on resonance-enhanced harmonics. The first observation was by Toma et al., in which the 13th harmonic of a Ti:sapphire laser (787 nm, 140 fs) was enhanced for pump intensities greater than 85 TW cm^{-2} [5]. Under these pump conditions, the intensity of the 13th harmonic was found to exceed that of the 11th harmonic by several tens of percent. Enhancement was restricted to the 13th harmonics only, and no resonance was observed for the 15th order and higher. This enhancement was identified to be due to the resonance with a 13-photon transition to the 5g state of Ar. The second observation was at a completely different condition [6], where mid-infrared laser pulses (2.5 ~ 4.5 µm, 300 fs) were used to investigate harmonics generated in alkali-metal atoms. Resonance enhancements were observed at pump intensities below 1.0 TW cm^{-2}. This intensity was enough to drive a multiphoton resonance between the 4s ground state and the 4p excited state, due to the long wavelength of the pump laser used. In this case, resonance enhancement was observed for multiple harmonics.

The lack in multiple harmonic enhancement for Ti:sapphire pump laser can be understood by comparing the differences with the harmonics generating by the much longer wavelength mid-infrared pulses. For harmonics generated by Ti:sapphire pulses, the resonant states are only weakly coupled to the ground state, and lie within one to two photons of the ionization threshold. Therefore, most of the harmonics that can be driven by resonance enhancement lie well above the ionization threshold. On the other hand, multiphoton ionization dominates for alkali-metal atoms pumped by mid-infrared lasers, which is characterized by non-perturbative behavior of the nonlinear polarization, at intensities where tunneling ionization is negligible. Therefore, one can conclude that to observe multi-harmonic resonance, it is crucial to work in the multiphoton ionization regime, and use wavelengths and atoms such that there are many harmonics above resonance, and that are still below or close to the ionization threshold.

4.1.2 Chirp-Induced Enhancement of Harmonic Generation from Indium-Containing Plasmas

Tuning of the harmonic wavelength by changing the oscillator spectrum is not practical because the adjustment of the oscillator spectrum cannot be directly transferred to the final laser spectrum due to gain narrowing and gain saturation processes, as well as necessity of readjustment of stretcher and compressor, which is a cumbersome procedure. Thus, it is not very practical to tune harmonic wavelengths by changing the laser spectrum. A much simpler approach to tune the harmonic wavelength without modifying the driving laser spectrum is by controlling the chirp of the fundamental radiation [7–9].

Indium-containing molecules may possess appropriate electron transitions, which could be used for the resonance enhancement of nonlinear optical response. While the enhancement of a single harmonic at the plateau region in the case of indium plasma may be further optimized by means of coherent control of the driving pulse, a variety of indium-containing semiconductor materials (InSb, InGaP, and InP) can be used for harmonic generation. In this subsection, we analyze the HHG from indium-containing plumes using the spectral tuning of the high-order harmonics by controlling the chirp of the driving laser radiation. We show that the chirp control allows further optimization of the 13th harmonic generation. A 200-fold ratio between the intensities of the 13th harmonic (61 nm) and the neighboring harmonics was reported in those studies [10].

The pump laser used in these studies was a chirped-pulse amplification Ti:sapphire laser system, operating at a 10 Hz pulse repetition rate. A portion of the uncompressed radiation (pulse energy $E = 30$ mJ, pulse duration $\tau = 300$ ps, central wavelength $\lambda = 793$ nm) was split from the main beam by a beam splitter and used as a heating pulse (Figure 4.4). This pulse beam was focused by a spherical lens on an indium-containing target kept in a vacuum chamber. The focal spot diameter of the heating pulse beam at the target surface was adjusted to be ~600 μm. The laser intensity on the target surface was varied between 5×10^9 and 9×10^{10} W cm^{-2}. This beam produced an ablation plume predominantly consisting of neutrals and singly charged ions. After some adjustable delay (20–50 ns), the driving laser beam ($E = 120$ mJ, $\tau = 48$ fs, $\lambda = 793$ nm, spectral width 19 nm) was focused on the indium-containing plasma from a direction parallel to the target surface, using a spherical lens of 500 mm focal length.

The maximum intensity of the femtosecond laser beam at the focal spot was 4×10^{17} W cm^{-2}. This intensity considerably exceeded the barrier suppression intensity of singly charged ions. The care was taken, by adjusting the position of the laser focus to be either in front of the laser plume or behind it, to optimize the high-harmonics output. The intensity of the driving laser pulse

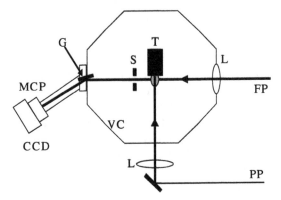

Figure 4.4 Schematic of the experimental setup on high-order harmonic generation from indium-containing plasmas. VC, vacuum chamber; T, target; S, slit; G, grating; L, lenses; MCP, microchannel plate; CCD, charge-coupled device; FP, femtosecond pulse; PP, picosecond heating pulse. *Source:* Ganeev et al. 2006 [10]. Reproduced with permission from American Physical Society.

(i.e. the femtosecond pulse) at the preformed plasma was thus varied between 2×10^{14} and 8×10^{15} W cm^{-2}. The high-order harmonics were analyzed by a grazing incidence XUV spectrometer with a Hitachi 1200 grooves/mm variable line spacing, flat field grating. The XUV spectrum was detected by a MCP with a phosphor screen and recorded by a CCD camera. Polished, 2 mm wide strips of In, InSb, InP, and InGaP were used as the targets. The spectral properties of target plasma in the visible and ultraviolet ranges were analyzed by a fiber optics spectrometer.

4.1.2.1 Preparation of the Optimal Plasmas

Efficient HHG from laser plumes can be achieved at some specific parameters of the ablated plume. The indium-containing plumes were prepared by adjusting the excitation of targets and by analyzing the plasma emission spectra from the low-excited plasmas. Figure 4.5 shows the visible/UV spectra from In, InSb, InGaP, and InP plumes in the cases of weak and tight focusing. In the studies presented below the plumes were prepared at the conditions of weak focusing, when the intensity of heating pulse did not exceed 3×10^{10} W cm^{-2}. In that case the low-excited "optimal" plasma was produced, which was suitable for efficient HHG. The term "optimal" refers to the conditions when maximum conversion efficiency and highest cutoff energy in the medium can be achieved.

The emission spectra of the plasma plumes were recorded during a single shot of the picosecond pulse, without further excitation of the plasma by the femtosecond pulse. This was done to obtain gross information about the plasma composition prior to its interaction with the driving beam. Of course, as these measurements were timeintegrated ones, one cannot say exactly what plasma conditions existed at the instant of the propagation of the femtosecond

Figure 4.5 Spectral measurements of the In, InSb, InGaP, and InP plasmas produced at the tight and weak focusing conditions of the heating pulse radiation. *Source*: Ganeev et al. 2006 [10]. Reproduced with permission from American Physical Society.

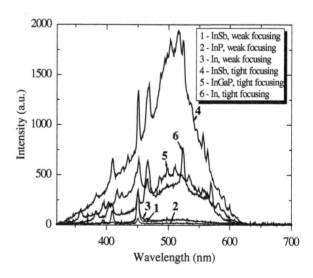

beam through the plume. Hence, one cannot assume that the main laser pulse interacted with the same mixture of excited neutrals and ions for the various delays used in these experiments. Next, the time-resolved, spectrally integrated studies of the surface plasma using fast photodiode showed that the plasma emission disappears during a period of 10–20 ns at the optimal heating pulse intensity.

It is seen from Figure 4.5 that, in the case of weak focusing on the target, the plasma consists mostly of neutral In atoms and small amount of In II [as seen from the relative intensities of the spectral lines belonging to the neutral In (525 nm) and singly ionized In (383 and 468 nm)]. Along with spectral analysis in the visible and UV ranges (using the fiber optic spectrometer), this plasma composition was also confirmed by the observations in the XUV range (using the XUV spectrometer). However, during the interaction of this plasma with the femtosecond pulse, an increase in the concentration of the singly charged ions as well as the generation of multiply charged ions was observed. These ions perhaps appear during the propagation of the leading part of the femtosecond laser pulse, since the ionization of neutral indium atoms occurred at considerably lower intensities ($\sim 5 \times 10^{13}$ W cm^{-2}) compared to the peak intensity of the driving pulse. The ionized medium, with a higher electron density in the beam center than in the outer regions, acts as a negative lens, leading to defocusing of the laser beam in the plasma. Some ionization of the medium takes place at the intensities exceeding the barrier suppression intensity. Moreover, a high electron density can limit the shortest wavelength generated and restrict the yield at higher laser intensities. When the intensity of the laser reaches the saturation intensity for singly ionized particles, a significant fraction of these species gets depleted and the HHG essentially stops. This saturation intensity may be defined as the minimum peak intensity for which the singly charged ions have been completely depleted at the end of the pulse. Such depletion leads to an abrupt end of the nonlinear optical process giving rise to the HHG. The harmonic generation was most efficient when the plume consisted of neutral indium atoms and singly charged ions.

The atom and ion densities in the interaction zone are determined by the relations describing the thermodynamic parameters of plasma. The calculations of the particle densities in the laser plume were analogous to those presented in Ref. [11]. The plasma density in the interaction zone calculated using these relations varied from 7×10^{16} to 9×10^{17} cm^{-3} depending on the plasma sample, the delay, and the distance between the target and the laser beam.

A majority of the HHG studies from the laser plumes were, so far, focused on the single-atom species, though the HHG from diatomic and multi-atomic molecules could offer some advantages compared to the single atoms due to their elongated structures. While the single atoms with their low ionization potentials are likely to see ionization saturation clamping of the HHG to low orders, this is not obvious for multi-electron and molecular species possessing rather high-ionization saturation intensities. Further, these multi-atomic clusters, being excited and evaporated from the solid surface, could exhibit some properties of nanosized structures with enhanced nonlinear optical response due to local field enhancement. Moreover, such structures may possess appropriate electronic transitions, which could be useful for the resonance enhancement of the nonlinear optical response for certain harmonics.

In order to get some information on the structure of ablated plumes, the films of the ablation plasma deposited on different substrates were studied. These substrates (glass, silicon wafer, copper, aluminum, and silver strips) were placed at a distance of 50 mm from the targets. The deposition of the target materials was carried out at different focusing conditions of the heating pulse. The tight and weak focusing conditions for the preparation of the laser plume were compared. The formation of nanoparticles is governed by the thermodynamic conditions at the target surface. After the creation of the nanoparticles at the target surface, their spatial properties do not change during the evaporation and deposition on the substrates. The presence of the nanoparticles in the plumes may be inferred by analyzing the structural characteristics of the deposited material.

The spatial characteristics of the deposited films were analyzed by scanning electron microscopy (SEM), transmission electron microscopy (TEM), and total reflection X-ray fluorescence (TXRF) spectrometry. Those studies have shown that, in the case of tight focusing, the deposited films contain a lot of nanoparticles with variable sizes. Figure 4.6 shows the SEM images of the deposited indium nanoparticles on the surface of a silicon wafer. In the case of weak focusing, the concentration of nanoparticles is considerably smaller compared to the tight focusing conditions. In the weak focusing condition, the deposited film was almost homogeneous, while in the tight focusing condition, nanoparticles with size ranging from 30 to 100 nm appeared in the SEM images. Analogous results were observed using the TEM and TXRF. In particular, the TXRF of indium deposited on a glass slide at weak

Figure 4.6 SEM images of indium film deposited on the silicon wafer at (a) weak and (b) tight focusing of the heating pulse radiation. *Source:* Ganeev et al. 2006 [10]. Reproduced with permission from American Physical Society.

focusing conditions showed a thin film-like deposition without the inclusion of considerable amount of nanoparticles.

The above structural studies showed that the "optimal" plumes prepared at weak focusing conditions consisted of single particles, and not clusters. The formation of nanoparticles in laser plumes requires a much more intense laser pulse on the target surface. At these intensities, the plasma is far from the conditions for the efficient HHG. One may have to explore other opportunities for the creation of the low-excited plumes containing nanoparticles applicable for the observation of the local field-induced enhancement of harmonic generation.

4.1.2.2 Optimization of High Harmonic Generation

In the first set of HHG studies, harmonic generation from the indium plume was carried out. The high-order harmonics up to the 39th order (at $\lambda = 20.3$ nm) were observed in these experiments and showed a plateau-like pattern, with the harmonics in the range of 11th–33rd orders appearing at nearly equal intensity, excluding the 13th harmonic. The 13th harmonic ($\lambda = 61$ nm) was observed to be extraordinarily strong (Figure 4.7), analogous to the studies discussed in subsection 4.1.1. Various characteristics of the HHG were systematically studied in order to maximize the yield of the harmonics from the indium plasma. The optimal laser plasma was created by weak focusing of the heating laser beam. The influence of the time delay between the heating and driving pulses on the harmonic yield was investigated. As seen in Figure 4.8, the harmonic output considerably increased when the delay exceeded 20 ns and then remained approximately constant up to the maximum used delay (57 ns). Also, a decrease in the conversion efficiency was observed with an increase of the heating pulse intensity (at $I_{pp} > 5 \times 10^{10}$ W cm^{-2}). This effect was attributed to the observed generation of multiply charged ions at higher heating pulse intensities and the ionization-induced defocusing of the femtosecond pump due to generation of a large amount of free electrons in the plasma plume.

The position of the focus of the driving laser beam was adjusted by placing it either in front of the laser plume or behind it (at the same distance from the target) to optimize the high harmonics output. It was observed that the latter position produces more intense harmonics than the former. Shifting the focus of the main beam toward the center of the plasma only led to the diminishing

H23 H19   H13   H11

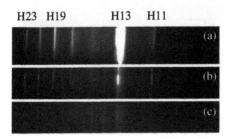

Figure 4.7 Harmonic distribution images in the case of indium plasma at the conditions of (a) chirp-free 48 fs pulses, (b) negatively chirped 95 fs pulses, and (c) negatively chirped 250 fs pulses. *Source*: Ganeev et al. 2006 [10]. Reproduced with permission from American Physical Society.

Figure 4.8 The 13th harmonic intensity from indium plume as a function of the time delay between the heating and the driving pulses. *Source*: Ganeev et al. 2006 [10]. Reproduced with permission from American Physical Society.

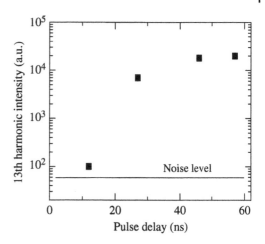

of the number of singly charged ions available for harmonic generation due to their further ionization at the high laser intensities. The harmonic yield as a function of the focus-plume position was different for low- and high-power laser intensity. For low laser intensities, a single peak at $z = 0$ was observed, while for high laser intensity, a two-peak pattern with distinct difference between their intensities appeared. No harmonics were generated at $z = 0$ in the latter case. This pattern remained the same, independent of the harmonic order in the plateau region. Such a behavior was reported by several authors in laser–gas jet experiments [12, 13], the explanation of which was based on the depletion, phase-mismatching, and self-defocusing effects. In discussed case, the latter factor seems to be the main reason of the observed dependence. A role of laser beam self-defocusing during the HHG has already been analyzed by many authors (see Ref. [14] and references therein). A simple model taking into account the role of the free electrons in the plume on the defocusing of fundamental beam predicts spatial phase modulation of the beam. The modification of the beam propagation through the ionized plume limits the laser intensity in the ionized medium and affects the conversion efficiency.

Since the sizes of nonlinear medium are comparable (due to the plasma spreading prior to the interaction with the driving radiation) with the Rayleigh length of the femtosecond laser beam ($L_R \sim 1$ mm), the above results can be related with the inhomogeneity of the plume and variations of the concentration of neutral atoms and singly charged ions. A deviation in the optimal position of the laser beam waist away from the $z = 0$ position, was due to the increase of plasma concentration and excitation as well as a considerable growth of free electrons. The increase in the ionization of plasma resulting in a higher abundance of free electrons leads to the defocusing of driving and harmonic radiation and its absorption, as well as the asymmetric dependence of harmonic output respectively to the position of the plume.

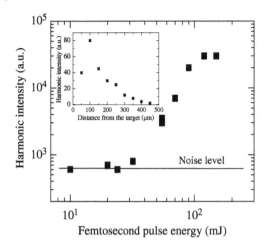

Figure 4.9 Dependence of the 19th harmonic yield on the driving pulse energy in the case of indium plasma. Inset: Variation of the 21st harmonic intensity as a function of the distance between the target surface and the axis of the driving radiation. *Source*: Ganeev et al. 2006 [10]. Reproduced with permission from American Physical Society.

The intensity of the high-order harmonics saturated at high intensities of the driving laser radiation. Figure 4.9 shows the dependence of the 19th harmonic intensity with the main pulse energy. One can see that the intensity of the 19th harmonic generated from the indium plume gets saturated at the energy of femtosecond pulse of 90 mJ, corresponding to $I_{fp} \approx 5 \times 10^{14}$ W cm^{-2}. The intensity of femtosecond pulse was well above the one [$(1-5) \times 10^{13}$ W cm^{-2}] at which the neutral atoms from various plumes are ionized. This observation could be considered as an additional indication of the crucial role of the singly charged ions in the generation of the high-order harmonics. Previously reported studies have also revealed the singly charged ions as the medium responsible for harmonic generation [15].

A discussion about the role of the multiply charged ions on the HHG may proceed through the growing influence of the free electrons appearing simultaneously with these ions as they do not participate directly in this process. The role of the free electrons is crucial due to their influence on the propagation of femtosecond beam through the ionized medium. They play a decisive role both in the restriction of cutoff energy and conversion efficiency [16, 17]. The optimal conditions for the HHG were found to be at the distance of 100–150 μm from the target surface, depending on the harmonic order. The inset of Figure 4.9 shows the dependence of the 21st harmonic yield on the distance between the optical axis of the driving beam and the target surface. The optimal distance, at which maximum harmonic yield was observed, was changed for different delays between the heating and driving pulses.

4.1.2.3 Chirp Control
All the above-discussed studies were performed with chirp-free 48 fs driving laser pulses. The resonance-induced enhancement of the single harmonic in

indium plasma using the modification of laser spectrum was first reported in Ref. [1]. We now analyze the process of harmonic wavelength tuning by changing the chirp of the fundamental radiation, without modifying the driving laser spectrum. The chirp of the driving laser pulse was varied by adjusting the separation of the gratings in the pulse compressor. A reduction in the grating separation from the chirp-free condition generated positively chirped pulses, and an increase of the grating separation provided negatively chirped pulses.

At high intensities of the driving radiation, the spectral structure of high-order harmonics should contain a positive chirp due to the self-phase modulation (SPM) of the driving laser pulse propagating through an ionizing medium as well as a dynamically induced negative chirp. Since the harmonic chirp results in a broadening of the harmonic spectrum, it should be appropriately compensated for to achieve sharp harmonics.

The change of the laser chirp resulted in a considerable variation of the intensity distribution of harmonics in the vicinity of the 13th harmonic, while other components of harmonic spectrum remained nearly unchanged. In the chirp-free case, a very strong 61 nm radiation and a plateau-like shape of the neighboring harmonics were observed (Figure 4.7a). The intensity of the 13th harmonic exceeded the intensity of neighboring ones by about two hundred times. However, in the case of negative chirp, a considerable decrease of this ratio was observed (see Figure 4.7b,c in the cases of 95 fs and 250 fs negatively chirped pulses). The 13th harmonic radiation shifted toward the short-wavelength side, out from the resonance conditions. In that case, the 13th harmonic yield was considerably decreased compared to that of the neighboring ones (Figure 4.10). In the case of positively chirped pulses, the same strong 13th harmonic was observed as in the case of chirp-free 48 fs pulses. The variation of harmonic spectrum toward the long-wavelength side led to the closeness of the harmonic wavelength with the resonance transition of In II ions

Figure 4.10 Variation of the harmonic spectrum from indium plume with the pulse chirp and pulse width. (a) chirp-free 48 fs pulses, (b) negatively chirped 95 fs pulses, and (c) negatively chirped 250 fs pulses. Each curve is shifted vertically to avoid overlap for visual clarity. *Source*: Ganeev et al. 2006 [10]. Reproduced with permission from American Physical Society.

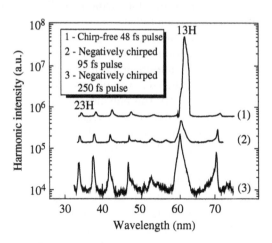

(62 nm). In that case, the 13th harmonic yield was governed by the competition of the resonance-induced amplification and the absorption processes.

The influence of the length of nonlinear medium on the 13th harmonic yield was also investigated. Most of the discussed studies were performed at a plasma plume length of 0.6 mm. The length of the plume was increased by using a cylindrical lens for focusing the heating pulse radiation on the surface of the indium target. The increase in the plume length led to a considerable decrease of the 13th harmonic yield compared to that of the neighboring harmonics. The ratio between the intensities of the13th harmonic and the neighboring harmonics decreased from 200 (for 0.6 mm long plasma) to 30 (for 3 mm long plasma) and 12 (for 6 mm long plasma), probably due to a considerable absorption of the 13th harmonic radiation in the vicinity of the resonance transition (62 nm).

It is important to calculate the absorption length, which is defined as the propagation distance after which the harmonic radiation intensity gets attenuated by a factor of e. This length is given by $l_{abs} = -L_{pl}/\ln(T)$, where L_{pl} is the plasma length and T is the transmission at a given wavelength and a given plasma density. In those studies, the absorption length was measured to be 2.6 mm. It was shown that a strong reabsorption restricts the possibility of using long indium plasma for increasing the 13th harmonic yield.

Analogous studies were carried out with the InP, InSb, and InGaP semiconductor targets as one can expect to observe a strong 13th harmonic yield from these samples as well. All these targets showed some peculiarities of harmonic spectra. Figure 4.11 shows the harmonics distribution in the range between the 9th and 29th harmonics for four indium-containing plumes. A strong 13th harmonic was observed in most of these plumes, though with different enhancement ratios over the neighboring harmonic orders. The maximum value of this ratio was observed for the In and InP plumes. The InGaP target showed a monotonic decrease of the intensity of the higher harmonics, while in the case of the InSb target a strong 21st harmonic radiation was observed at the appropriate chirp of the driving radiation.

The variation of the intensity of the 21st harmonic for different chirps of fundamental radiation was analyzed for InSb plasma. As in the case of the 13th harmonic from indium plume, it was found that in the case of positively chirped 140 fs pulses, the 21st harmonic intensity exceeded that of the neighboring ones by a factor of 7–10. This enhancement of a single harmonic intensity considerably diminished in the case of negatively chirped pulses (Figure 4.12). Moreover, in this target, the intensity of the 11th harmonic exceeded that of 13th harmonic, unlike in other indium-containing targets.

4.1.2.4 Discussion

The harmonic peaks shift to the longer wavelength side for positive chirp, because the leading edge of the driving laser pulse has more red component than the trailing edge [7–9]. The initial lower intensity portion of the pulse

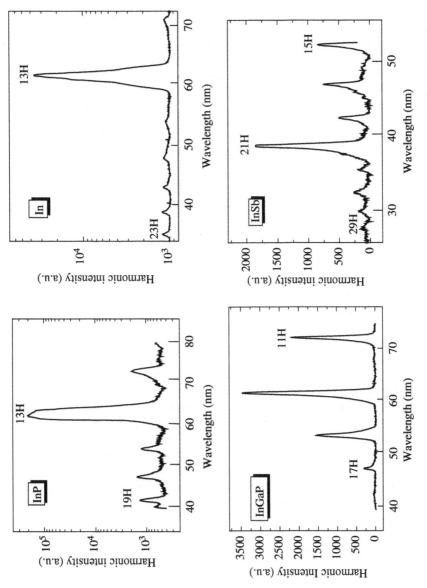

Figure 4.11 Harmonic spectra from InP, In, InGaP, and InSb plumes. *Source:* Ganeev et al. 2006 [10]. Reproduced with permission from American Physical Society.

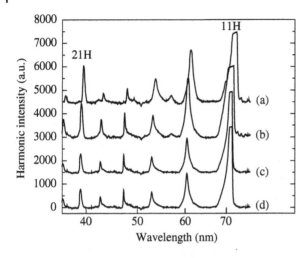

Figure 4.12 Variation of the 21st harmonic yield with chirp of the driving pulse: (a) positively chirped 80 fs pulses, (b) chirp-free 48 fs pulses, (c) negatively chirped 90 fs pulses, (d) negatively chirped 160 fs pulses. Each curve is shifted vertically to avoid overlap for visual clarity. *Source*: Ganeev et al. 2006 [10]. Reproduced with permission from American Physical Society.

creates harmonics. As the pulse intensity reaches its peak, the condition for HHG gets spoilt. Thus it is the leading edge of the pulse, which contributes to the HHG. The harmonics produced with positively chirped laser pulses were red-shifted because the harmonics produced in the leading edge of the laser pulse come from the red part of the laser spectrum. The same can be said about the blue-shifted harmonics produced by negatively chirped pulses. The variation of the position of the 13th harmonic (in the case of In plume) and 21st harmonic (in the case of InSb plume) leads to the resonance enhancement of single harmonics (at the corresponding wavelengths).

To date, no observations of the spectra of studied In-containing compounds below 100 nm have been reported. The only study in this spectral range was related with the analysis of photo-absorption lines of In I, In II, and In II* spectra between 37.5 and 70 nm [2]. The information from the past work, however, helped to identify the transitions, which can be responsible for the strong variations of harmonic yield distribution in discussed experimental conditions, when the controlled chirp of fundamental pulse was applied for achieving the resonance-like enhancement of a single harmonic.

The data on inner shell transitions of In ions are relevant not only for research in atomic structure, but also for applications in nonlinear optics, in particular resonance-induced enhancement of high-order harmonic yield. At $Z = 49$, indium occupies a position in the periodic table where filling of the 5p subshell has just begun and the nf ($n \geq 4$) wave functions are held out from the atomic core by a strong barrier. Thus the 4d excitation spectrum is dominated by broadband intense $4d \rightarrow np$ ($n \geq 5$) transitions below 4d threshold [18].

The indium plasma emission observed in the range of interest is due to radiative transitions between the ground state ($4d^{10}5s^{2}\,^{1}S_{0}$) and the lowly lying state ($4d^{10}5s5p$) of In II. Among them, the transition at 19.92 eV (62.24 nm), corresponding to the $4d^{10}5s^2\,{}^1S_0 \rightarrow 4d^95s^25p\,({}^2D)\,{}^1P_1$ transition of In II, is exceptionally strong. The absorption oscillator strength gf of this transition has been calculated to be 1.11 [2], which is more than 12 times larger than other transitions from the ground state of In II. This transition can be driven into resonance with the 13th harmonics (61 nm, 20.33 eV) by the AC-Stark shift, thereby resonantly enhancing its intensity. It is difficult to calculate the AC-Stark shift accurately. But according to Ref. [19], the AC-Stark shifts can be several electron volts in magnitude.

The enhancement of the 13th harmonic (×200), generated from In plasma, considerably exceeded that of the 21st harmonic (×10), generated from InSb plasma. Probably, the reason of such a difference is related with the difference in oscillator strengths of the transitions involved. Here we have to underline that one of goals of this study was to demonstrate that the enhancement of harmonics can be realized not only near the beginning of the plateau, as it was shown previously [1], but also at the middle of the plateau region as well. It gives some hope for further findings in the enhancement of the intensity of the higher-order harmonics in the short-wavelength range using different targets.

The generation of arbitrarily shaped spectra of HHG by adaptive control of the pump laser pulse in laser–gas jet experiments was demonstrated in Ref. [20]. This allowed the selection of some spectral ranges of the generated harmonics. Their approach, though very promising for the engineering of a whole harmonic spectrum by enhancement and suppression of a group of harmonics, did not provide a single-harmonic output in the plateau region, unlike the discussed results, where a significant intensity enhancement was observed for a single harmonic by more than 2 orders of magnitude over that of the nearest neighboring harmonics.

Most of the resonance-related HHG studies analyzed the harmonic spectrum as a function of laser intensity with regard to the occurrence of enhancements for particular intensities. Another early approach involved the variation of the driving radiation spectrum to tune the harmonic wavelength and adjust it to the atomic or ionic resonances of the nonlinear medium [21, 22]. The discussed approach is close to the latter one, though the spectrum of the fundamental radiation was not changed as a whole but instead changed the spectral distribution inside the pulse by controlling the chirp of laser radiation. As it was shown earlier, this led to the tuning of the harmonic wavelength and thus allowed achieving the resonance enhancement of harmonic yield from some plumes.

The question arises as to why a multiphoton resonance with some excited state of ions leads to a pronounced resonance in the harmonic spectrum while a multiphoton resonance with other excited states does not? One can assume that the competition between the reabsorption, phase matching (mismatching), and growth of harmonic intensity, as well as the population of the excited states and the transition life-time have to be at "optimal" conditions to show the resonance-induced growth of single harmonic in the plateau region.

It follows from the preceding discussion that the origin of the strong yield of the single harmonics in the plateau region is associated with the resonance-induced growth of nonlinear optical frequency conversion. Therefore, let us examine the resonance-induced growth mechanism in a little more detail. The role of atomic resonances in harmonic generation had been a main subject of discussion in the early studies of low-order harmonic generation [5, 23, 24]. The harmonic enhancement was attributed to the existence of oscillating electron trajectories that revisit the origin more than once after having been ejected, via tunneling, from the atom or ion [25]. Since such trajectories start from the resonantly populated excited state, with a nonzero initial kinetic energy, they still have nonzero instantaneous kinetic energy when they return to the origin. As usual, the recombination results in the emission of harmonics, but due to the relatively low probabilities of recombination, the population in the laser-driven wave packets increases continuously and the probability for harmonic emission grows with the number of allowed recollisions. This multiple recollision is predicted to enhance the harmonics under atomic and ionic resonances [25].

In some reports, the structures appearing in the harmonic yield versus intensity graphs were attributed to the resonance effects [5, 23]. The fact that the production of high-energy electrons is enhanced at specific intensities suggests the existence of a resonant process that increases the number of collisions by generating more free electrons or by directing them more efficiently to the nucleus. It has been demonstrated that the resonances, together with multiple recollision processes, which have been shown to be at the origin of the enhancements of the magnitudes of above-threshold ionization peaks in the high-energy range, can also play a determining role on the magnitudes of harmonic lines within the plateau. Since the HHG is a result of similar impacts on the core, followed by recapture, it is likely that an increased number of electron–ion collisions due to resonant ionization will also affect the harmonic spectrum.

The increase of harmonic yield has also been explained in the frames of the resonance amplification model based on the analysis of one-dimensional (1D) time-dependent Schrödinger equation for simple model potentials [4]. However, the calculations using this model lead to the enhancement of a group of harmonics, together with the harmonic for which the resonance conditions are fulfilled [6], contrary to observations of single harmonic enhancement. The amplification of a single harmonic in the case of electrons with nonzero initial energy was shown in Ref. [25] at certain conditions. According to estimations, such an enhancement of 13th harmonic yield could be observed at an intensity of the driving radiation of $\sim 1.2 \times 10^{14}$ W cm^{-2} and $\sim 2.3 \times 10^{14}$ W cm^{-2} in the cases of harmonic generation using In I and In II cores, respectively. The enhancement of the 21st harmonic yield could be observed at an intensity of the driving radiation of $\sim 1.2 \times 10^{14}$ W cm^{-2} in the cases of harmonic generation at In II core. The results described in Ref. [25] show that the growth of single harmonic intensity considerably depends on the wavelength and intensity of driving radiation. In the discussed case, the resonance-induced enhancement of single harmonic was observed in a broad range of laser radiation intensities. This points toward the possibility of the tuning of the resonance amplification during laser pulse. A detailed analysis taking into consideration the influence of the electrons of inner shells (together with the electrons of outer shells) needs to be carried out for the explanation of the observed peculiarities. This stipulation emanates from the presence of strong transitions from the inner shell $4d^{10}5s^2\,{}^1S_0 \rightarrow 4d^95s^2$ np, mf of In II, some of which coincide or are close to the frequencies of amplified harmonics [2].

Some theoretical simulations predict that in the strong-field regime, a single multiphoton resonance can enhance many harmonics of the fundamental laser radiation simultaneously [6]. This regime is characterized by the nonperturbative response of a medium to an intense driving field, when the nonlinear polarization giving rise to the qth harmonic increases with the laser intensity I_{fp} much more slowly than I^q. However, with the exception of some cases in which a single harmonic was enhanced [3, 5], atomic resonance effects have been observed in very few studies of the HHG in the rare gases. Even in the above-mentioned studies the enhancement was less than an order of magnitude, which is insufficient for most applications.

The near absence of resonance enhancement of the high-order harmonics in previous laser–gas jet experiments can be understood by considering the difference between the excitation spectra of atoms and ions of available solid targets and rare gases. In the case of plasma from various targets, there is a high probability to find a proper target for which the fulfillment of multi-resonance conditions in vacuum ultraviolet (VUV) range can lead to the enhancement of the harmonic yield.

The characteristic range of the conversion efficiencies achieved in the plateau region in the case of HHG from gas jets did not considerably exceed 10^{-6} in early experiments. The conversion efficiency has been subsequently improved up to 10^{-5} by proper phasematching [26]. In the case of laser–solid surface experiments, wherein both odd and even harmonics are generated in the direction of specular reflection, the conversion efficiency is higher and comparable to that of saturated collisional XUV lasers, and is much higher than that for other XUV sources. The HHG from laser plasma in conventional conditions shows a conversion efficiency comparable with the one achieved in the case of HHG from gas jets. However, as it is shown in the present work, the resonance conditions can considerably increase the efficiency of a single harmonic in the plateau region to exceed the one for HHG from laser–solid surface experiments in the XUV region.

4.2 Harmonic Generation from Different Metal Plasmas

4.2.1 Chromium Plasma: Sample for Enhancement and Suppression of Harmonics

The superb temporal and spatial coherence of high-order harmonics most commonly generated using noble gases make them attractive for application in XUV interferometry [27] and spectroscopy [28], and their short duration has opened the way toward attosecond pulse generation. Another interesting opportunity for the development of coherent XUV sources arises from the HHG after the propagation of short laser pulses through the laser plasma produced at the surface of solid targets [29–34]. However, the maximum observed order of harmonics reported in those studies was limited to the 9th [31], 11th [29, 32], 13th [30], 21st [33], and 27th [34], respectively, due to some concurred effects in high-excited plasma. No plateau pattern for high harmonics was reported in these studies.

In this subsection, we analyze studies on harmonic generation from the femtosecond pulse propagated through the low-excited chromium plasma created by heating pulse radiation. The harmonics up to the 33rd order ($\lambda = 24.12$ nm) were observed and the influence of various parameters on conversion efficiency was analyzed. A steep decrease of conversion efficiency for low-order harmonics (up to the 15th order) was followed by a long plateau with the conversion efficiency of 10^{-7}. We discuss the conditions when some of the harmonics (in particular, the 27th order) disappear from the plateau spectrum [35].

HHG scheme was similar to the one described in the previous subsection. The heating pulse was focused on a 4 mm thick Cr slab target by a spherical lens. After some delay, the femtosecond main pulse (8 mJ, 150 fs) was focused on the chromium plasma from the orthogonal direction, using a 200 mm focal length lens. The generated harmonics were analyzed by an XUV spectrometer.

The harmonic intensity increased sharply when the delay between the heating and main pulses was varied from 0 to 10 ns, and did not show large changes up to the maximum delay of 94 ns used in these experiments (Figure 4.13a). The efficient high-harmonic emission was observed in the case when only strong neutral Cr lines appeared from the laser ablation of target (Figure 4.13b). High harmonics up to the 33rd order (24.12 nm) were observed in these experiments (Figure 4.14a). The heating pulse intensity in this case was measured to be $I_{hp} = 3 \times 10^{10}$ W cm^{-2}. The harmonics generated from the chromium plume appeared to be similar to those observed in gas harmonics, with characteristic shape of plateau for harmonics exceeding 15th order (Figure 4.14c). The conversion efficiency in the plateau region was measured to be 10^{-7}. The plateau disappeared when the heating pulse energy was increased, that led to the generation

Figure 4.13 (a) The 19th harmonic intensity as a function of the delay between heating and main pulses, and (b) emission spectrum of chromium plasma in the visible range at optimal conditions of frequency conversion. $I_{hp} = 4 \times 10^{10}$ W cm^{-2}. *Source:* Ganeev et al. 2005 [35]. Reproduced with permission from AIP Publishing LLC.

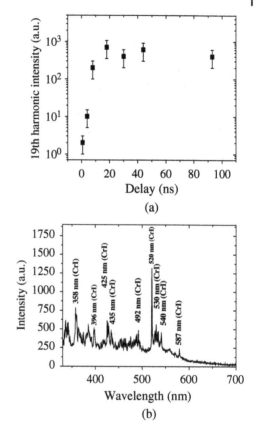

of multiple-charged ions in Cr plume (Figure 4.14b). In that case the heating pulse intensity was 1×10^{11} W cm^{-2}.

The relatively low cutoff harmonic energy can be caused by free-electrons-induced self-defocusing. The self-defocusing of the main beam induced by free electrons can considerably decrease the effective peak intensity within the nonlinear medium. In that case, cutoff energy is no longer determined by a peak intensity, but by a laser intensity at which the ionization of nonlinear medium occurs.

The harmonic intensity scaled as $(I_{hp})^4$ for small heating pulse energies. The nonlinear growth of the harmonic intensity with heating pulse intensity can be explained as a result of the increase of particle density with growth of I_{hp}. It has been shown that the harmonic yield has a quadratic dependence on the particle density in the plateau region [36, 37]. With a growth of heating pulse energy above 10 mJ, corresponding to the intensity $I_{hp} > 5 \times 10^{10}$ W cm^{-2}, a considerable decrease of harmonic intensity was observed. A decrease of conversion efficiency at the conditions when further growth of heating pulse intensity leads

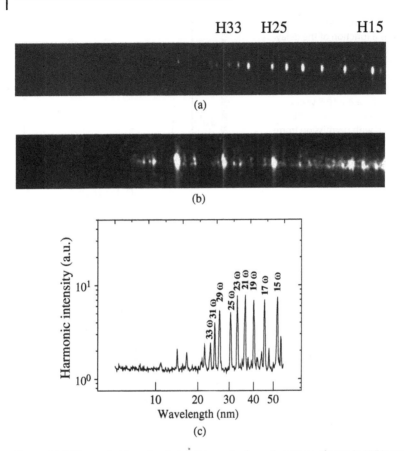

Figure 4.14 Harmonic (a) and emission (b) spectra from chromium plasma in XUV range. (c) High-order harmonic intensity distribution in the plateau region. *Source:* Ganeev et al. 2005 [35]. Reproduced with permission from AIP Publishing LLC.

to the abundance of multiple charged ions can be attributed to processes such as phase mismatching and ionization-induced self-defocusing of the main beam, both of which are due to the appearance of free electrons in the plume.

The peculiarity of those studies, alongside the enhanced 29th harmonic, was a considerable suppression of the 27th harmonic when compared to the neighbor harmonics. Such a feature was observed in the cases of different focusing geometry using both 200 and 100 mm focal length lenses. In the latter case, a flat gold mirror was used instead of cylindrical one in grazing-incidence XUV spectrometer in order to analyze the spatial characteristics of harmonic radiation. Figure 4.15a presents the harmonic spectrum from the chromium plasma in the case of 100 mm focal length lens. In that case the harmonics as high as 41st one were observed, though their appearance was unstable. The absence

Figure 4.15 Comparison of harmonic spectra from (a) chromium and (b) boron plumes. *Source*: Ganeev et al. 2005 [35]. Reproduced with permission from AIP Publishing LLC.

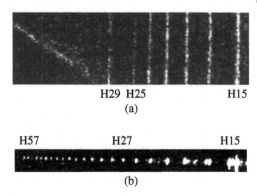

H29 H25   H15

(a)

H57   H27   H15

(b)

of the 27th harmonic was observed in a broad range of main pulse intensities. A comparison of harmonic spectra generated from the chromium and boron plasmas (Figure 4.15b) has shown that in the latter case a plateau comprised all harmonics, up to the 57th order.

The uniformity of laser plume can affect conversion efficiency. The plume became nonuniform only at high intensities of heating pulse radiation. Such conditions were avoided due to the high plasma density led to the growth of reabsorption of harmonic radiation, appearance of a large amount of free electrons, self-defocusing, plasma rippling, etc., that prevented achieving the efficient conversion in the XUV region.

Those studies were aimed to show that the target material plays an important role in achieving some specific conditions for separate harmonic. In particular, in the case of Cr plume, a considerable decrease of conversion efficiency for 27th harmonic can be related with both propagation effects and single-atom or single-ion response. These findings were further realized in analogous experiments with other plumes.

4.2.2 Studies of Resonance-Induced Single Harmonic Enhancement in Manganese, Tin, Antimony, and Chromium Plasmas

The investigations of HHG from plasma plumes infer an interesting approach. The method capitalizes on the efficient harmonic generation from low-ionized laser plasma generated on the surfaces of various solid-state targets. By using plumes generated from specific materials, coincidental overlap between the harmonic wavelength and a strong radiative transition of neutrals and singly charged ions can lead to considerable enhancement of the harmonic yield. By using solid target atoms for HHG, there is the possibility to investigate resonance enhancements with materials that were not accessible in the past. The observation of intensity enhancement of a single harmonic in the plateau region has been reported. In particular, the 80-times intensity enhancement of the 13th harmonic of Ti:sapphire laser pump was demonstrated, using indium plasma as the nonlinear medium, and by varying the spectrum of the pump laser. Plasma plume of GaAs and InSb also showed enhancement of single harmonics at different harmonic orders.

Currently, intensity enhancement of single harmonics has been limited to relatively low- to middle-orders. Therefore, an important direction would be to further extend the photon energy at which such intensity enhancements can be realized. Such studies would pave the way for the creation of intense, quasimonochromatic source of coherent XUV radiation. In this subsection, we discuss the active control of intensity enhancement of single harmonics using plumes of various materials, by varying the chirp of the pump laser [38]. Such enhancement always occurred when the wavelength of the harmonic was spectrally in the vicinity of a strong radiative transition with large oscillator strength.

Various targets were analyzed to identify promising materials that demonstrate the enhancement of specific harmonics in the plateau region. Among them, Sb, Mn, Sn, and Cr, showed the highest enhancement of harmonics, apart from indium plasma. Those studies were performed by varying the chirp of the main pump laser pulse, to tune the harmonic wavelengths to the wavelength of the ionic transitions with strong oscillator strengths. Varying the laser chirp resulted in a considerable change in the harmonic spectrum generated from the laser plasma.

The main pump laser was a chirp-free 35 fs duration pulse. All the targets used in those experiments showed some intensity enhancement (or reduction) of some specific harmonic order under these conditions. One method of varying the harmonic spectrum distribution in the plateau is by tuning the central wavelength of the main pump laser. However, this is not practical because the adjustment of the oscillator spectrum cannot be directly transferred to the final laser spectrum due to gain narrowing and gain saturation processes. One also needs to readjust the stretcher and the compressor, making the whole alignment process very difficult and cumbersome. A much simpler approach to tune the harmonic wavelength without modifying the driving laser spectrum is by controlling the chirp of the fundamental radiation. In the following, we analyze some peculiarities of HHG from several above-mentioned plasma plumes using this technique.

4.2.2.1 Manganese Plasma

Initially, the harmonic generation from manganese plasma was observed up to the maximum cutoff of H29. The harmonic spectrum showed a conventional plateau pattern for higher orders of harmonics. The intensity of the subnanosecond heating pulse that produces the plasma plume for this case was $I_{hp} = 1 \times 10^{10}$ W cm^{-2}. However, by further increasing the subnanosecond pulse intensity on the manganese target surface, a considerable increase in the harmonic cutoff was achieved. An interesting observation was the emergence of a plateau pattern at higher orders (from the 33rd to 93rd harmonic), which was followed by a steep drop of harmonic intensity. This second plateau appeared in place of a harmonic plateau between 15th and 29th orders, which were observed for moderate irradiation of Mn target by the subnanosecond pulse [39].

In those HHG experiments with 800 nm driving radiation, no considerable enhancement of a specific single harmonic was observed, although slight increase of several harmonics between the 33rd and 41st orders compared to other harmonics was clearly seen in harmonic spectra. A different pattern was observed in the case of 400 nm driving pulses. The maximum harmonic order (21st) in this case was considerably lower compared to the case of 800 nm pump (101st). However, enhancement of a single harmonic was observed for this experimental configuration (Mn plasma pumped by 400 nm pulse) (Figure 4.16). The intensity of the 17th harmonic was five times more intense than those of the neighboring harmonics. Interestingly, the wavelength of this harmonic ($\lambda = 23.5$ nm) was close to the wavelength of the 33rd harmonic ($\lambda = 24.3$ nm) in the case of 800 nm pump, which also showed some enhancement with respect to the neighbor harmonics, although much less pronounced.

The harmonic wavelength in the case of 400 nm pump was tuned by varying the chirp of the 800 nm laser. However, the intensity of the 17th harmonic remained strong, since it was not detuned out of resonance. This behavior can be explained by the narrow bandwidth of the 400 nm pulses (~8 nm), which only allowed the tuning of the 17th harmonic within a narrow spectral range (0.25 nm). This level of spectral tuning seems to be insufficient to detune from

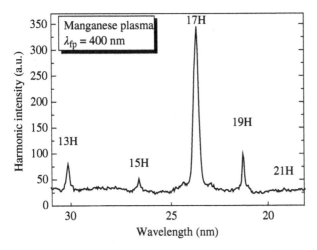

Figure 4.16 Harmonic spectra from manganese plasma in the case of 400 nm driving radiation. *Source:* Ganeev et al. 2007 [38]. Reproduced with permission from American Physical Society.

the resonance line responsible for the enhancement of the 17th harmonic. Note that, in the case of 800 nm laser, the variation of the laser chirp allowed a considerable change in the enhancement of specifics harmonic in previous works [1, 40–42].

The plasma characteristics were analyzed by simulations using the HYADES code [43]. The expansion of the manganese target interacting with the heating pulse laser was simulated to determine the electron density, ionization level, and ion density of this plume as a function of the heating pulse intensity, at a distance of 300 μm from the target surface. These simulations showed that, at $I_{hp} = 3 \times 10^{10}$ W cm^{-2}, the ionization level of Mn plume achieves 1, and the concentration of singly charged ions and electrons becomes equal to 3.3×10^{17} cm^{-3}. What is especially important here for HHG is that the ion density increases considerably with an increase in the heating pulse intensity. As a result, the harmonics, especially near the cutoff, will increase nonlinearly in intensity, thus helping the detection of these harmonics.

The enhancement of 23.5 nm radiation ($E_{H17(400)} = 52.9$ eV) in the case of Mn plasma could be associated with the resonance-induced growth of conversion efficiency in the presence of ionic lines with strong oscillator strengths. The Mn III and Mn II lines in the range of 51–52 nm were studied in past works and proved to possess strong oscillator strengths [44, 45]. These results indicate that the influence of some of these transitions led to the growth of the 17th harmonic yield.

4.2.2.2 Chromium Plasma

In previous studies, chromium plasma showed resonance-induced intensity enhancement and suppression properties for some harmonic order. In Ref. [35], a considerable decrease of the 27th harmonic of 796 nm radiation ($\lambda = 29.5$ nm, $E_{H27} = 42.2$ eV) with regard to the neighboring ones was reported and attributed to some ionic transitions demonstrating strong absorption oscillator strength. In the Cr harmonic spectrum, the ratio of the intensity of the 27th and that of the neighboring harmonics changed from almost zero to a value close to 1 by using harmonic tuning [41]. Although the observation of the extinct 27th harmonic generated from Cr plume has previously been reported in Ref. [35] (see subsection 4.2.1), no spectral variations of the harmonic wavelength were performed in those experiments to confirm the crucial role of the resonance-induced variation of the 27th harmonic yield.

At the same time, in the case of Cr, a strong 29th harmonic of the 795 nm laser ($\lambda = 27.3$ nm) belonging to the mid-plateau region has been reported [41]. This pattern was observed in the case of chirp-free pulses. The variation of the chirp of the pump laser led to the decrease of the 29th harmonic yield compared to the neighboring harmonics. The maximum ratio of the intensity of the 29th harmonic compared to that of the 31st harmonic was measured to be 23.

The discussed studies using 800 nm, 35 fs pulses, confirmed previously, reported peculiarities of the harmonic spectra generating from chromium plasma and revealed new features of harmonics in the case of 400 nm driving pulses. In particular, an enhanced 29th harmonic ($\lambda = 27.6$ nm, $E_{H29} = 45.1$ eV) approximately coincided with the short-wavelength wing of the strong spectral band of the $3p \rightarrow 3d$ transitions of Cr II ions. Moreover, the observed enhancement of the 15th harmonic of 400 nm driving radiation also can be attributed to the enhancement of nonlinear susceptibility of this harmonic induced by the influence of the same spectral band, though not so pronounced as in a case of 29th harmonic of 800 nm radiation.

Previous studies of photoabsorption and photoionization spectra of Cr plasma in the range of 41–42 eV have demonstrated the presence of strong transitions, which could be responsible for such a suppressed pattern of harmonic spectrum [46–48]. In particular, the region of the "giant" $3p \rightarrow 3d$ resonance of Cr II spectra was analyzed in Ref. [48] and the strong transitions, which could both enhance and diminish the optical and nonlinear optical response of the plume, were revealed. The neutral and ionized Cr spectra, previously believed to be completely different, were shown here to be rather similar.

Those and other studies used the photoabsorption/photoionization spectra to identify the areas of strong absorption. A considerable effort has been made to explain the spectral structure previously observed in the 3p excitation region of both neutral and ionized Cr, but the attempts have not been successful. In any case, the reported data explained entirely the nonlinear optical response of Cr plasma in the case of harmonic generation using 800 nm main pump laser, which was revealed in previous [35, 41] and present studies.

In the first set of experiments with Cr plume, both the extinction of the 27th harmonic and enhancement of the 29th harmonic of the 800 nm pump laser were observed (Figure 4.17). Variation of the chirp of the main laser pulse led to further enhancement of the 29th harmonic yield (Figure 4.18). The optimal conditions, at which this harmonic showed maximum conversion efficiency, corresponded to positively chirped 135 fs pulses. In this case, the harmonic output was approximately two times stronger compared to the case of chirp-free pump laser. The enhancement factor of 29th harmonic compared with the neighboring harmonics (18-times enhancement) was slightly less than recently reported data (23-times enhancement, [41]), probably due to broader spectrum of the pump laser for the former case (40 nm, as compared with 20 nm in Ref. [41]).

The experiments with 400 nm pump laser showed similar enhancement of a single harmonic, whose wavelength was close to previously observed harmonics that showed intensity enhancement using the 800 nm pump. Figure 4.19 shows the harmonic spectrum obtained for 400 nm main pump interacting with chromium plasma. The observed enhanced 15th harmonic (5-times enhancement, $\lambda = 26.7$ nm, $E_{H15} = 46.7$ eV) almost coincides with a

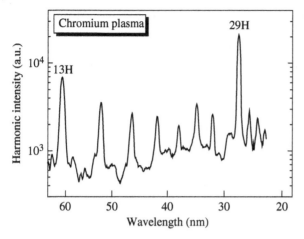

Figure 4.17 Harmonic spectra from Cr plasma in the case of 800 nm, 35 fs chirp-free main pulses. *Source*: Ganeev et al. 2007 [38]. Reproduced with permission from American Physical Society.

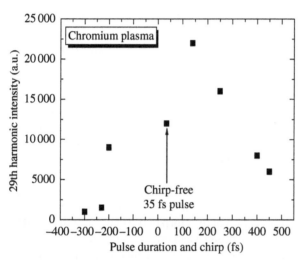

Figure 4.18 Variation of 29th harmonic intensity at different chirps of 800 nm radiation. Chromium plasma. *Source*: Ganeev et al. 2007 [38]. Reproduced with permission from American Physical Society.

broad spectral emission of singly charged Cr, which was also responsible for the enhancement of the 29th harmonic of the 800 nm pump laser in previous experiments. The chirp of the 400 nm radiation was changed by varying that of the 800 nm pump, but as in the case with Mn plasma, this did not lead to the tuning of harmonic wavelength and correspondingly did not show the relative change in the 15th harmonic intensity compared to its neighbors.

The calculations of *gf* values in the photon energy range of 40–60 eV presented in Ref. [46] show a group of the transitions in the 44.5–44.8 eV

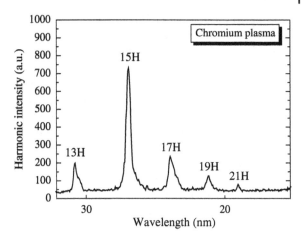

Figure 4.19 Harmonic spectrum from Cr plasma in the case of 400 nm driving pulses. *Source*: Ganeev et al. 2007 [38]. Reproduced with permission from American Physical Society.

region possessing very strong oscillator strengths (with *gf* varying between 1 and 2.2), which considerably exceeded those of other transitions in the range of 40–60 nm. These transitions were assumed to be responsible for the observed enhancement of the 29th harmonic. At the same time, the strong photoabsorption lines within the 41–42 eV region reported in the above work could considerably decrease the yield of 27th harmonic.

4.2.2.3 Antimony Plasma

The HHG of the antimony atom has previously been investigated using InSb plume [10], as well as with pure Sb plume excited on the surface of solid antimony target [49]. Past works have shown that for InSb plasma, in the case of positively chirped laser pulses, the intensity of the 21st harmonic of 795 nm laser considerably exceeded that of the neighboring harmonics [10], with enhancement factor of 10. On the other hand, for chirp-free and negative chirped laser pulses, the 21st harmonic intensity was only slightly higher than that of the neighboring ones. The role of Sb in 21st harmonic enhancement was confirmed by the studies of pure In plume, where no enhancement was observed for this harmonic. A confirmation of this conclusion was also obtained in [49]. The intensity enhancement of a single high-order harmonic at a wavelength of 37.67 nm was demonstrated using low ionized antimony laser-ablation plume. The conversion efficiency of this harmonic was reported to be 2.5×10^{-5} and the output energy was 0.3 µJ. Such an enhancement of single-harmonic was caused by the multiphoton resonance with the strong radiative transition of the Sb II ions. The intensity of the 21st harmonic at the wavelength of 37.67 nm was 20 times higher than that of the 23rd and the 19th harmonics.

The Sb I spectrum is dominated by two peaks, a broad one centered near 31.24 eV with a bandwidth close to 1 eV full width at half maximum (FWHM) and a narrower one centered near 32.22 eV. The spectrum also contains a large population of Sb II, which gives rise to peaks at 32.4 and 32.7 eV. However, the strongest transitions among these Sb I lines calculated and measured in [50], was the $4d^{10}5s^25p^3\,{}^2D_{5/2} - 4d^95s^25p^4\,({}^3P)\,{}^2F_{7/2}$ transition ($E_{phot} = 31.5$ eV). The gf value of this transition is calculated to be 1.54, which is few times higher than those of other transitions in this spectral range. At the same time, among the calculated gfs for the $4d^{10}5s^25p^2 - 4d^95s^25p^3$ transitions of Sb II ion, the ${}^3P_2 - ({}^2D)\,{}^3D_3$ transition ($E_{phot} = 32.8$, $gf = 1.36$) also shows a strong oscillator strength, which could influence the nonlinear optical response of plasma due to proximity of the harmonic wavelength with this transition.

These transitions can be driven into resonance with the 21st harmonic (37.8 nm, 32.9 eV) by the AC Stark shift, thereby resonantly enhancing its intensity. It is difficult to calculate the AC Stark shift accurately, but according to Ref. [19] the AC-Stark shifts can be several electron-volts in magnitude.

In the discussed experiments, enhancement of the 21st harmonic generated from the antimony plume was observed using main pump lasers with different pulse durations and chirp. Figure 4.20 presents the variations in the intensity of the 21st harmonic at different chirps of the main pump. One can see that the optimum conversion efficiency is achieved with a negative chirp and pulse duration of 210 fs. For antimony, no enhanced harmonics were observed for 400 nm pump.

The difference in the results obtained in the discussed work with those of Ref. [10] is attributed to the different central wavelength and the spectral bandwidth of the pump laser. The enhancement factor for the 21st harmonics in these studies (8×) was less than those in the refereed works (10× [10] and 20× [49]). This could be explained by the different methods that were used

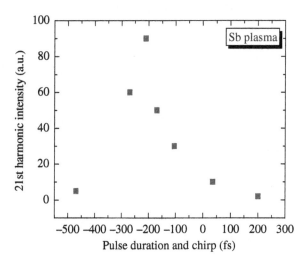

Figure 4.20 Variations of 21st harmonic yield from Sb plasma at different chirps of driving 800 nm radiation. *Source*: Ganeev et al. 2007 [38]. Reproduced with permission from American Physical Society.

for tuning the harmonic wavelength to the ionic/neutral transitions of Sb. In Ref. [49], wavelength tuning of the pump laser was realized by shifting the master oscillator wavelength close to the resonance lines of ions, while in Ref. [10] the chirp was varied, but using pump laser with a narrower bandwidth. These results infer that the use of pump laser with narrower bandwidth is preferable to increase the enhancement of the harmonics.

4.2.2.4 Tin Plasma

Strong enhancement of the 17th harmonic of 800 nm pump (47.1 nm, 26.5 eV) was observed in the Sn plume (Figure 4.21). This process was studied by shifting the harmonic wavelength in the range of 0.5 nm relative to the chirp-free position of the harmonic wavelength. The results of these experiments are presented in Figure 4.22. One can see the variation of this harmonic yield, which was optimized by a negatively chirped 70 fs pump laser. The origin of this phenomenon is similar to previously presented data on resonance-induced enhancement of single harmonics from specific plumes. In the case of tin plasma, the 15-times increase in the harmonic yield at specific chirp of the pump laser is attributed to the proximity of the 17th harmonic wavelength to ionic transitions possessing strong oscillator strength.

Such an enhancement has previously been reported and optimized by tuning the central wavelength of the master oscillator of the laser system [40]. In discussed work, the observation of strong single HHG at the wavelength of 46.76 nm by using tin laser-ablation plume was reported. The intensity of the 17th harmonic at the wavelength of 46.76 nm was 20 times higher

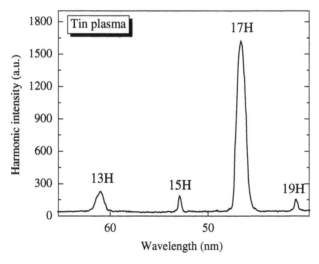

Figure 4.21 Harmonic distribution in the case of Sn plume. $\lambda = 800$ nm. *Source*: Ganeev et al. 2007 [38]. Reproduced with permission from American Physical Society.

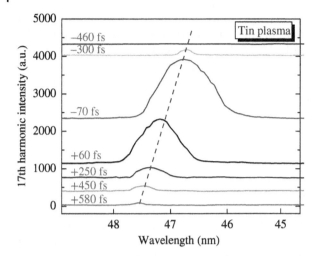

Figure 4.22 Tuning of 17th harmonic at different chirps of driving radiation. Tin plasma. *Source:* Ganeev et al. 2007 [38]. Reproduced with permission from American Physical Society.

than its neighboring harmonics. The output energy of the 17th harmonic was measured to be 1.1 µJ. The origin of this enhancement was attributed to resonance with strong radiative transition of the Sn II ion, produced within the laser-ablated plume.

In past work, the Sn II ion has been shown to possess a strong transition $4d^{10}5s^25p^2P_{3/2}$ – $4d^95s^25p^2$ (^1D) $^2D_{5/2}$ at the wavelength of 47.20 nm ($E_{phot} = 26.24$ eV) [51]. The gf value of this transition has been calculated to be 1.52, and this value is five times larger than other transitions from the ground state of Sn II. Therefore the enhancement of the 17th harmonic with the 800 nm wavelength laser pulse can be explained as being due to resonance with this transition.

4.2.2.5 Discussion of Harmonic Enhancement
The photon fluxes of enhanced harmonics considerably differ each from other. Previously, highest conversion efficiency ($\sim 10^{-4}$) has been reported in the case of the 13th harmonic generated from the indium plume. The pulse energy of this radiation (61.5 nm) was estimated to be 2 µJ. In the case of 29th harmonic ($\lambda = 27.6$ nm) produced from chromium plasma, the conversion efficiency was estimated to be 2×10^{-5}. Thus the energy of 27.6 nm radiation was assumed to be 0.4 µJ. These studies showed that the yield of enhanced harmonic strongly depended on the tuning of harmonic wavelength.

The tuning of harmonics can be realized not only by the artificial change of the chirp of driving radiation, but also by the dynamically induced chirp during interaction of strong laser field with the ionic medium. In particular, the multiphoton ionization of the medium occurring in the laser focus imposes a change of the refractive index due to the free electron contribution. In that case, the refractive index decreases, resulting in a blueshift of harmonics. This phenomenon was reported by different groups (see, for example, [52]). The self-induced redshift of harmonic wavelength was rarely observed in these studies. In particular, in Ref. [53], it was shown that the chirp-induced shifts of harmonics generated in Xe gas jet in the case of relatively long (300 fs) laser pulses exhibit a trend either toward the blue or the red, depending on particles density. The redshift was observed at low gas densities. This process might be related to Kerr effect, cluster formation in the gas jet, electron–ion recombination, and ion expansion. The former mechanism was found consistent with the experimental finding reported in Ref. [53]. The chirp-induced redshift and blueshift were explained by the model including both atoms and ions. The neutrals assumed to be responsible for the blueshift, while the ions were the main reason of redshift.

The ionic medium imposes both the phase and spatial change of the propagated radiation. Whether the refraction of the beam is significant in the plasma or not depends on the intensity of the radiation. Free electrons themselves do not lead to the self-defocusing when their concentration is insignificant and they distributed homogeneously in the area of laser–plasma interaction. The main peculiarity here is a spatial inhomogeneity of the focused radiation. The creation of additional free electrons along the propagation axis (due to considerably higher intensity compared to the wings of the beam) leads to the self-defocusing caused by the gradient of free electron concentration along the beam waist radius. This process also leads to further phase mismatch between the harmonic and fundamental waves. The studies of the plasma nonlinearities leading to the self-defocusing at the conditions of HHG are presented in Ref. [54].

There is a primary difference between observation of the enhancement of single harmonics in the discussed work, using Sb-, Sn-, Cr-, and Mn-containing plumes, with the previous studies on resonance enhancement of harmonic generation in alkali metals [6], where multiharmonic enhancement was reported, and in rare gases [3, 5], when the single harmonic intensity was enhanced a few times compared to those of neighboring harmonics. Chirp control of the pump laser allows one to achieve an optimal relation between the quasiresonance conditions, reabsorption, and induced self-defocusing, leading to a considerable enhancement of the single harmonic yield.

The very few demonstrations of resonance enhancement of the high-order harmonics in previous laser–gas HHG experiments can be understood by considering the difference between the excited spectra of atoms and ions of available solid targets and few rare gases. In the case of plasma from various targets, there is a higher probability to find a proper target for which the fulfillment of multi-resonance conditions in XUV range can lead to the enhancement of the harmonic yield.

4.2.3 Enhancement of High Harmonics from Plasmas Using Two-Color Pump and Chirp Variation of 1 kHz Ti:Sapphire Laser Pulses

4.2.3.1 Advances in Using High Pulse Repetition Source for HHG in Plasmas

Enhancement of HHG in gaseous media as a result of atomic and ionic resonances has been the subject of extensive theoretical work over the last 10 years [5, 6, 25, 55, 56]. The most significant experimental results in this field were, however, reported for HHG in plasma plumes. Weakly ionized plasmas from some solid targets displayed resonances in the excitation of neutrals at certain harmonic wavelengths. These experiments also showed that singly charged ions can enhance the yield of specific harmonics. As a much wider range of solids are available compared with gaseous target materials, HHG studies in plasma plumes dramatically increase the chance of finding an ionic transition that is resonant with a harmonic wavelength.

Resonant enhancement in HHG from metal and semiconductor targets ablated by picosecond pulses [39, 57] has been previously studied using a single pump pulse at 800 nm. However, when HHG is driven by a two-color pump scheme, emission spectra can display both odd and even harmonics, increasing the chance of a spectral overlap with an ionic resonance. Another promising route toward locating additional resonances is to tune the harmonic wavelength. This could be achieved by various methods: tuning the fundamental wavelength of laser pulse; chirping the laser radiation; altering the laser intensity to control the ionization rate of the nonlinear medium; and adaptive pulse-shaping.

Most of previous studies of resonance-induced and two-color pump-induced enhancement of harmonics in laser plasmas were demonstrated using relatively low (10 Hz) pulse repetition rate laser sources [58]. Increasing the pulse repetition rate to 1 kHz considerably increases the average power of high-order harmonics [59], and allows improved statistics in experimental applications of the emission.

An attractive feature of resonant enhancement is that it can increase the conversion efficiency of a specific harmonic by more than an order of magnitude [57]. In this subsection, we analyze resonant enhancement with two-color pumping at 1 kHz pulse repetition rate to generate strong harmonics in different spectral ranges [60]. This unique source will be ideal for various applications in physics, chemistry, and biology, and for advancing nonlinear X-ray optics and attosecond physics.

Those experiments were performed using a 1 kHz Ti:sapphire chirped pulse amplification (CPA) laser delivering 2.5 mJ pulses of 40 fs duration at 780 nm. In these experiments, a portion (1.5 mJ, 20 ps) of the uncompressed pulse was split from the beam line prior to the laser compressor stage. This pulse was focused to an intensity of $I_{ps} = 5 \times 10^9 - 3 \times 10^{10}$ W cm^{-2} on the target.

The compressed pulse (30 fs, 1 mJ) was focused into the plasma in a direction orthogonal to that of the picosecond pulse. The position of the focus with respect to the plasma plume was chosen to maximize the harmonic signal. The intensity at the focus of the femtosecond pulse was estimated to be $I_{fs} = 5 \times 10^{14}$ W cm^{-2}. The delay between plasma initiation and femtosecond pulse was varied in the range of 6–57 ns using an optical delay line. The high-harmonic spectrum was analyzed using XUV spectrometer.

To drive HHG using two colors, the second harmonic (2ω) of the fundamental (ω) femtosecond pulse was generated using a 0.5 mm thick BBO crystal in a type-I phase-matching scheme. Group velocity dispersion between the ω and 2ω pulses in nonlinear crystal was compensated for using a calcite plate. The second harmonic conversion efficiency was 4%. HHG was enhanced by the presence of the 2ω field despite the 25 : 1 energy ratio between the ω and 2ω pulses.

The chirp of the femtosecond laser pulse was tuned by adjusting the separation of the two gratings in the CPA compressor. Only the wavelength at the leading edge of the pulse contributes significantly to HHG because at the intensity used the strong-field-induced plasma grows increasingly with time, eventually preventing HHG; thus the wavelengths of the high-harmonic comb from a chirped pulse can be controlled through this ionization gating effect [8, 61]. The chirp was calibrated by measuring the spectrum, spectral phase, and pulse duration of the laser pulses as a function of grating separation using spectral shearing interferometry for direct electric field reconstruction (SPIDER) [62].

4.2.3.2 Comparison of Plasmas Allowing Generation of Featureless and Resonance-Enhanced HHG Spectra

In this subsection, we analyze the results of single- and two-color HHG studies using three different target materials: silver, chromium, and vanadium. The optimum delay between heating and driving pulses was found to be 40 ns for these three targets. Through the chirping technique, resonant enhancements of single odd, and in some cases even, harmonics were achieved.

Figure 4.23a,b shows harmonic spectra generated in silver plasma using the 780 nm pump. Harmonics above the 50th order were routinely generated. Variation of the laser chirp allowed a considerable tuning of harmonic wavelengths (2.8 nm in the case of 17th harmonic, Figure 4.23c), while the relative intensities of harmonics in the plateau region remained approximately the same over a broad range of the driving laser chirps, which led to variation of pulse duration in the range of −97 to +110 fs (positive and negative values of pulse duration correspond to positively and negatively chirped pulses). Note that effective tuning of harmonic wavelengths through chirp can be achieved only for broadband pulses, where a pronounced difference in the wavelength components at the leading and trailing parts of the pulse can be produced. The pulses used in the discussed experiments had a bandwidth of ~40 nm.

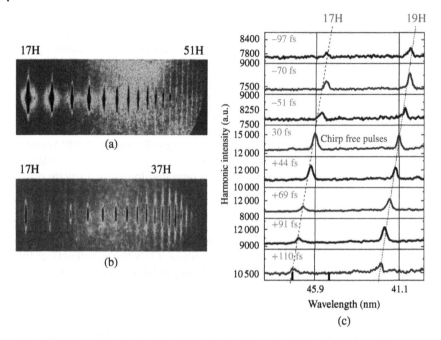

Figure 4.23 Harmonic spectra from Ag plasma in the cases of (a) apertureless and (b) apertured single color pump (780 nm). (c) Tuning of 17th and 19th harmonics by changing the distance between the gratings in the compressor stage. Positive and negative values of pulse duration correspond to positively and negatively chirped pulses. Dotted lines show the tuning of 17th and 19th harmonics with different chirps. Black lines show the wavelengths of these harmonics at chirp-free conditions. Thick black lines on the left side of bottom graph show the tuning range of 17th harmonic (2.8 nm). *Source:* Ganeev et al. 2012 [60]. Reproduced with permission from Optical Society of America.

The introduction of a second-harmonic field increased the conversion efficiency of odd harmonics, in agreement with previous studies in gaseous [63–68] and plasma targets [69]. In the case of the silver plasma, strong even harmonics were also observed at the same intensity as the odd ones (Figure 4.24a), which extended into the cutoff region in some cases, as seen in Figure 4.24b. Tight focusing ($f = 200$ mm) led to much stronger odd harmonics than even harmonics, as seen in Figure 4.24c, while the application of the 50 mm local length lens for the focusing of driving radiation in to the plasma increased the intensity of the even harmonics (Figure 4.24d).

An interesting feature of the spectrum in Figure 4.24a is the enhanced even (20th) harmonic compared with neighboring even harmonics, although this enhancement (2×) was not as pronounced as those previously reported in studies of single odd harmonics from some plasmas [57]. This lower enhancement of harmonics in the vicinity of resonance line is probably due to a weaker influence

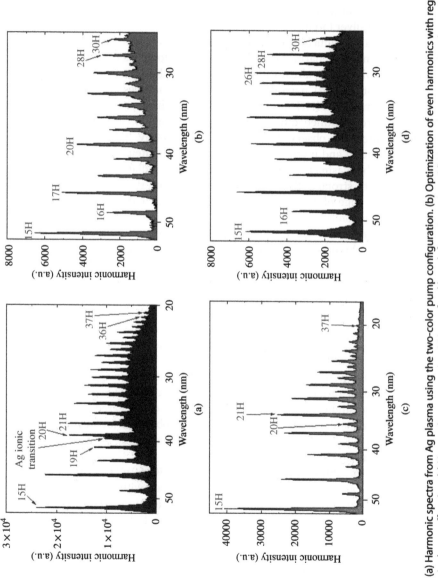

Figure 4.24 (a) Harmonic spectra from Ag plasma using the two-color pump configuration. (b) Optimization of even harmonics with regard to the odd ones in the cutoff region. (c) Harmonic spectra using 200 mm focal length focusing mirror. (d) Harmonics spectra using the 500 mm focal length lens. *Source:* Ganeev et al. 2012 [60]. Reproduced with permission from Optical Society of America.

of this ionic resonance on the nonlinear optical response of 20th harmonic. No line was identified from the NIST tables in this part of the spectrum, but the behavior was similar to resonance enhancements in other metal plasmas. One can note that this ionic line appeared along with harmonic spectra at stronger excitation of silver target.

The resonance-induced growth of a single even harmonic motivates to search for improvement of harmonic yield in two additional media, chromium and vanadium, which have already displayed, enhanced properties using low pulse repetition rate (10 Hz), narrowband (10 nm) laser sources [35, 70].

Two regimes for HHG in chromium plasma displaying different spectral features were identified during those studies. For a weak excitation of the target, a cutoff was observed at 31.2 nm ($E = 39.74$ eV, 25H). With increased target excitation and increased femtosecond pulse intensity inside the plume by moving the plasma toward the focus of 780 nm radiation, a second plateau appeared with strong 27th and 29th harmonics, and the cutoff was extended toward the range of $E = 58.81$ eV (37th harmonic). For zero-chirp the 27th and 29th harmonics were approximately equal in intensity. Varying the laser chirp changed the relative intensities of these two harmonics, while the intensities of the other harmonics remained approximately the same. As the chirp was varied and the pulse duration was changed from +114 to −92 fs, the intensity of the 27th harmonic considerably increased (15×), while the 29th harmonic became weaker, as seen in Figure 4.25a. This is because the 27th harmonic ($\lambda = 28.88$ nm, $E = 42.91$ eV for zero-chirp) is shifted toward the ionic resonance. The presence of such a resonance is confirmed by overexciting the chromium plasma; the ionic line can be seen at the blue side of the 27th harmonic on Figure 4.25b.

Those results were consistent with other HHG studies in Cr plasma, which also displayed a variation in the 27th harmonic intensity for different chirps [35]. In those studies, the 27th harmonic almost disappeared from the harmonic spectra, and a strong 29th harmonic was observed for zero-chirp. The experiments in Ref. [40] were performed using 800 nm laser pulses, so the 29th harmonic was closest to the resonance.

Calculated gf values at photon energies range of 40–60 eV ($\lambda = 20.66$–31 nm) clearly show a group of transitions from 44.5–44.8 eV with very strong oscillator strengths (gf between 1 and 2.2) [46]. These transitions are much stronger than those in the range of 40–60 nm, and are likely to be responsible for the enhancement of the 27th and 29th harmonics. Furthermore, strong photoabsorption lines within the 41–42 eV region reported in Ref. [46] could decrease the yield of the 25th harmonic.

Two-color pumping revealed that the resonance ionic line similarly impacts the efficiency of even harmonic generation. Figure 4.26a shows that even harmonics between the 24th and 26th orders (38.15–41.33 eV) almost disappeared when chirp-free pulses were used, while the 28th harmonic (44.50 eV)

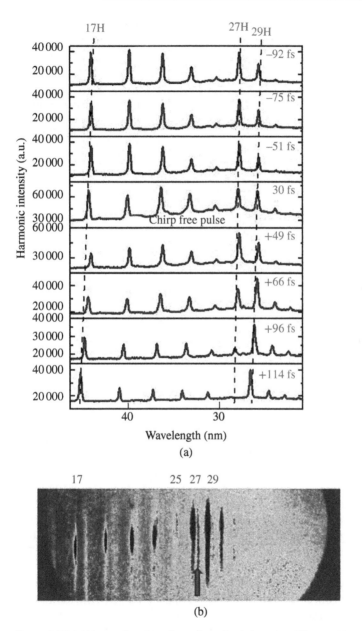

Figure 4.25 (a) Harmonic spectra from chromium plasma at different chirps of laser radiation. Positive and negative values of pulse duration correspond to positively and negatively chirped pulses. (b) Harmonic spectrum at overexcited conditions of Cr plasma formation, with ionic lines appearing close to the enhanced 27th and 29th harmonics. The arrow shows one of these lines close to the 27th harmonic. *Source*: Ganeev et al. 2012 [60]. Reproduced with permission from Optical Society of America.

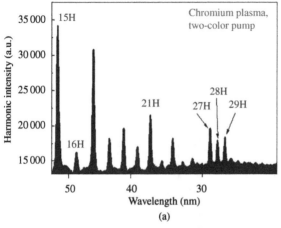

(a)

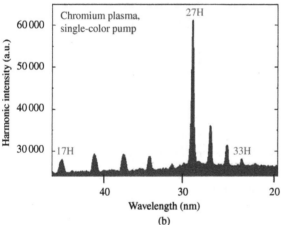

(b)

Figure 4.26 (a) Two-color pump-induced spectra of harmonics from Cr plasma and (b) the spectra obtained at analogous experimental conditions by removing the second harmonic crystal from the path of 780 nm radiation. *Source*: Ganeev et al. 2007 [60]. Reproduced with permission from Optical Society of America.

was much stronger (5×) than the lowest order harmonics and almost equal in intensity to the enhanced odd ones. The efficiency with which some high harmonics were generated was two times greater for two-color pumping than for single-color pumping, as seen in Figure 4.26a,b in the case of the 15th–17th harmonics.

Finally, we analyze the results of HHG studies using vanadium plasma. In previous studies, the conversion efficiency in the plateau range of the harmonics of 800 nm radiation was 1.6×10^{-7} [70]. Those plasma emission spectra measurements and calculations of the ionization conditions and harmonic cutoffs in the laser-ablation plume showed that the higher harmonics originated from the interaction of the femtosecond laser pulses with doubly charged vanadium ions, which allowed generation of harmonics up to the 71st order.

In the discussed case, the dynamics of vanadium harmonics variations was the same as in the case of chromium plasma once the chirp and corresponding redistribution of laser spectrum along the pulse caused the tuning of harmonic wavelength. Only low-order harmonics (up to the 23rd order) were obtained when the V target was weakly excited ($I_{ps} = 6 \times 10^9$ W cm^{-2}, Figure 4.27a). With a stronger excitation ($I_{ps} = 1 \times 10^{10}$ W cm^{-2}), the spectrum extended to higher energies, with the appearance of a "second plateau" beginning with a strong 27th harmonic, as shown in Figure 4.27b. Two-color pumping led to an increase in the harmonic yield (Figure 4.27c), accompanied by the appearance of an enhanced (4×) 26th harmonic ($\lambda = 30$ nm, $E = 41.33$ eV), which is attributed to the influence of a strong ionic transition at this energy. Neither of the ionic transitions were identified to be responsible for this enhancement. The NIST database contains one relatively strong transition ($3p^6 3d - 3p^5$ ($^2P_0$) $3d^2$ (^1G); $\lambda = 31.2$ nm) in this spectral region for quadruple ionized vanadium, but it is unlikely that such a high level of ionization was achieved at the used laser-ablation parameters.

4.2.3.3 Discussion

These two-color HHG studies confirm previously reported spectral features in the vicinity of the 3p – 3d transitions of the Cr II ions [35] and reveal a strong even harmonic generated close to those transitions. The observed enhancement of the 28th harmonic of the 780 nm pump (corresponding to the 14th harmonic of 390 nm radiation) is attributed to the influence of these transitions, although it is not as pronounced compared with the 29th harmonic of 800 nm radiation. The enhancement of the 27th and 28th harmonics from vanadium plasma can also be attributed to an overlap with ionic transitions. Although those ionic transitions were not identified, the ionic lines were seen by overexciting the target.

Calculations of harmonic enhancement for some plasmas, in particular for the chromium ion, can be found in Refs. [71, 72]. The coincidence of the enhanced harmonics observed in the experiments with the giant 3p – 3d resonance implicates the involvement of this transition in the enhancement. The resonant enhancement can also be considered from a macroscopic perspective. At resonance conditions, when the harmonic frequency is close to the atomic transition frequency, the variation in the wave number of a single harmonic could be large, and the influence of dispersion from free-electrons can be compensated by the atomic dispersion for a specific harmonic order [73]. In this case, improvement of the phase-matching conditions for single harmonic can be achieved.

HHG from laser-produced plasmas using three techniques (resonance-induced and two-color pump-induced enhancement, together with the application of a high repetition rate laser source) allowed to improve the plasma harmonic yield. Together with an increased HHG output, an enhanced yield of even harmonics compared with odd ones was observed. Resonance-induced enhancement in the intensity of some single even harmonics when the ratio between the 780 and 390 nm pulse energies was 25 : 1.

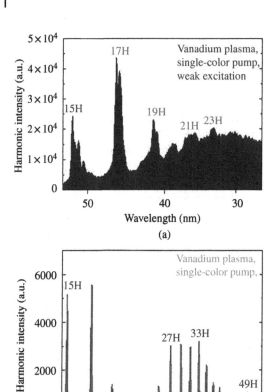

(a)

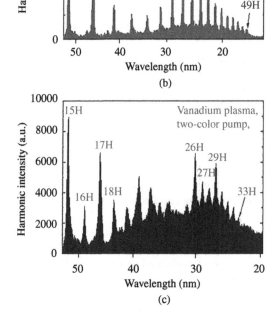

(b)

(c)

Figure 4.27 Variations of harmonic spectra at (a) weak excitation of V target ($I_{ps} = 6 \times 10^9$ W cm^{-2}), (b) stronger excitation of target ($I_{ps} = 1 \times 10^{10}$ W cm^{-2}), and (c) application of two-color pump. *Source*: Ganeev et al. 2007 [60]. Reproduced with permission from Optical Society of America.

This novel high repetition rate source for resonance- and two-color-induced plasma HHG improved the average power of the emitted XUV radiation by 2 orders of magnitude compared with previous plasma HHG studies using 10 Hz lasers with equal pulse energy. This improvement will be useful both for investigating the temporal characteristics of the high-harmonics from laser plasma and for studying quantum path interference of long and short electron trajectories, both of which require large numbers of laser shots.

During those experiments, several new resonance-enhanced processes were revealed, mostly due to the modification of electron trajectories in HHG by the presence of a weak second harmonic field. This led to the generation of enhanced odd and even harmonics in different regions of the plateau. In particular, an enhanced 20th harmonic in silver plasma (2×), 26th harmonic in vanadium plasma (4×), and 28th harmonic in chromium plasma (5×) were observed. These results support theoretical predictions that the involvement of autoionizing states of atoms and ions will enhance nonlinear optical processes in the vicinity of a resonance. Crucially, using broadband (40 nm) pulses for plasma HHG permitted tuning of the spectral position of the harmonic comb over a broad range (2.8 nm in the case of the 17th harmonic), thus allowing us to ensure that a particular harmonic coincided with the resonance wavelength of an autoionizing state.

4.3 Peculiarities of Resonant and Nonresonant Harmonics Generating in Laser-Produced Plasmas

4.3.1 Spatial Coherence Measurements of Nonresonant and Resonant High-Order Harmonics Generated in Different Plasmas

4.3.1.1 Introduction

HHG in gases driven by intense, femtosecond laser pulses provides a useful source of short-wavelength coherent radiation in the XUV region. HHG can be described by a semi-classical three-step recollision model [74], in which an electron wavepacket is launched into the continuum and accelerated by the strong electric field of the laser field before recombining with the parent ion within an optical cycle of its launch, leading to the emission of XUV radiation. Extensive research has been conducted to enhance the efficiency of the process. As part of this effort, a variety of alternative nonlinear media for HHG have been investigated, such as molecules, microdroplets, and laser-produced ablation plumes. An important feature of the HHG process is that since the harmonic radiation builds up coherently in the generating medium, the high coherence of the drive laser beam can be transferred to the harmonic radiation, which can thus exhibit near full spatial coherence [75]. This makes high-order harmonics an attractive table-top, short-wavelength source for applications, such as diffraction imaging, holography and, more generally, for short-wavelength interferometry using wave-front division.

Studies of the spatial coherence of high harmonic radiation have proved useful in helping to elucidate the underlying physics of the HHG process [76], particularly in identifying mechanisms that can degrade the coherence. Previous measurements in gas targets showed that for laser intensities high enough to cause significant ionization of the neutral medium, the rapid production of free electrons can degrade the coherence [77, 78]. The refractive index of the free electrons imparts a rapidly varying phase on the harmonic. This can lower its spatial coherence if the rate of formation of free electrons varies at different points across the focused beam leading to decorrelation of the time-dependent fields. This can be caused by density or laser-intensity variations across the laser focus.

Ablation targets for HHG differ from conventional (neutral) gas targets in a number of ways. First, they are usually preformed plasmas with a density distribution that is determined by the ablation dynamics and subsequent evolution of the plasma, which depend on the target composition and ablation pulse parameters in a complex way. Secondly, the HHG process can be resonantly enhanced in the plasma.

The differences between ablation and conventional gas target for HHG raise interesting questions about the degree of spatial coherence of high harmonics from ablation plumes. It is known that these harmonics are emitted as low divergence, near Gaussian beams (around 2 mrad FWHM in this experiment). In the following, we analyze the results of direct measurement of their spatial coherence [79]. It was found that the spatial coherence of both nonresonant and resonant high harmonics from carbon, zinc, and indium ablation targets generated with few-cycle pulses is reasonably high (in the range 0.6–0.75) and somewhat higher than for harmonics generated in argon gas under similar experimental conditions. This finding confirms that high-order harmonics from ablation plumes can be used in applications requiring high spatial coherence. For such applications, other features of the harmonic radiation from ablation targets may be useful, such as the resonant enhancement of particular harmonic orders that give rise to quasi-monochromatization of the radiation, thus reducing the requirements for spectral filtering.

4.3.1.2 Measurements of the Spatial Coherence of Harmonics

The experimental set-up is shown in Figure 4.28. A Ti:sapphire laser, which provided pulses of 30 fs duration and energies of up to 2.5 mJ at a repetition rate of 1 kHz, was used in the discussed studies. These pulses were focused into a 1 m long differentially pumped hollow core fiber (250 μm inner core diameter) filled with neon. The spectrally broadened pulses at the output of the fiber system were compressed by 10 bounces of double-angle technology chirped mirrors. High-intensity few-cycle pulses (775 nm central wavelength, 0.5 mJ,

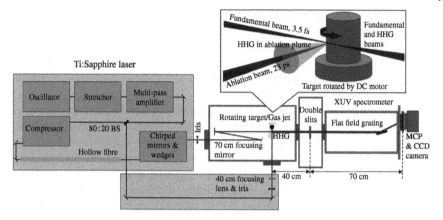

Figure 4.28 Experimental setup for measuring the spatial coherence of high harmonics generated in ablation plasma plumes and gas targets. A chirped pulse amplification system with a hollow fiber pulse compressor was used to produce the few-cycle pulses to drive HHG. A beam-splitter (BS) was used to pick off part of the stretched laser pulse for ablation. A rotating target set-up allows operation at 1 kHz pulse repetition rate. The harmonics were analyzed using a spatially resolving XUV spectrometer with microchannel plate (MCP) detector. Double slits could be introduced into the harmonic beam to produce an interference pattern on the detector, from which the coherence of the radiation could be determined. *Source*: Ganeev et al. 2014 [79]. Reproduced with permission from AIP Publishing LLC.

4 fs) were typically obtained. The compressed pulses were characterized spatially and temporally with a spatially encoded arrangement for spectral shearing interferometry for direct electric field reconstruction (SEA-F-SPIDER) [80]. A part of the uncompressed beam of the CPA Ti:sapphire laser (central wavelength 800 nm, pulse energy 120 µJ, pulse duration 23 ps, pulse repetition rate 1 kHz) was split from the beam line prior to the laser compressor stage and was directed into the vacuum chamber to create an ablation plume from the target. The picosecond pulses were focused by a 400 mm focal length lens to an intensity on the target surface of $I_{ps} \approx 8 \times 10^9$ W cm^{-2}. The length and density of the ablation plume was estimated to be ~0.5 mm and ~3×10^{18} cm^{-3}, respectively. Ablation plumes from carbon, zinc, and indium were produced in these experiments. The 15 mm diameter rods of these materials rotating at 30 rotations per minute were used at the targets for laser ablation. Rotating targets considerably improve the stability of harmonics compared with fixed targets in the case of high pulse repetition rate (1 kHz). In particular, it was shown that once the target rotation is stopped, the harmonic efficiency from the plasmas decreased for more than 1 order of magnitude within 1–2 seconds [81]. Harmonics were generated with 4 fs laser pulses that were focused into the ablation plumes using a 700 mm focal length spherical mirror. The delay between the picosecond and the femtosecond pulse was set to 33 ns. The focal position of

the femtosecond pulse with respect to the ablation plume was chosen to maximize the harmonic signal, and the intensity in the ablation plume was estimated to be $I_{fs} \approx 3 \times 10^{14}$ W cm^{-2}. This estimate is consistent with the 70 eV high harmonic cutoff observed in neon. Neon provides a more reliable corroboration of intensity than the HHG cutoff in argon due to the Cooper minimum in argon at 48 eV that masks the real cutoff position.

The harmonics were analyzed using a spatially resolving XUV spectrometer. The spatially resolved spectra of the generated harmonics on the phosphor screen were recorded with a CCD camera. The ablation target could also be replaced by a gas target placed at the same position to generate harmonics in an Ar, or Ne (density $\approx 10^{18}$ cm^{-3}, $L \approx 1.5$ mm). The spatial coherence of the high harmonics was measured using double slit interference. A pair of slits was mounted on a translation stage and placed 40 cm from the targets and 70 cm from the MCP. The slits were made in a tungsten foil and had 50 μm spacing, 6 μm width, and were 10 mm long. Each interference pattern measured in this experiment was integrated over ≈ 1000 laser shots (≈ 1 s). A series of measurements were made for different targets (laser-produced plasmas and Ar gas) in the following energy ranges: 15–30 eV for C, Zn, and In; and 15–40 eV for Ar. The relatively low cutoff observed for C is in agreement with earlier work [82].

For each fringe pattern, the fringe visibility $V = (I_{max} - I_{min})/(I_{max} + I_{min})$ was determined, where I_{max} and I_{min} are the maximum and minimum intensities of the interference pattern. With equal intensity at both slits, as was the case in the measurements, the fringe visibility at the center of the fringe pattern is equal to the modulus of the complex coherence factor (CCF), μ_{12}, where $\mu_{12} = \gamma_{12}(\tau = 0)$. Here $\gamma_{12}(\tau) = \langle E_1(t + \tau)E_2^*(t)\rangle / \sqrt{\langle|E_1|^2\rangle\langle|E_2|^2\rangle}$ is the complex degree of coherence between the fields E_1 and E_2 at each slit, and the angled brackets denote a time average over the pulse duration. Figures 4.29–4.32 show the double-slit interference data for high harmonics from C, Zn, and In plasmas, and from Ar gas targets, respectively. In each figure the spatially resolved HHG spectrum is shown in part (a), the spatially integrated spectrum is shown in part (b), and a spatial lineout of a single harmonic (interference pattern) is shown in part (c). Taking an average of the visibility at the center of the fringe pattern over 20 shots, the measured visibilities, which are equal to $|\mu_{12}|$, for the different targets are as follows: C_{H13} (13th harmonic) $V = 0.63 \pm 0.03$; indium $V = 0.66 \pm 0.06$; Zn $V = 0.74 \pm 0.04$; Ar_{H15} $V = 0.47 \pm 0.08$. The HHG spectra from Zn and In show resonantly enhanced harmonics due to the overlap of harmonic radiation with plasma resonance lines. In the case of Zn (Figure 4.30), the emission spectrum is dominated by a peak at ≈ 18 eV, which was identified as the overlap of H11 (photon energy ≈ 17 eV, FWHM bandwidth ≈ 1 eV) with the 18.3 eV Zn III transition ($3d^{10} \rightarrow 3d^9$ (^{2}D) 4p), which has an oscillator strength significantly greater than other lines in this spectral region. For In (Figure 4.31), an enhanced feature at ~20 eV is observed, which was attributed to the overlap of H13 (photon energy ≈ 20 eV, bandwidth ~1 eV) with the 19.9 eV ground to autoionizing

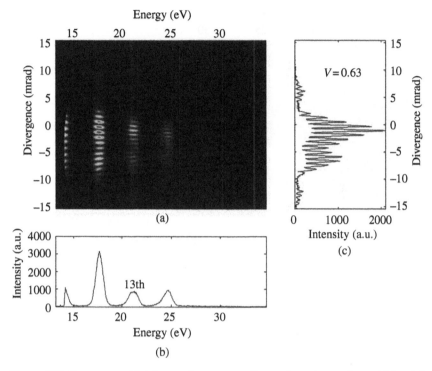

Figure 4.29 Nonresonant high harmonics generated in a carbon plasma plume. (a) Spatially resolved HHG spectrum showing interference fringes from the double slits. (b) Spectral line out. (c) Spatial line out of the interference pattern for the 13th harmonic. The average visibility of fringes near the center of the pattern is $V = 0.63$. *Source*: Ganeev et al. 2014 [79]. Reproduced with permission from AIP Publishing LLC.

state transition in In^+ ($4d^{10}5s^2\,^1S_0 \rightarrow 4d^9 5s^2 5p\,(^2D)\,^1P_1$), which has an oscillator strength $\approx 10\times$ larger than other transitions.

The maximum visibility measured in the discussed experiment ($V = 0.74$ for the Zn target) is lower than that reported in some earlier work, but is consistent with a small departure from full spatial coherence for the driving laser field. In the absence of other effects that can degrade the coherence, the visibility of the fringe pattern for the qth harmonic is approximately given by $V_q \approx 1 - q(1-V_1)$, where V_1 is the visibility for the driving laser field. Hence, a value of $V_1 = 0.98$ for the laser beam would already completely account for the visibility $V_{11} = 0.74$ measured for harmonics from the Zn target. Although femtosecond laser beam exhibits essentially full spatial coherence, one cannot rule out such a small decrease in coherence of the radiation delivered to the target, for example due to nonlinear propagation effects in air and the vacuum chamber window coupled with a small intensity asymmetry of the transverse beam profile.

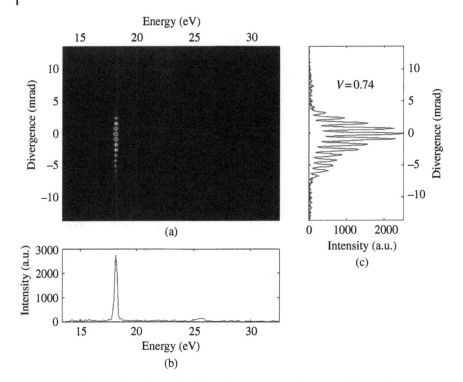

Figure 4.30 Resonantly enhanced high harmonics generated in a zinc plasma plume. (a) Spatially resolved HHG spectrum showing interference fringes from the double slits. (b) Spectral line out showing enhancement of the 11th harmonic due to its overlap with a plasma transition line in Zn^+ at 18.3 eV marked by solid line in (a). (c) Spatial line out of the resonant harmonic. The average visibility of fringes near the center of the pattern is $V = 0.74$. *Source*: Ganeev et al. 2014 [79]. Reproduced with permission from AIP Publishing LLC.

However, the significantly lower visibility ($V = 0.44$) observed for harmonics from Ar gas, compared to the plasma targets (average visibility 0.68) was attributed to increased free electron production during the drive laser pulse in Ar compared to the other targets. One can note that only rapid variation of the free electron density during the HHG process can degrade the harmonic spatial coherence. Preexisting free electrons in the ablation plumes should not degrade the coherence, since their density is effectively static on the timescale of the few-cycle pulse. The tunnel ionization of the targets by a 4 fs laser pulse was simulated using ionization rates from Ammosov–Delone–Krainov (ADK) theory. In the absence of details of the ionic composition of the ablation plasmas, one can assume that the plasma targets are fully singly ionized. ADK rates are known to be reasonably accurate for rare gases such as Ar, but for multi-electron atoms they significantly overestimate the ionization rate. Therefore, the calculations for C, Zn, and In provide an upper limit for the

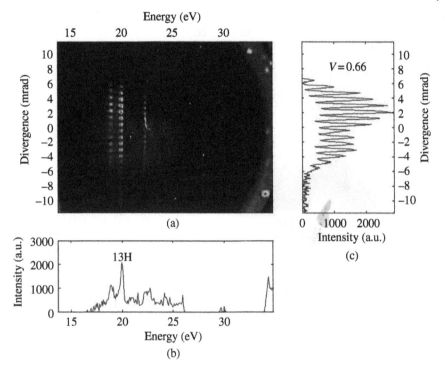

Figure 4.31 Resonantly enhanced high harmonics generated in an indium plasma plume. (a) Spatially resolved HHG spectrum showing interference fringes from the double slits. (b) Spectral line out showing enhancement of the 13th harmonic due to its overlap with a plasma transition line in In⁺ at 19.9 eV marked by solid line in (a). (c) Spatial line out of the resonant harmonic. The average visibility of fringes near the center of the pattern is $V = 0.66$. *Source*: Ganeev et al. 2014 [79]. Reproduced with permission from AIP Publishing LLC.

ionization fraction in those atoms. For a peak intensity of 3×10^{14} W cm⁻², the ionization fraction at the peak of the pulse was calculated to be ∼0.1 for Ar ($I_p = 15.75$ eV), compared with ∼10⁻⁵ for C⁺ ($I_p = 24.4$ eV), 0.004 for In⁺ ($I_p = 18.87$ eV) and ∼0.01 for Zn⁺ ($I_p = 17.96$ eV). Using the target densities given earlier, this implies a free electron density of ∼10¹⁷ cm⁻³ produced during HHG in the Ar target, which is at least 3×, 8×, and 3000× higher than for Zn, In, and C targets, respectively. The measured visibility for the case of Ar is consistent with earlier work in which free electron densities in the region of 10¹⁷ cm⁻³ produced during harmonic generation were shown to lead to a reduction in harmonic visibility to the range 0.4–0.6 for similar harmonic orders [77]. One can note that for carbon, an intensity of 3×10^{14} W cm⁻² is close to the ionization saturation intensity. Therefore, a small fraction of neutral carbon atoms in the ablation plume ($I_p = 11.26$ eV) would increase the electron density significantly compared to the value calculated above. This

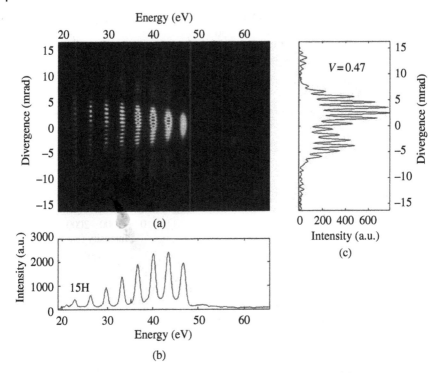

Figure 4.32 High harmonics generated in argon gas. (a) Spatially resolved HHG spectrum showing interference fringes from the double slits. (b) Spectral line out. (c) Spatial line of the 15th harmonic. The average visibility of fringes near the center of the pattern is $V = 0.44$. *Source*: Ganeev et al. 2014 [79]. Reproduced with permission from AIP Publishing LLC.

could explain the lower visibility for harmonics generated in carbon compared to the other plasma targets.

4.3.2 Demonstration of the 101st Harmonic Generation from Laser-Produced Manganese Plasma

4.3.2.1 Low Cutoffs from Plasma Harmonics

Intensive research in HHG has resulted in the development of unique coherent soft X-ray sources with high brightness, extremely short pulse duration, and high spatial coherence. The majority of such studies have been performed using gas as the nonlinear medium, with the successful demonstration of harmonics with wavelength in the range of few nanometers and pulse duration as short as hundreds of attoseconds [83]. Furthermore, recent results have shown that the interaction of relativistically intense lasers with solid surfaces can be an alternative method for HHG with wavelengths extending into the spectral water-window region [84]. However, typically observed conversion efficiencies of HHG are on the order of 10^{-6}, which can be a disadvantage when considering application of harmonics.

HHG from plasma plume generally results in a relatively low cutoff. Initially, the highest order observed so far from singly charged ions of laser plasma was the 65th harmonic ($\lambda = 12.2$ nm) produced from boron plasma [85]. This photon energy observed from plasma HHG (101 eV) have fallen short compared with the well-known cutoff rule $E_c \approx I_p + 3.17U_p$ (where I_p is the ionization potential of the atom, and U_p is the ponderomotive potential) [74]. Rather, overionization processes of the plasma plume have had deterministic effects for the highest observable harmonic energy. Such overionized plasma defocuses the main pump laser, limiting the maximum intensity, and also creates a phase mismatch between the main pump laser and the harmonics. The influence of these two processes greatly reduces the harmonic conversion efficiency, or even fully stops HHG.

Previous studies show that harmonic emission from plasma plume originates mostly from singly charged ions [40, 42, 86]. In this case, there is a linear correlation between the cutoff energy of the harmonics and the second ionization potential of the target atom [86]. As a result, there is a maximum cutoff energy that one can obtain with this method, since even the highest second ionization potentials ($\sim$25 eV) will limit the cutoff to about the 60th harmonics [16, 85].

At the same time, harmonic emission from doubly charged ions, as opposed to singly charged ions, can in theory extend the cutoff up to much higher energies, since the barrier suppression intensity in this case is higher. However, extending the cutoff in such cases can be achieved only if one can compensate for the negative effects of self-defocusing and phase mismatch of the main pump laser, due to the presence of higher free electrons density. Such a possibility of using harmonic emission from higher charged ions to generate higher order plasma harmonics has yet to be experimentally realized for HHG from plasma plume, while, in the case of gas-jet harmonics, it has been demonstrated, for example, with ionized Ar [87].

In the following parts of this subsection, we show that, by optimizing the conditions of the manganese plasma plume, one can extend the cutoff of the HHG from plasma plume to the 101st order (wavelength $\lambda = 7.9$ nm). An important technique for cutoff optimization was the analysis of time-resolved ultraviolet spectra of the plasma emission. From those data one can conclude that harmonics near the cutoff were generated from the interaction of the femtosecond pump laser with doubly charged manganese ions, which allowed the generation of 157 eV photons [88].

4.3.2.2 Experimental Arrangements and Initial Research
The generation of harmonics from plasma plume requires two pump lasers, first, a long (subnanosecond) pulse to create the plasma, and a second, short (femtosecond) pulse to generate the harmonics from the plasma plume.

Two temporally synchronized laser beams from a high-power Ti:sapphire laser system are necessary to independently control the characteristics of radiation (heating and main pulses) and to study the plasma conditions for efficient harmonic generation. For this purpose, the 10 Hz, multi-terawatt, 35 fs beamline was used. The output of this beamline was configured into two beams before compression, with each beam having a maximum energy of 200 mJ and pulse duration of 210 ps. Each beam is equipped with a variable energy controller, which can independently vary the pulse energy using a computer. One of the two beams is sent to a vacuum compressor, which compresses the 210 ps pulse to 35 fs, with maximum energy of 150 mJ. Two beams used in the discussed study are referred as line 1 (subnanosecond heating pulse; 210 ps pulse duration, 200 mJ maximum pulse energy) and line 2 (femtosecond pulse; 35 fs pulse duration, 150 mJ maximum pulse energy). To create ablation, the heating pulse from line 1 was focused on to a manganese target placed in vacuum chamber, by using a plano-convex lens (focal length $f = 150$ mm). The focal spot diameter of this pulse beam on the target surface was adjusted to be ~600 μm. The intensity of the subnanosecond pulse, I_{pp}, on the target surface was varied between 7×10^9 W cm^{-2} and 4×10^{10} W cm^{-2}. This intensity range was defined from previous studies of HHG from different targets [89]. After some delay, part of the femtosecond pulse (energy $E = 8$–25 mJ, temporal duration $t = 35$ fs, central wavelength $\lambda = 800$ nm, bandwidth 40 nm FWHM) from line 2 was focused to the position of the plasma plume, by using a MgF$_2$ plano-convex lens ($f = 680$ mm). The maximum intensity of the femtosecond pulses was $I_{fp} = 3 \times 10^{15}$ W cm^{-2}, above which the conditions for efficient HHG were degraded. The harmonic spectrum was spectrally dispersed by a homemade XUV spectrometer. The 2 mm thick, 3 mm long Mn plates were used as the targets. The harmonic generation from Mn plume was also compared with the one from other plasma samples (Ag, In, Au, V, Ge, Pt, Cr, Ga, Bi, Cu, Al, and Pb).

A strong correlation between the cutoff harmonics (H) generating from different targets and the ionization potentials (I_i) of the particles participating in HHG was observed during those studies, which was close to the data previously reported in Refs. [86, 90]. For most targets, a linear relation was observed between the second ionization potential and the cutoff energy, implying the important role of singly charged ions for harmonic generation from plasma plume. This relation was confirmed by comparing the reported harmonic cutoffs using main pump lasers with different pulse duration (35 fs (discussed study), 48 fs [90], and 150 fs [86]). This dependence shows a linear relation between the H and I_i for targets, where a harmonic plateau was observed (Figure 4.33).

The above dependence is represented by the empirical relation of H (harmonic cutoff) $\approx 4I_i$ eV $- 32$. From this relation, one can draw the next conclusion. When singly charged ions are strongly involved in the HHG from plasma

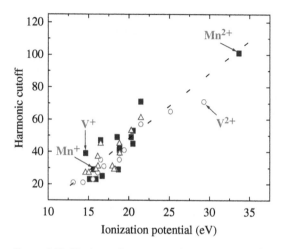

Figure 4.33 Maximum harmonic order observed as a function of the ionization potential of singly charged and doubly charged ions from various targets. The averaged line shows an empirical relation H (harmonic cutoff) $\approx 4I_i$ eV − 32. Filled squares are the results obtained in discussed work, open circles are the results obtained from Refs. [70, 86], and open triangles are the results of Ref. [90]. *Source:* Ganeev et al. 2007 [88]. Reproduced with permission from American Physical Society.

plume, maximum harmonics are generated with targets that have higher second ionization potentials. This means that the generation of additional free electrons by further increasing the main pump intensity, due to the ionization of singly charged ions, leads to the saturation of the H (I_{fp}) dependence, thus restricting the generation of higher order harmonics.

The empirical relation was found exclusively for the specific wavelength range of the driving radiation (Ti:sapphire laser, 800 nm). This relation (H (harmonic cutoff) $\approx 4I_i$ eV − 32) has a steeper slope than the one (H (harmonic cutoff) $= 2.5I_i$ eV), which was reported in the case of 248 nm driving pulses [33]. This seems to be consistent since a short-wavelength laser makes less significant ponderomotive energy shift. Further studies have to carry out to get insight into the mechanisms that alter the standard cutoff law ($I_p + 3.17U_p$) to the empirical one by considering the role of doubly charged ions.

One could not definitely verify the participation of doubly charged ions in the HHG from plasma plume. The three-step model predicts that, in such case, the HHG should occur through the ionization of doubly charged ions, followed by a recombination of the electron accelerated in the laser field with the triply ionized core. In most cases, there is no need to include doubly charged ions for explaining the cutoff of the HHG from plasma plume, since their appearance indicates a higher free electrons density in the plume.

Surprisingly, the discussed studies of the HHG from manganese plasma plume have revealed new features related to this process. As in previous

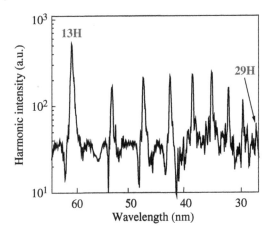

Figure 4.34 Harmonic spectrum from manganese ablation obtained at $I_{fp} = 5 \times 10^{14}$ W cm^{-2} and $I_{pp} = 1 \times 10^{10}$ W cm^{-2}. *Source*: Ganeev et al. 2007 [88]. Reproduced with permission from American Physical Society.

cases with other targets, harmonic generation from this material up to the maximum cutoff $H = 29$ was observed, which well coincides with the empirical rule, taking into account the second ionization potential of Mn ($I_{2i} = 15.64$ eV) (see Figure 4.33). The harmonic spectrum showed a conventional plateau pattern (Figure 4.34). The intensity of the subnanosecond pulse that produces the plasma plume for this case was $I_{pp} \approx 1 \times 10^{10}$ W cm^{-2}. However, by further increasing the subnanosecond pulse intensity on the manganese target surface, as well as the main pulse intensity, a considerable increase in the harmonic cutoff was achieved. Harmonics as high as the 101st order were clearly identified in this case, though the conversion efficiency for most harmonic orders was smaller compared with the case of smaller heating pulse intensities (Figure 4.35). An interesting observation was the emergence of a "second" plateau pattern at higher orders (from the 33rd to 93rd harmonic) with further steep drop of harmonic intensity up to 101st order ($\lambda = 7.9$ nm). This second plateau appeared in place of a harmonic plateau between the 15th and 29th orders that were observed for moderate irradiation of Mn target by the subnanosecond pulse. The newly observed cutoff well coincided with the empirical $H(I_i)$ dependence, taking into account the involvement of doubly charged ions and the third ionization potential of manganese ($I_{3i} = 33.67$ eV) (Figure 4.33).

The similar pattern of harmonic distribution (i.e. the appearance of second plateau) was discussed in Ref. [91]. They considered collective (plasmon) electron oscillations in laser–matter interaction and proposed following scenario of electron movement: (i) the electron is ejected from the ground state and recombines into the same ground state, with HHG according to the three-step model, (ii) the electron is ejected from plasmon excited state and recombines in the ground state with HHG, thus being responsible for second cutoff, and (iii) the electron is ejected from plasmon excited state and recombines in the

Figure 4.35 A lineout of high-order harmonic spectrum obtained at $I_{fp} = 3 \times 10^{15}$ W cm^{-2} and $I_{pp} = 3 \times 10^{10}$ W cm^{-2}. Inset: A region of resolved harmonic distribution from the 67th to 101st orders. *Source*: Ganeev et al. 2007 [88]. Reproduced with permission from American Physical Society.

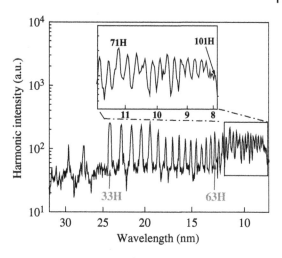

same plasmon excited state with HHG. Then, the electron makes an emitting transition into ground state, thus making resonance enhancement of a single harmonic possible.

4.3.2.3 Analysis of Cutoff Extension

Some theoretical simulations predict that the multiplateau structure of harmonic spectra is characteristic of the weakly relativistic regime [91]. The appearance of the complex pattern in the harmonic distribution has been discussed in Ref. [92]. The second and even the third plateau appeared in their calculations of harmonics in the case of laser-driven tightly bound systems. They suggested that, while for neutral atoms, the low-energy part of harmonics is too short to develop any evident structure, for multiply charged ions, this part of the HHG spectrum is extensive and exhibits its own characteristic structure. However, these calculations were based on the interaction of the weakly relativistic laser intensities (10^{18} W cm^{-2}) with the very tightly bound hydrogen-like ions (such as O^{7+}, $I_p \approx 871$ eV), which are quite far from used experimental conditions of the moderate laser intensities and low charges of the ions (10^{15} W cm^{-2}, Mn^{2+}, $I_p \approx 34$ eV). Their conditions even predicted the plateau appearance not from the "tunneling-recombination" mechanism [74, 93, 94], but from the inner atomic dynamics when a laser-driven bound electronic wavepacket sweeps over the center of the binding potential. In the discussed case (moderate laser intensities and low-charged particles), the doubly charged manganese ions appearing during the propagation of a leading part of the pulse, as well as during further ionization of plasma by the enhanced heating pulse, participate in the harmonic generation alongside with the singly changed ions, but with higher cutoff energy, which defines by the conventional three-step model [74, 94]. The involvement of the higher-charged

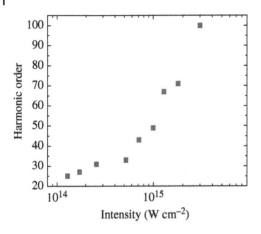

Figure 4.36 Harmonic cutoff as a function of main pulse intensity. *Source*: Ganeev et al. 2007 [88]. Reproduced with permission from American Physical Society.

species (compared to the singly charged ones) leads to the increase of the binding potential and consequently to the growth of the harmonic cutoff of the prolonged ("second") plateau.

Systematic investigations of this phenomenon were performed by first varying the main pump intensity. No saturation in the harmonic cutoff with an increase in the main pulse intensity was observed up to 3×10^{15} W cm^{-2} (Figure 4.36). This inferred the possibility of obtaining even higher harmonics from manganese plasma plume, by using higher main pulse intensities. However, further growth of pump intensity led to the appearance of strong plasma emission in the XUV range caused by a considerable tunneling ionization, which overlapped harmonic emission. One should note that the highest harmonic was restricted by the spectral resolution of the spectrometer, as well as the continuum emission from the plasma in the range of 5–10 nm.

Next, the time-resolved spectral studies of the emission from the Mn plasma in the UV range were performed. These results showed that, for conditions when higher order harmonics were generated, the main laser pulse interacted with the singly charged ions, for the maximum delays used in these experiments. The doubly charged ions appeared after the leading part of the main laser started to irradiate the plume, since the main pulse intensity exceeded the barrier suppression intensity for singly charged ions. Further interaction of main laser radiation with these ions led to their ionization, acceleration of electrons in laser field, and their recombination with the core, producing the emission of higher harmonics.

The time gate for each UV spectral measurement was 20 ns. Measurements were taken each 10 ns from the beginning of the irradiation of Mn target, up to 150 ns. These time-resolved measurements were performed at both the "optimal" and "non-optimal" conditions for HHG. The change in the Mn plasma emission was made in the vicinity of the UV spectral lines related with the excitation of the singly charged ions (253–263 nm), where various lines

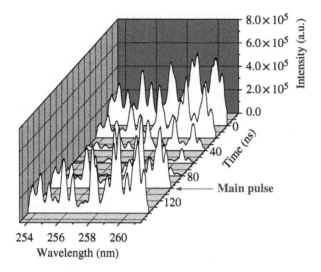

Figure 4.37 Time-resolved UV spectra of the "optimal" plasma produced on the surface of manganese target. $I_{pp} \approx 3 \times 10^{10}$ W cm^{-2}. *Source*: Ganeev et al. 2007 [88]. Reproduced with permission from American Physical Society.

could be compared within a narrow spectral range. The "optimal plasma" was generated, at which both the plateau-like distribution and the maximum cutoff were observed [at $I_{pp} \approx (2-3) \times 10^{10}$ W cm^{-2}]. The term "optimal plasma" is referred to the conditions when maximum HHG conversion efficiency can be achieved by adjusting the charge state of plasma, together with the pulse delay and focusing geometry of femtosecond pulse. For 210 ps heating pulse, the average plasma density was $(3-4) \times 10^{17}$ cm^{-3} at abovementioned pulse intensities and 100 ns delays, which was estimated from the calculations using the HYADES code [43]. Figure 4.37 shows the result of the change in such time-resolved spectra during the first 150 ns, for a heating pulse to main pulse delay of 100 ns.

The decay times of ionic lines at optimal conditions for harmonic generation were about 60 ns. One can see a fast decrease in the intensities of all Mn$^+$ lines, which were excited in these conditions. In these time-resolved spectra, the main pulse arriving 100 ns after the beginning of plasma formation excites exclusively the ionic lines in the spectrum, while no increase in the intensity of the neutral lines can be observed after the interaction of the main pump laser with the plasma.

A different behavior of the UV spectrum dynamics was observed when one increases the heating pulse intensity at the surface of the manganese target. An insignificant increase in the heating pulse intensity from 3×10^{10} to 4×10^{10} W cm^{-2} led to a considerable increase in the emission from the manganese plume. Apart from an approximately fivefold growth of the intensities of the ionic lines, some additional lines (associated predominantly with other ionic transitions) appeared and showed longer decay times. The time-resolved UV spectra at 90 and 100 ns showed a considerable difference in the excitation of ionic lines by the femtosecond main pump for optimal and nonoptimal conditions of the HHG. Similar behavior could be expected in the XUV range as well, though no measurements of the time-resolved spectra of plasma lines in the range of plateau-like harmonics were performed. However, their appearance in the time-integrated XUV spectra led to a decrease or even almost disappearance of the "low-order" harmonics (between the 13th and 29th orders), as well as to the worsening of harmonic generation conditions at the shorter-wavelength range.

On the basis of the above analysis, further optimization of HHG was performed by choosing the appropriate intensities of the heating pulse and main pulse. This time-resolved study of manganese plasma conditions allowed the precise identification of the range of heating pulse and main pulse intensities, at which optimal harmonic generation is achieved, without the prevalent influence of free electrons on the phase mismatch and self-defocusing. In that case the highest order harmonics were generated from the Mn plume.

In the discussed studies using other plasma materials, the opposite influence of the heating pulse intensity on HHG was observed, when plasma conditions were optimized for achieving highest harmonic order and best conversion efficiency. In particular, in the case of Au plume, the plateau pattern of harmonic distribution deteriorated when the heating pulse energy was increased above 17 mJ, corresponding to the intensity of $I_{pp} \approx 3 \times 10^{10}$ W cm^{-2} at the target surface. The strong spectral lines from multiply ionized gold atoms appeared in the XUV spectrum when the heating pulse energy was further increased, and no any harmonics with wavelengths shorter than 33 nm were distinguished.

Laser-ablation itself is a complex phenomenon, especially at relatively low laser intensities used for plasma harmonics. The behavior of laser ablation will change considerably, depending on the equation of state, ionization potential, and cohesive energy of the material. The processes that determine harmonic generation from plasma plume are complex, and may involve various factors that are not considered for gas harmonics. For example, the nonlinear medium is already weakly ionized for plasma harmonics, whose level depends on the pulse intensity. If the free electrons density is too high, it can induce phase mismatch between the pump laser and the harmonics, or defocusing of the pump laser. Both these effects can reduce or stop HHG.

To understand this different behavior of the HHG process for different plasma materials, simulations using the HYADES code were performed. The HYADES code is a 1D, three-geometry, Lagrangean hydrodynamics and energy transport code. The electron and ion components are treated separately in a fluid approximation and are loosely coupled to each other. The diffusion

Table 4.1 Simulations of Mn and Au plasma characteristics at different heating pulse intensities using the HYADES code.

Intensity (10^{10} W cm^{-2})	Target	1	2	3	5
Electron density (10^{17} cm^{-3})	Mn	1.3	3.25	3.77	6.13
	Au	7.32	14.2	18.5	25.2
Ion density (10^{17} cm^{-3})	Mn	1.3	3.25	3.77	4.50
	Au	4.7	6.03	6.8	7.52
Ionization level	Mn	0.62	1.0	1.0	1.36
	Au	1.56	2.35	2.72	3.35

Source: Ganeev et al. 2007 [88]. Reproduced with permission from American Physical Society.

approximation is used for modeling all energy transport phenomena. The degree of ionization is determined by a Saha, Thomas-Fermi, or other models.

The expansion of the manganese and gold targets interacting with the heating pulse was simulated, which allowed determining the electron density, ionization level, and ion density of these plumes as a function of the heating pulse intensity, at a distance of 300 µm from the target surface. Results of these calculations for 100 ns delay are summarized in Table 4.1. One can see from these data that, already at 1×10^{10} W cm^{-2}, there is a considerable difference in the ionization states of Mn and Au plasmas, which can lead to the difference in nonlinear optical response. At this intensity, the ionization level of Au plume becomes considerably higher than 1, which leads to the appearance of additional free electrons, due to the ionization of singly charged ions. The ratio between the electron density and ion density in gold plasma continues to increase at higher pulse intensities, and at $I_{pp} = 5 \times 10^{10}$ W cm^{-2}, their ratio becomes higher than 4. The increase in the free electron density for Au plume prevents efficient harmonic generation and extension of harmonic cutoff, due to the growing phase mismatch between the harmonic and pump laser waves.

The characteristics of manganese plasma under the same conditions are considerably different from those of Au. The ionization level of the Mn plume is considerably lower than that of Au under the same heating pulse intensity, and doubly charged ions can only be expected at $I_{pp} > 3 \times 10^{10}$ W cm^{-2}. Therefore, the effect of free electrons on the HHG is smaller. What is especially important for HHG is that the ion density increases considerably with an increase in the heating pulse intensity. As a result, the harmonics, especially near the cutoff, will increase nonlinearly in intensity, thus helping the detection of these harmonics. These combined features of relatively low electron density and high ion density in manganese plasma allowed the demonstration of the highest harmonics ever observed from plasma plume.

The results show that new findings in the extension of the harmonic cutoff toward the soft X-ray region may be possible, by searching for the optimal HHG conditions of the plasma plume through simulation of plasma characteristics and the time-resolved analysis of the UV spectral dynamics. Some indirect confirmation of this approach was recently found for harmonic generation from vanadium plasma. Assuming that singly charged ions ($I_{2i} = 14.65$ eV) are the highest ionization states participating in HHG, the cutoff for vanadium is estimated to be at about the 30th harmonics, according to the previously mentioned empirical relation. In fact, the 39th harmonic from this plume was observed, which is in the range of uncertainty of the empirical relation presented in Figure 4.33. However, some studies have shown that the HHG from vanadium plume at specific plasma conditions has a cutoff at considerably shorter wavelength (with harmonics demonstrated up to the 71st order [70]). Hydrodynamic simulations showed that doubly ionized atoms were present within the vanadium plasma. If one uses the ionization potential of the doubly charged vanadium ions ($I_{3i} = 29.31$ eV), the maximum expected cutoff would be about the 79th harmonic. This is in a reasonable agreement with the experimentally observed cutoff (71st harmonic), especially since the estimation of the cutoff was done using the laser intensity when focusing into the vacuum. In the experiment, however, defocusing of the incoming laser beam due to plasma build-up and increase of the free electrons density is expected, thereby decreasing the effective intensity, and thus the achievable cutoff. These two results, as well as the data on harmonic generation from singly and doubly charged manganese ions, are underlined in Figure 4.33 assuming their corresponding ionization potentials.

4.3.3 Isolated Subfemtosecond XUV Pulse Generation in Mn Plasma Ablation

4.3.3.1 Application of a Few-Cycle Pulses for Harmonic Generation in Plasmas: Experiments with Manganese Plasma

In the course of the ongoing efforts to enhance the low XUV harmonic intensities and to study fundamental aspects of HHG, alternative media such as molecules, microdroplets, and solids have been employed. In the following, we discuss study focused on HHG from transition metal plasmas. These are very promising targets in view of the giant resonances found in the photoionization cross-sections. For example, the Mn^+ cross-section is $\sim$40 Mb at 50 eV photon energy, whereas rare gas atoms have cross-sections between 1 and 8 Mb at this photon energy. Photorecombination – the third step in the recollision model – is the inverse process of photoionization and therefore HHG and photoionization must exhibit the same resonances. This has been confirmed not only by previous resonance-induced experiments with laser-produced transition metal plasmas but also in the study of HHG from xenon gas [95].

Resonance-induced enhancement of a single harmonic or, in some cases, a group of harmonics of the laser radiation allowed considerable improvement of harmonic efficiency in some specific XUV spectral ranges related with high oscillator strengths of ionic states of metals. This was confirmed in multiple studies following the initial observation of this phenomenon in indium plasma. In particular, the strong enhancement of a single harmonic was reported in Cr [42], and Mn [88] plasmas. The Mn plasma is of special interest since it shows the highest harmonic cut-off energy observed in plasma plumes (101st harmonic, see previous subsection). In early studies, multicycle (30 [88] and 140 fs [96]) laser pulses were employed and the generation of all harmonics in the plateau was observed together with a strongly enhanced single harmonic.

The progress in the generation of few-cycle pulses allowed the observation of various new effects including the realization of isolated attosecond-pulse generation in gas media [97–99]. In this connection, it is interesting to analyze resonance-induced processes observed in an ablation plume using the shortest available drive laser pulses. In this subsection, we analyze the observation of resonance enhancement in manganese plasmas using sub-4 fs pulses. The most interesting feature observed in those experiments was the suppression of almost all neighboring harmonics in the vicinity of a resonantly enhanced small spectral range of about 2.5 eV bandwidth covering only a single harmonic at the photon energy of ∼50 eV. In the following, we discuss HHG studies in Mn plasma at various pulse durations of the drive laser pulse (3.5–25 fs) and compare these results with those obtained using 40 fs pulses. Theoretical modeling suggests that the resonantly enhanced harmonic emission constitutes a near isolated subfemtosecond pulse. We also analyze the influence of carrier envelope phase (CEP) on the harmonics from manganese plasma and show some CEP-dependent pictures of Mn HHG spectra [100].

The experimental setup was similar to the one described in subsection 4.3.1.2. The harmonic spectrum in the case of the manganese plasma was strikingly different compared with other plasma samples (for example, Ag plasma) analyzed in separate experiments. While all other samples that were studied showed a relatively featureless harmonic spectrum with extended cutoff (Figure 4.38, upper curve showing the spectrum of harmonics generating in the silver plasma), the Mn plasma allowed the generation of a strong single harmonic substantially enhanced compared with the neighboring ones (Figure 4.38, bottom curve).

One can note that in earlier work the harmonic spectra from manganese plasmas for 30 and 140 fs pulses also showed enhanced harmonics around 50 eV. The assumption of the resonance nature of the enhancement of harmonics of the ∼800 nm radiation of Ti:sapphire lasers in this spectral region is supported by the presence of a strong giant resonance in the vicinity of 50 eV confirmed by experimental [45, 101] and theoretical [44] studies. The enhancement of a single harmonic can be attributed to the broadband resonances of the ions of

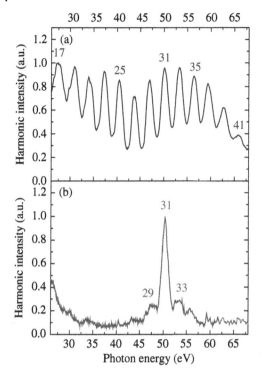

Figure 4.38 Harmonic spectra from the silver plasma (a) and manganese plasma (b). *Source*: Ganeev et al. 2012 [100]. Reproduced with permission from Optical Society of America.

few metals, such as V, In, Cd, Cr, Cd, and Mn. These "giant" resonances have been experimentally confirmed in the literature [44, 45, 48] and discussed in a few theoretical studies [71, 72, 102–104].

However, in previous studies using multicycle drive pulses, the intensity of the enhanced harmonics was only a few times higher than neighboring harmonic orders. It was possible to reproduce this behavior in the discussed experiments using 40 fs pulses from another Ti:sapphire laser at similar intensity inside the laser plasma (4×10^{14} W cm^{-2}). The raw image of the harmonic spectrum presented in Figure 4.39a shows several enhanced harmonics starting from the 31st order followed by an extended second plateau. The extension of the harmonic cutoff exceeding the 71st order is attributed to the involvement of doubly charged Mn ions as the sources of HHG. This feature of Mn plasma harmonics has already been reported earlier [88]. Here a typical Mn harmonic spectrum in the case of 3.5 fs pulses is also presented (Figure 4.39b). No second plateau, which was seen in the case of multicycle (40 fs) pulses, is observed for the few-cycle pulse. Most strikingly is the observation of a single very strong, broadband (2.5 eV), 31st harmonic. Only two weak neighboring harmonics (around the strong emission) are seen in the 30–65 eV spectral range. The ratio between the intensities of the enhanced harmonic to the weak neighboring

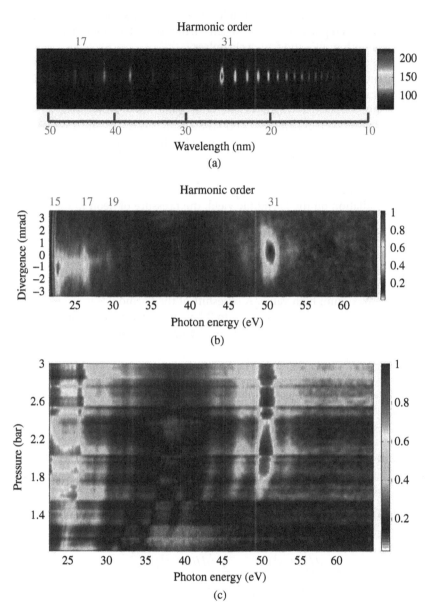

Figure 4.39 Raw images of harmonic spectra from manganese plasma in the case of (a) 40 fs and (b) 3.5 fs probe pulses obtained at the same intensity. (c) Raw images of harmonic spectra from Mn plasma at different pressures of neon in the hollow fiber obtained at the same energy of probe laser pulses. *Source*: Ganeev et al. 2012 [100]. Reproduced with permission from Optical Society of America.

harmonics exceeds 1 order of magnitude. One can note that at a lower intensity of the femtosecond pulse ($<2 \times 10^{14}$ W cm^{-2}), the strong harmonic disappeared when using both multi- and few-cycle pulses.

The distinctive structure of the harmonic spectra, both for 40 and 3.5 fs pulses clearly points to the involvement of Mn resonances centered around 50–51 eV. The same can be said about the photoionization or photoabsorption characteristics of Mn + plasma, which is due to the "giant" 3p → 3d resonance [101]. The laser polarization dependence of this emission was varied and it was found that the 50 eV radiation abruptly disappears with the change of the polarization state of the femtosecond probe pulses from linear to circular, which is a clear signature of the emission being due to HHG.

To analyze the effect of the spectro-temporal characteristics of the femtosecond radiation on the harmonic yield, the pressure of neon in the hollow fiber system was varied, thus changing the duration of the harmonic drive pulse [105]. The spectral and intensity variations of manganese harmonic spectra in the range of 22–62 eV as the functions of neon pressure in the hollow fiber are shown in Figure 4.39c. One can clearly see that, with change of pressure (from 1 to 2.3 bar), the single 31st harmonic intensity varies from almost zero to its maximum high value. A blueshift of the harmonics is also evident. Further increase of neon pressure up to 3 bar, at which the experiments with the 3.5 fs pulses were carried out, did not change the harmonic distribution.

4.3.3.2 Theoretical Calculations and Discussion

In the following, we analyze the results of numerical simulations within a 1D model. One can assume that the main contribution to the resonant peak in the spectrum comes from Mn$^+$ ions. Note that the ionization potential of Mn^{2+} ions (33.7 eV) is more than twice higher than the ionization potential of Mn$^+$ ions (15.6 eV). The time-dependent Schrödinger equation was solved by means of the split-operator method [106]. The Mn$^+$ target is modeled using a potential supporting a metastable state by a potential barrier [71, 104]. The shape of the potential is (see Figure 4.40a)

$$V(x) = -a + \frac{a}{1 + \exp\left(\dfrac{x+b}{c}\right)} + \frac{a}{1 + \exp\left(\dfrac{-x+b}{c}\right)}$$
$$+ \frac{d/(e+x^2)}{1 + \exp\left(\dfrac{x+b}{c}\right)} + \frac{d/(e+x^2)}{1 + \exp\left(\dfrac{-x+b}{c}\right)} \tag{4.1}$$

where a, b, c, d, and e are parameters. They were chosen to be 1.672, 1.16, 0.216, 8.95, and 0.63, respectively, so that the width of the resonance and the energy gap between the ground and the resonant states resemble the experimental data [101]. The metastable state of this model potential is at 51.8 eV above the ground state. The laser field is $E(t) = E_0 f(t) \cos(\omega_0 t + \varphi)$, where $f(t)$ is the pulse

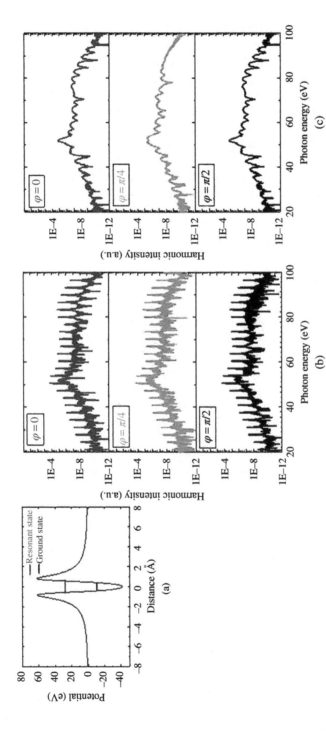

Figure 4.40 (a) Potential used for the numerical simulations. Calculated HHG spectra using (b) a long (40 fs) pulse and (c) few-cycle pulses at different values of the CEP ($\varphi = 0$, $\pi/4$, and $\pi/2$). *Source:* Ganeev et al. 2012 [100]. Reproduced with permission from Optical Society of America.

envelope, φ denotes the CEP, and ω_0 is the laser frequency corresponding to the central wavelength $\lambda = 760$ nm. The laser intensity is $I_0 = 4 \times 10^{14}$ W cm^{-2}. A CEP of $\varphi = 0$ means that the maximum of the envelope corresponds to a maximum of $\cos(\omega_0 t)$.

HHG spectra were calculated for pulse shapes with different lengths and for different values of φ (Figure 4.40b,c). A sin^2 envelope with a total length of four full cycles was used to model the 3.5 fs pulse, while an envelope with 4 cycles sin^2 switch-on/off, 13 cycles of constant intensity and 21 cycle total duration was used to model the 40 fs case. The long pulse led to a HHG spectrum that shows well-defined peaks at the odd harmonic orders and that is weakly dependent on the CEP (Figure 4.40b). Figure 4.40c shows the dependence of the harmonic spectrum on the CEP in the case of the short, few-cycle pulse. In all cases, the resonance dominated the spectrum. The most intense emission occurs around 51 eV, where the metastable state is located. This enhancement has been explained in terms of a four-step model [71], according to which the recombination step is split into two steps: trapping of the returning electron in the metastable state followed by radiative transition into the ground state. An alternative view is that, according to the quantitative rescattering theory [107], the emission spectrum is proportional to the photoionization cross-section, which exhibits a peak at the resonance energy. Note that the mechanism is different from the type of harmonic generation described in Ref. [108], where the initial state before ionization is already a superposition of ground and excited states. Some difference between harmonic spectra for $\varphi = 0$, $\pi/4$, and $\pi/2$ is found, though we note that the CEP dependence is strongest for the spectrum outside the region of the resonance. For random CEP the substructure of the spectrum will average out as confirmed by numerical averaging over 20 values of the CEP in the range from 0 to π. The shape of resonance peak depends on the CEP, and the case of $\varphi = \pi/4$ appears to be special since a dip due to trajectory interference seems to coincide with the resonance peak.

The experiments described above were carried out without CEP stabilization (that is, for random CEP values). The HHG experiments with Mn plasma using 3.5 fs pulses with stabilized CEP ($\varphi = 0$ and $\pi/2$) were also carried out, which led to some differences in that case (Figure 4.41), in particular the variation of harmonic distribution observed for the lower order harmonics (compare the middle and bottom curves of Figure 4.41; see also, following discussion). The spectral shapes of the 31st harmonic emission were approximately similar for these two fixed values of CEP, while a considerable difference in harmonic spectra was maintained when comparing to longer pulse duration and lower intensity of the driving pulses. Figure 4.41 shows measurements for 25 fs pulses (upper panel) and 3.5 fs pulses (middle and bottom panels) of the same energy. One can clearly see the absence of harmonic extension and resonance-induced HHG in the case of low-intensity, 25 fs pulses, which has

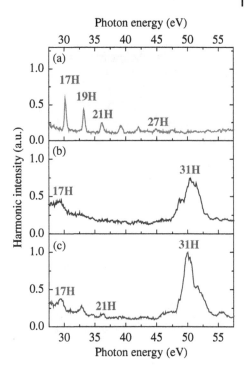

Figure 4.41 Experimental harmonic spectra generated from manganese plasma in the case of the absence of gas in the hollow fiber compressor ($t = 25$ fs) and random CEP (a), and at 3 bar pressure ($t = 3.5$ fs) at fixed CEP ($\varphi = 0$ (b); $\varphi = \pi/2$ (c)). *Source*: Ganeev et al. 2012 [100]. Reproduced with permission from Optical Society of America.

been reported in earlier studies of plasma harmonics from manganese ablation using low intensity pulses [88, 96].

In order to investigate temporal characteristics of the harmonic emission in numerical simulations, a Gabor transformation was performed [109]:

$$G(\omega, t) = \frac{\int d\tau \, a(\tau) \, \exp(i\omega\tau) \, \exp\left(-\frac{(t - \tau)^2}{2\sigma^2}\right)}{\sqrt{2\pi\sigma^2}}, \tag{4.2}$$

where $a(\tau)$ is the dipole acceleration from the simulation, σ is a parameter taken to be $\sigma = 1/(3\omega_0)$. The modulus squared, $|G(\omega,t)|^2$, is the time–frequency distribution. The temporal intensity profile of the XUV emission is calculated as the square of the time-dependent dipole acceleration after filtering out the photon energies below 1.2 au (corresponding to 32.7 eV). In fact, the emission profile is not affected much by the filtering since the spectrum is strongly dominated by the resonance. The results are shown in Figure 4.42. Comparing the short- and long-pulse regimes, one can notice that, whereas in Figure 4.42b–d emission of the resonance occurs at the end of the few-cycle pulse, Figure 4.42a shows that the resonance is repopulated and decaying each half cycle of the multicycle pulse. For the short pulse, the emission takes the form of a short

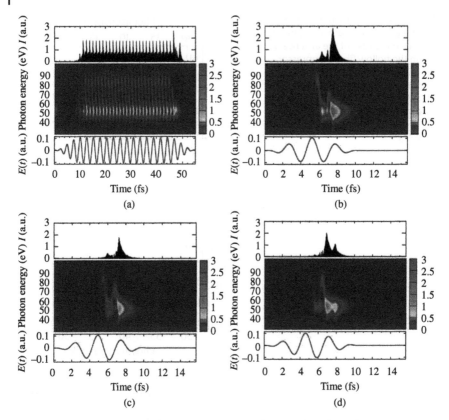

Figure 4.42 Calculated results for HHG driven by (a) a long (40 fs) pulse with CEP $\varphi = \pi/4$ and few-cycle pulses with CEPs of (b) $\varphi = 0$, (c) $\varphi = \pi/4$, and (d) $\varphi = \pi/2$. The top panels show the HHG temporal intensity profile obtained as the square of the time-dependent dipole acceleration after high-pass filtering above 32.7 eV. The middle panels show the time-frequency diagrams. The solid curves in the bottom panels show the time dependence of the electric field of the driving laser pulse. *Source*: Ganeev et al. 2012 [100]. Reproduced with permission from Optical Society of America.

burst confined to one or two half cycles of the driving laser field. In few-cycle pulse for the recollision following the last strong field peak the resonance is only weakly perturbed by the field. Thus the enhanced recombination associated with the resonance is largely confined to this last cycle.

For most CEPs, the emission can be viewed as an isolated subfemtosecond XUV pulse if the pulse length is defined, in the usual way, as the FWHM. This main emission burst is either preceded or followed by a small side peak. Similar emission profiles are found for three values of the CEP (0, $\pi/4$, and $\pi/2$, Figure 4.42b–d). The time of maximum emission varies in a range of less than 1 fs with CEP. The calculations show that usually subfemtosecond XUV pulses or at least ~1 fs XUV pulses, for different values of CEP, can be achieved. This is in sharp contrast with the usual strong CEP dependence of isolated attosecond pulse generation [110, 111]. This suggests that resonance-induced HHG driven by few-cycle pulses provides a route to isolated XUV attosecond pulse generation with reduced requirements for CEP stabilization.

As it was pointed out, the discussed experimental and theoretical studies of harmonic spectra from Mn plasma plumes showed some dependence on the CEP of the driving, few-cycle pulses. In Figure 4.41 it is evident that, when varying the CEP of the driving 3.5 fs pulse, the 30–40 eV region of the harmonic spectra is changed, with two additional peaks (19th and 21st harmonics) appearing when the CEP varies from 0 to $\pi/2$ rad (middle and bottom panels). The structure of the enhanced spectral windows around the 31st harmonic modifies its shape, the shoulder shifting from the left to the right side of the main peak, and the overall FWHM also decreasing. Some CEP-induced changes in harmonic spectra are also confirmed by the simulations, which point out a different temporal structure of the isolated pulse for various CEP values (Figure 4.42). Note that while the shape of the attosecond pulse changes with CEP, the emission time with respect to the IR driving pulse varies little.

The fact that weak CEP dependence of the harmonic spectra in the case of 3.5 fs pulses was observed could be attributed to the presence of a significant density of free electrons in the manganese plasma, which might diminish the difference between the HHG spectra recorded for different values of experimental CEP. The same can be said about other HHG experiments using silver and brass plasmas, which did not show significant differences in harmonic spectra when comparing few-cycle pulses with fixed and random CEP. The comparative studies were also carried out using gas media under similar experimental conditions and found a characteristic strong dependence of the HHG spectra on the CEP. Thus the weak influence of the CEP on the harmonic pattern generated by few-cycle pulses from the ablation plumes appears to be a common feature of plasma HHG.

References

1 Ganeev, R.A., Suzuki, M., Baba, M. et al. (2006). *Opt. Lett.* 31: 1699.
2 Duffy, G. and Dunne, P. (2001). *J. Phys. B* 34: L173.
3 Xu, H., Tang, X., and Lambropoulos, P. (1992). *Phys. Rev. A* 46: R2225.
4 Figueira de Morisson Faria, C., Copold, R., Becker, W., and Rost, J.M. (2002). *Phys. Rev. A* 65: 023404.
5 Toma, E.S., Antoine, P., de Bohan, A., and Muller, H.G. (1999). *J. Phys. B* 32: 5843.
6 Gaarde, M.B. and Schafer, K.J. (2001). *Phys. Rev. A* 64: 013820.
7 Chang, Z., Rundquist, A., Wang, H. et al. (1998). *Phys. Rev. A* 58: R30.

8 Kim, H.T., Lee, D.G., Hong, K.-H. et al. (2003). *Phys. Rev. A* 67: 051801.

9 Kim, H.T., Kim, I.J., Lee, D.G. et al. (2004). *Phys. Rev. A* 69: 031805.

10 Ganeev, R.A., Singhal, H., Naik, P.A. et al. (2006). *Phys. Rev. A* 74: 063824.

11 Ganeev, R.A., Kulagin, I.A., Suzuki, M. et al. (2005). *Opt. Commun.* 249: 569.

12 L'Huillier, A. and Balcou, P. (1993). *Phys. Rev. Lett.* 70: 774. M. Bellini, C. Corsi, and M. C. Gambino, *Laser Part. Beams* 20, 277 (2002).

13 Bouhal, A., Salières, P., Breger, P. et al. (1998). *Phys. Rev. A* 58: 389.

14 Bellini, M., Corsi, C., and Gambino, M.C. (2002). *Laser Part. Beams* 20: 277.

15 Ganeev, R.A. and Kuroda, H. (2005). *Opt. Commun.* 256: 242.

16 Ganeev, R., Suzuki, M., Baba, M. et al. (2005). *Opt. Lett.* 30: 768.

17 Ganeev, R.A., Baba, M., Suzuki, M., and Kuroda, H. (2005). *Phys. Lett. A* 339: 103.

18 Connerade, J.P. (1977). *Proc. R. Soc. London, Ser. A* 352: 561.

19 Pont, M. (1989). *Phys. Rev. A* 40: 5659.

20 Flettner, A., Pfeifer, T., Walter, D. et al. (2005). *Appl. Phys. B* 80: 277.

21 Puell, H. and Vidal, C.R. (1976). *Phys. Rev. A* 14: 2225.

22 Diels, J.-C. and Georges, A.T. (1979). *Phys. Rev. A* 19: 1589.

23 Balcou, P. and L'Huillier, A. (1993). *Phys. Rev. A* 47: 1447.

24 Miyazaki, K. and Sakai, H. (1992). *J. Phys. B* 25: L83.

25 Taieb, R., Veniard, V., Wassaf, J., and Maquet, A. (2003). *Phys. Rev. A* 68: 033403.

26 Gibson, E.A., Paul, A., Wagner, N. et al. (2003). *Science* 302: 95.

27 Descamps, D., LyngÅ, C., Norin, J. et al. (2000). *Opt. Lett.* 25: 135.

28 Larsson, J., Mevel, E., Zerne, R. et al. (1995). *J. Phys. B* 28: L53.

29 Ganeev, R.A., Redkorechev, V.I., and Usmanov, T. (1997). *Opt. Commun.* 135: 251.

30 Akiyama, Y., Midorikawa, K., Matsunawa, Y. et al. (1992). *Phys. Rev. Lett.* 69: 2176.

31 Krushelnick, K., Tighe, W., and Suckewer, S. (1997). *J. Opt. Soc. Am. B* 14: 1687.

32 Theobald, W., Wülker, C., Schäfer, F.R., and Chichkov, B.N. (1995). *Opt. Commun.* 120: 177.

33 Kubodera, S., Nagata, Y., Akiyama, Y. et al. (1993). *Phys. Rev. A* 48: 4576.

34 Wahlström, C.-G., Borgström, S., Larsson, J., and Pettersson, S.-G. (1995). *Phys. Rev. A* 51: 585.

35 Ganeev, R.A., Suzuki, M., Baba, M., and Kuroda, H. (2005). *Appl. Phys. Lett.* 86: 131116.

36 Reintjes, J.F. (1984). *Nonlinear Optical Parametric Processes in Liquids and Gases*. New York: Academic Press.

37 Altucci, C., Starczewski, T., Mevel, E. et al. (1996). *J. Opt. Soc. Am. B* 13: 148.

38 Ganeev, R.A., Elouga Bom, L.B., Kieffer, J.-C., and Ozaki, T. (2007). *Phys. Rev. A* 75: 063806.

39 Ganeev, R.A. (2009). *Phys. Usp.* 52: 55.

40 Ganeev, R.A., Singhal, H., Naik, P.A. et al. (2006). *J. Opt. Soc. Am. B* 23: 2535.

41 Suzuki, M., Baba, M., Ganeev, R. et al. (2006). *Opt. Lett.* 31: 3306.

42 Ganeev, R.A., Naik, P.A., Singhal, H. et al. (2007). *Opt. Lett.* 32: 65.

43 Rubenchik, A.M., Feit, M.D., Perry, M.D., and Larsen, J.T. (1998). *Appl. Surf. Sci.* 129: 193.

44 Dolmatov, V.K. (1996). *J. Phys. B* 29: L687.

45 Kilbane, D., Kennedy, E.T., Mosnier, J.-P. et al. (2005). *J. Phys. B* 38: L1.

46 McGuinness, C., Martins, M., Wernet, P. et al. (1999). *J. Phys. B* 32: L583.

47 McGuinness, C., Martins, M., van Kampen, P. et al. (2000). *J. Phys. B* 33: 5077.

48 West, J.B., Hansen, J.E., Kristensen, B. et al. (2003). *J. Phys. B* 36: L327.

49 Suzuki, M., Baba, M., Kuroda, H. et al. (2007). *Opt. Express* 15: 1161.

50 D'Arcy, R., Costello, J.T., McGuinness, C., and O'Sullivan, G. (1999). *J. Phys. B* 32: 4859.

51 Duffy, G., van Kampen, P., and Dunne, P. (2001). *J. Phys. B* 34: 3171.

52 Macklin, J.J., Kmetec, J.D., and Gordon, C.L. III (1993). *Phys. Rev. Lett.* 70: 766.

53 Brandi, F., Giammanco, F., and Ubachs, W. (2006). *Phys. Rev. Lett.* 96: 123904.

54 Ganeev, R.A., Suzuki, M., Baba, M., and Kuroda, H. (2006). *J. Opt. Soc. Am. B* 23: 1332.

55 Zeng, Z., Li, R., Cheng, Y. et al. (2002). *Phys. Scr.* 66: 321.

56 Bartels, R., Backus, S., Zeek, E. et al. (2000). *Nature* 406: 164.

57 Ganeev, R.A. (2007). *J. Phys. B* 40: R213.

58 Ganeev, R.A., Singhal, H., Naik, P.A. et al. (2010). *Phys. Rev. A* 82: 053831.

59 Ganeev, R.A., Hutchison, C., Siegel, T. et al. (2011). *J. Mod. Opt.* 58: 819.

60 Ganeev, R.A., Hutchison, C., Zaïr, A. et al. (2012). *Opt. Express* 20: 90.

61 Tosa, V., Kim, H.T., Kim, I.J., and Nam, C.H. (2005). *Phys. Rev. A* 71: 063808.

62 Iaconis, C. and Walmsley, I.A. (1998). *Opt. Lett.* 23: 792.

63 Cormier, E. and Lewenstein, M. (2000). *Eur. Phys. J. D* 12: 227.

64 Kim, I.J., Kim, C.M., Kim, H.T. et al. (2005). *Phys. Rev. Lett.* 94: 243901.

65 Mauritsson, J., Johnsson, P., Gustafsson, E. et al. (2006). *Phys. Rev. Lett.* 97: 013001.

66 Pfeifer, T., Gallmann, L., Abel, M.J. et al. (2006). *Opt. Lett.* 31: 975.

67 Charalambidis, D., Tzallas, P., Benis, E.P. et al. (2008). *New J. Phys.* 10: 025018.

68 Kim, I.J., Lee, G.H., Park, S.B. et al. (2008). *Appl. Phys. Lett.* 92: 021125.

69 Ganeev, R.A., Singhal, H., Naik, P.A. et al. (2009). *Phys. Rev. A* 80: 033845.

70 Suzuki, M., Baba, M., Kuroda, H. et al. (2007). *Opt. Express* 15: 4112.

71 Strelkov, V. (2010). *Phys. Rev. Lett.* 104: 123901.

72 Frolov, M.V., Manakov, N.L., and Starace, A.F. (2010). *Phys. Rev. A* 82: 023424.

73 Kulagin, I.A. and Usmanov, T. (2009). *Opt. Lett.* 34: 2616.

74 Corkum, P.B. (1993). *Phys. Rev. Lett.* 71: 1994.

75 Bartels, R.A., Paul, A., Green, H. et al. (2002). *Science* 297: 376.

76 Sauères, P., L'Huillier, A., Antoine, P., and Lewenstein, M. (1999). *Adv. At. Mol. Opt. Phys.* 41: 1.

77 Ditmire, T., Gumbrell, E.T., Smith, R.A. et al. (1996). *Phys. Rev. Lett.* 77: 4756.

78 Ditmire, T., Tisch, J.W.G., Gumbrell, E.T. et al. (1997). *Appl. Phys. B* 65: 313.

79 Ganeev, R.A., Abdelrahman, Z., Frank, F. et al. (2014). *Appl. Phys. Lett.* 104: 021122.

80 Witting, T., Frank, F., Arrell, C.A. et al. (2011). *Opt. Lett.* 36: 1680.

81 Hutchison, C., Ganeev, R.A., Witting, T. et al. (2012). *Opt. Lett.* 37: 2064.

82 Elouga Bom, L.B., Pertot, Y., Bhardwaj, V.R., and Ozaki, T. (2011). *Opt. Express* 19: 3077.

83 Tsakiris, G.D., Eidmann, K., Meyer-ter-Vehn, J., and Krausz, F. (2006). *New J. Phys.* 8: 19.

84 Dromey, B., Zepf, M., Gopal, A. et al. (2006). *Nat. Phys.* 2: 456.

85 Ganeev, R.A., Baba, M., Suzuki, M., and Kuroda, H. (2006). *J. Appl. Phys.* 99: 103303.

86 Ganeev, R.A., Suzuki, M., Baba, M., and Kuroda, H. (2005). *Appl. Phys. B* 81: 1081.

87 Gibson, E.A., Paul, A., Wagner, N. et al. (2004). *Phys. Rev. Lett.* 92: 033001.

88 Ganeev, R.A., Elouga Bom, L.B., Kieffer, J.-C., and Ozaki, T. (2007). *Phys. Rev. A* 76: 023831.

89 Elouga Bom, L.B., Kieffer, J.-C., Ganeev, R.A. et al. (2007). *Phys. Rev. A* 75: 033804.

90 Ganeev, R.A., Singhal, H., Naik, P.A. et al. (2007). *Appl. Phys. B* 87: 243.

91 Zanghellini, J., Jungreuthmayer, C., and Brabec, T. (2006). *J. Phys. B: At. Mol. Opt. Phys.* 39: 709.

92 Milošević, D.B., Hu, S.X., and Becker, W. (2001). *Phys. Rev. A* 63: 013403.

93 Krause, J.L., Schafer, K.J., and Kulander, K.C. (1992). *Phys. Rev. A* 45: 4998.

94 Lewenstein, M., Balcou, P., Ivanov, M.Y. et al. (1994). *Phys. Rev. A* 49: 2117.

95 Shiner, A.D., Schmidt, B.E., Trallero-Herrero, C. et al. (2011). *Nat. Phys.* 7: 464.

96 Ganeev, R.A., Suzuki, M., Baba, M., and Kuroda, H. (2009). *Appl. Phys. Lett.* 94: 051101.

97 Hentschel, M., Kienberger, R., Spielmann, C. et al. (2001). *Nature* 414 (6863): 509.

98 Witting, T., Frank, F., Okell, W.A. et al. (2012). *J. Phys. B* 45: 074014.

99 Altucci, C., Tisch, J.W.G., and Velotta, R. (2011). *J. Mod. Opt.* 58: 1585.

100 Ganeev, R.A., Witting, T., Hutchison, C. et al. (2012). *Opt. Express* 20: 25239.

101 Kjeldsen, H., Folkmann, F., Kristensen, B. et al. (2004). *J. Phys. B* 37: 1321.

102 Milošević, D.B. (2010). *Phys. Rev. A* 81: 023802.

103 Redkin, P.V. and Ganeev, R.A. (2010). *Phys. Rev. A* 81: 063825.

104 Tudorovskaya, M. and Lein, M. (2011). *Phys. Rev. A* 84: 013430.

105 Robinson, J.S., Haworth, C.A., Teng, H. et al. (2006). *Appl. Phys. B* 85: 525.

106 Feit, M.D., Fleck, J.A. Jr., and Steiger, A. (1982). *J. Comp. Phys.* 47: 412.

107 Lin, C.D., Le, A.-T., Chen, Z. et al. (2010). *J. Phys. B* 43: 122001.

108 Milošević, D.B. (2006). *J. Opt. Soc. Am. B* 23: 308.

109 Gabor, D. (1946). *J. Inst. Electr. Eng.* 93: 429.

110 Chipperfield, L.E., Gaier, L.N., Knight, P.L. et al. (2005). *J. Mod. Opt.* 52: 243.

111 Goulielmakis, E., Schultze, M., Hofstetter, M. et al. (2008). *Science* 320: 1614.

5

Resonance Processes in Ablated Semiconductors

The application of elemental semiconductors (tellurium, arsenic, germanium, selenium, silicon, and antimony) for high-order harmonic generation (HHG) during propagation of the femtosecond pulses (FPs) through the laser plasmas produced on the surfaces of those species may reveal various attractive features related with the specific properties of semiconductor elements. This approach in the studies of the nonlinear optical properties of materials has been developed in the 1990s [1] and is currently applied for the analysis of the characteristics of various solids [2]. The metals were among mostly used targets for laser ablation and HHG, though some crystals (such as LiF, MgF_2, CaF_2, and NaF) and semiconductor compounds (such as GaAs and InSb) were also studied using this approach.

In the meantime, laser-ablation-induced HHG spectroscopy of semiconductors can reveal the resonance-induced enhancement of some harmonic orders in the extreme ultraviolet (XUV) range as well as the cluster-induced growth of harmonic yield. The latter assumption has been demonstrated in Ref. [3] where 3rd and 5th harmonic generation of an infrared (IR) (1064 nm) pulsed laser has been studied in ablation plasmas of the wide bandgap compounds CdS and ZnS. The study of the temporal behavior of the harmonic emission has revealed the presence of distinct compositional populations in these complex plasmas. Species ranging from atoms to nanometer-sized particles have been identified as emitters, and their nonlinear optical properties were studied separately due to strongly differing temporal behavior. It was found that, at short distances from the target, atomic species are mostly responsible for harmonic generation at early times from the beginning of ablation (<500 ns), while clusters mostly contribute at longer times (>1 µs). Harmonic generation thus emerges as a powerful and universal technique for ablation plasma diagnosis and as a tool to determine the nonlinear optical susceptibility of ejected semiconductor clusters or nanoparticles.

Such plasmas may contain ions with the transitions near the wavelengths of those harmonics. Thus the resonance-induced growth of single harmonic

Resonance Enhancement in Laser-Produced Plasmas: Concepts and Applications,
First Edition. Rashid A. Ganeev.
© 2018 John Wiley & Sons, Inc. Published 2018 by John Wiley & Sons, Inc.

could be realized in the plasma particles produced during laser ablation of semiconductors. Other methods of harmonic enhancement from the plasmas include the application of longer medium and two-color pump. Most of the HHG studies were carried out using the narrow plasmas ($l \leq 0.5$ mm). In the meantime, recent studies have revealed the usefulness of the application of the longer plasma media for efficient harmonic generation [4]. One has to maintain the extended plasma formation at the conditions when the length of nonlinear optical medium does not exceed the coherence length of harmonics. To resolve the phase mismatch problem, one can use the quasi-phase-matching (QPM) of harmonics by division of extended medium onto the group of separated plasma jets.

In this chapter, we analyze the generation of high-order harmonics of the ultrashort pulses of Ti:sapphire laser in the plasma plumes produced during laser ablation of various semiconductors (Te, Se, Si, As, Sb, and Ge). The resonance-induced enhancement of single harmonics was observed in the Ge, Se, Sb, As, and Te plasmas. The enhanced 35th harmonic obtained from selenium plasma was 12 times stronger than the neighboring harmonics.

Some other peculiarities in those plasmas are analyzed taking into account energy levels of ions. We discuss the enhancement of the intensity of the 27th harmonic radiation produced during the HHG from the GaAs plasma by controlling the chirp of the fundamental Ti:sapphire laser radiation. The influence of Ga and As ions on the enhancement of 27th harmonic radiation is also analyzed. The intensity enhancement of the same harmonic order (H27) at 29.44 nm by using the lowly charged tellurium ions in the laser-ablation plume is discussed as well. The enhancement of the single harmonic to the multiphoton resonance with a Te II transition possessing by the strong oscillator strength was obtained although the wavelength difference between the 27th harmonic and the strong oscillator strength transition was 1.42 nm. Further, we discuss the intensity enhancement of a single high-order harmonic at a wavelength of 37.67 nm using the low ionized antimony laser-ablation plume. The conversion efficiency of this harmonic was 2.5×10^{-5} and the output energy was 0.3 μJ. Such an enhancement of single harmonic was caused by the multiphoton resonance with the strong radiative transition of the Sb II ions. The cutoff energy of the harmonics generated in Sb plasma was 86 eV (55th harmonic). Finally, strong intensity enhancement or extinction of some single harmonics is discussed in the case of low-excited laser-produced plasmas of various materials (GaAs, Cr, InSb, stainless steel). The intensities of some of the harmonics in the mid- and end-plateau regions vary from ~23-fold enhancement to their almost disappearance compared to those of the neighboring ones. It is also shown that the observed intensity enhancement (or extinction) can be varied by controlling the chirp of the driving laser radiation.

5.1 High-Order Harmonic Generation During Propagation of Femtosecond Pulses Through the Laser-Produced Plasmas of Semiconductors

In this section, we analyze the HHG studies in the extended plasma plumes (EPPs) produced on the surfaces of elemental semiconductors (Te, Se, Si, As, Sb, and Ge) [5]. The objective of studies was to reveal the attractive properties of those plasmas. We discuss the results of HHG optimization in the plasma plumes using different approaches.

5.1.1 Optimization of HHG

The uncompressed radiation of Ti:sapphire laser (central wavelength 802 nm, pulse duration 370 ps, pulse energy $E_{hp} = 6$ mJ, 10 Hz pulse repetition rate) was used for extended plasma formation. These heating pulses were focused using a 200 mm focal length cylindrical lens inside the vacuum chamber that contained the semiconductor targets to create the EPPs (Figure 5.1). The ablation of 5 mm long tellurium, selenium, silicon, arsenic, antimony, and germanium targets was studied. The intensity of heating pulse on the target surface was 4×10^9 W cm^{-2}. The ablation beam cross-section on the target surface was 5×0.08 mm^2. The driving compressed pulse from the same laser with the energy of 3 mJ and 64 fs pulse duration was used, after 35 ns delay from the beginning of ablation, for harmonic generation in the plasma plumes. The driving pulse was focused using the 400 mm focal length spherical lens onto the extended plasma at a distance of 100 μm above the target surface. The intensity of driving pulse at the focus area was 1×10^{15} W cm^{-2}.

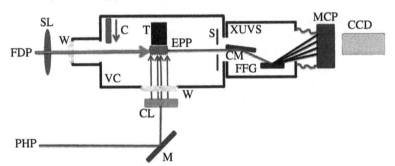

Figure 5.1 Experimental scheme for harmonic generation in the extended plasma plumes. FDP, femtosecond driving pulse; PHP, picosecond heating pulse; SL, spherical lens; CL, cylindrical lens; VC, vacuum chamber; W, windows of vacuum chamber; C, BBO crystal; T, target; EPP, extended plasma plume; S, slit; XUVS, extreme ultraviolet spectrometer; CM, cylindrical mirror; FFG, flat field grating; MCP, microchannel plate registrar; CCD, charge-coupled device camera. *Source:* Ganeev et al. 2015 [5]. Reproduced with permission from AIP Publishing LLC.

The plasma and harmonic emissions were analyzed by an extreme ultraviolet spectrometer (XUVS)that contained a cylindrical mirror and a 1200 grooves/mm flat field grating (FFG) with variable line spacing. The spectrum was recorded on a microchannel plate (MCP) detector with the phosphor screen, which was imaged onto a charge-coupled device (CCD) camera. The movement of MCP along the focusing plane of FFG allowed observation of harmonics in different regions of XUV.

Different methods were applied for the enhancement of the coherent XUV radiation generated in the semiconductor plasmas:

a) The presence of strong ionic lines in the vicinity of some harmonics was observed, alongside with the enhancement of those harmonics. Some of these results will be discussed in the Subsection 5.1.2.

b) The delay between heating and driving pulses is crucial for optimization of the harmonic generation in the plasma plumes. At the initial stages of semiconductor plasma formation, the concentration of particles (atoms and singly charged ions) is insufficient for harmonic generation. The increase of the delay between pulses (Δt) allowed the appearance of the plasma particles along the path of driving pulse, which led to the steady growth of harmonic yield from the semiconductor plasma. Further increase of delay led to the saturation of harmonic yield at ~30–40 ns and the gradual decrease of HHG conversion efficiency at longer (>70 ns) delays.

c) The influence of the distance between the semiconductor surface and the optical axis of propagation of the driving radiation at a fixed delay between pulses (35 ns) was analyzed. This distance (d) was varied by a manipulator, which controlled the position of the semiconductors with regard to the driving beam. A decrease of harmonic yield with the increase of d was observed, which corresponded to the $I_H \sim 1/d^{1.7}$ dependence.

d) Those studies were carried out using the EPPs ($l = 5$ mm). One can define the processes restricting HHG efficiency by analyzing the $I_H(l)$ dependence, which should follow the quadratic rule once the phase mismatch and absorption play insignificant roles. The discussed studies have shown the almost quadratic growth of the conversion efficiency of studied lower order harmonics up to $l \approx 4$ mm with further saturation of harmonic yield.

e) The increase of the energy of heating pulse on the semiconductor surface led to the growth of plasma concentration and harmonic yield. The dependences of harmonic yield on the energy of heating pulse were analyzed using different semiconductor plasmas. The common feature of those measurements was the observation of the maximums of the $I_H(E_{hp})$ dependences at $E_{hp} \approx 3$–4 mJ with the following saturation and decrease of harmonic yield in the case of stronger heating pulses. The decline of HHG was related with the appearance of large amount of free electrons in the extended plasma leading to the phase mismatch between the interacting waves.

The analysis of $I_H(\Delta t)$, $I_H(E_{hp})$, $I_H(l)$, and $I_H(d)$ dependences allowed maximizing the harmonic yield from the plasmas produced on the surfaces of semiconductors. In the following subsection, we analyze studies of the resonance-induced growth of single harmonics at the optimal conditions of experiments. In the case of most semiconductor targets, the following parameters were used for achieving the maximal harmonic yield, the efficient resonance enhancement, and two-color-pump-induced odd and even harmonic generation in the extended plasmas: 35 ns delay between the heating and driving pulses, propagation of the driving pulse 100 μm above the target surface, 0.8 J cm^{-2} fluence of the heating pulse, 8×10^{14} W cm^{-2} intensity of the focused driving radiation, and 0.3 mm thick beta barium borate (BBO) crystal for the two-color pump.

5.1.2 Resonance-Induced Enhancement of Harmonics

Figure 5.2 shows the harmonic spectra obtained from the plasmas produced on the germanium, tellurium, and antimony targets. The Ge, Te, and Sb plasmas showed the enhanced 21st, 27th, and 21st harmonics respectively, with the moderate enhancement factors of 2×–3.5× compared with the neighboring lower order harmonics. Although the data about the ionic transitions involved in the enhancement of these harmonics are scarce, some conclusions could be drawn from the studies reported in Refs. [6–8].

The enhancement factor for the 21st harmonic generated in the 5 mm long antimony plasma (2×) was smaller compared with the case of narrower plasma plume. Previous studies of the 3 mm long Sb plasma have shown the 10× growth of the 21st harmonic of 795 nm radiation compared with the 19th and 23rd orders [9]. The difference in the enhancement factors could be related with the use of different wavelengths of driving radiation (802 and 795 nm), which can lead to better conditions of enhancement in the latter case.

The appearance of some ionic transitions in the range of ~42 eV, which correspond to the enhanced 27th harmonic generating in the tellurium plasma, has been shown in the theoretical calculations presented in Ref. [7], while their experimental data were reported only for the photon energies above 50 eV. It is not clear whether those transitions enhance the nonlinear optical response of tellurium plasma during propagation of the 802 nm pulses. The observations of the enhanced 27th and neighboring even harmonics using both single-color and two-color pumps point out the influence of the group of transitions, rather than the single one, on the harmonic yield in the 29–32 nm spectral range.

Figure 5.3 shows the harmonic spectrum obtained from the selenium plasma. The enhancement factor of the single (35th) harmonic generated in the selenium plasma was considerably larger compared with the resonance enhancement in the Ge, Sb, and Te plasmas. It was 12 times stronger compared with the neighboring harmonics. This resonance-enhanced single harmonic has a

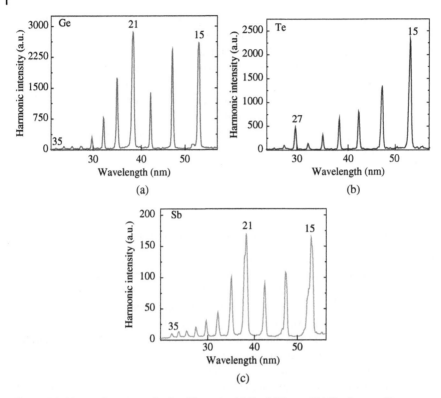

Figure 5.2 Harmonic spectra obtained from the (a) Ge, (b) Te, and (c) Sb plasmas. These plasmas were produced using the 4 mJ, 370 ps pulses. The intensity of driving pulses was 8×10^{14} W cm^{-2}. *Source:* Ganeev et al. 2015 [5]. Reproduced with permission from AIP Publishing LLC.

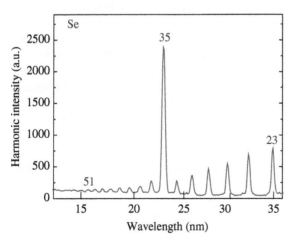

Figure 5.3 Harmonic spectrum obtained from the Se plasma. This plasma was produced using the 4 mJ, 370 ps pulses. The intensity of driving pulses was 8×10^{14} W cm^{-2}. *Source:* Ganeev et al. 2015 [5]. Reproduced with permission from AIP Publishing LLC.

shortest wavelength reported so far ($\lambda_{27H} = 22.91$ nm). The HHG experiments with circularly polarized femtosecond pulses did not reveal the strong emission in this spectral range.

The doubly ionized selenium atom has the ground configuration $4s^2 4p^2$ and the excited configurations $4s4p^3$, $4s^2 4pnd$ ($n > 4$), and $4s^2 4pns$ ($n > 5$). The studies of doubly charged Se were reported in Ref. [10] with the shortest wavelength of 40 nm. The photoionization cross-section measurements for the Se II were reported in Ref. [11]. The photon energy studied in those experiments was ranged between 18.0 and 31.0 eV. However, they also reported that the strong third-order line corresponded to the 54.62 eV energy, without the deliberation of the origin of this transition. Note that this transition is close to the enhanced 35th harmonic ($h\nu = 54.11$ eV).

The studies of high-order nonlinear processes through exploitation of intermediate resonances show that the proximity of the wavelengths of specific harmonic orders and the strong emission lines of ions does not necessarily lead to the growth of the yield of single harmonics. The response of the medium during propagation of intense pulse includes the resonance-induced enhancement of specific nonlinear optical processes, the absorption of emitted radiation, and the involvement of collective macroprocesses, such as phase matching between the interacting waves.

5.1.3 Two-Color Pump

The important goal of these studies of semiconductor plasma was the enhancement of harmonic yield by different means. Various approaches have previously been examined in the case of plasma harmonics to increase the conversion efficiency of coherent XUV radiation. Methods of harmonic enhancement from the plasmas, apart from resonance-induced approach, include the application of two-color pump and QPM. The former method is analyzed in this subsection, while the latter approach will be discussed in the following subsection.

The application of additional field for the HHG in isotropic medium can considerably modify the generating spectrum. This assumption has been confirmed earlier during numerous two-color pump HHG experiments using gases and plasmas as the small-sized ($l = 0.3$–0.5 mm) nonlinear media. Those studies have revealed both the enhancement of harmonic yield and the generation of strong even harmonics, alongside with the odd ones. The physical mechanism responsible for the observed enhancement of odd and even harmonics in the case of two-color pump at low ratios between the energies of second harmonic and fundamental pulses was attributed to the growth of ionization rate due to the influence of a weak shorter wavelength pump and the involvement of larger amount of accelerated electrons in harmonic generation. In the following, we show the results of studies of the two-color pump of the extended (5 mm long) plasmas produced on the semiconductor targets at the 5 : 95 ratio between the two pumps.

Part of the driving pulse was converted in the nonlinear crystal (BBO, type I, crystal length 0.3 mm, 401 nm conversion efficiency 5%) placed inside the vacuum chamber on the path of the focused driving beam, as shown in Figure 5.1. The driving and second harmonic beams were then focused inside the semiconductor plasma. The polarizations of these two pumps were orthogonal to each other.

Figure 5.4 shows the harmonic spectrum obtained from the extended Te plasma in the case of the two-color pump. One can distinguish the prevalence of even harmonics over the odd ones along the almost whole range of generation. Those studies have shown that the significant difference in the intensities of two pumps does not play any important role in the case of extended plasma. The advanced properties of extended plasma were emphasized while comparing with the odd and even harmonics generated in the 0.5 mm long tellurium plasma. In the latter case, the cutoff of even harmonics was significantly lower compared with the one of odd harmonics. Moreover, the intensity of even harmonics was three to four times smaller compared with the odd ones, contrary to the case of extended plasma.

The delay-dependent experiments were conducted using the BBO crystals of different lengths. The 0.02, 0.3, 0.5, 0.7, and 1.0 mm crystals were used for second harmonic generation. The group velocity dispersion in the BBO crystal leads to a temporal separation of two pulses. Owing to this effect in the type-I BBO crystal, the 802 nm pulse (ω) was delayed ($\Delta_{cryst} = d[(n^o_\omega)_{group}/c - (n^e_{2\omega})_{group}/c] \approx 57$ fs for the 0.3 mm long BBO) with

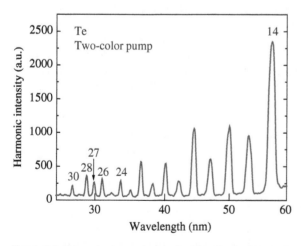

Figure 5.4 Two-color-pump-induced odd and even harmonic spectra obtained from the Te plasma. This plasma was produced using the 4 mJ, 370 ps pulses. The intensity of driving pulses was 8×10^{14} W cm^{-2}. The ratio of 401 and 802 nm pulse energies was 5 : 95. The polarizations of two pump waves were orthogonal to each other. *Source:* Ganeev et al. 2015 [5]. Reproduced with permission from AIP Publishing LLC.

respect to the 401 nm pulse (2ω) due to $n^{\circ}{}_{\omega} > n^{e}{}_{2\omega}$ in this negative uniaxial crystal. Here Δ_{cryst} is the delay between two pulses after leaving the crystal, d is the crystal length, $c/(n^{\circ}{}_{\omega})_{\text{group}}$ and $c/(n^{e}{}_{2\omega})_{\text{group}}$ are the group velocities of the ω and 2ω waves in the BBO crystal, c is the light velocity, and $n^{\circ}{}_{\omega}$ and $n^{e}{}_{2\omega}$ are the refractive indices of crystal at the wavelengths of the ω and 2ω pumps. The duration of second harmonic pulse is given by $t_{2\omega} \approx [(\Delta_{\text{cryst}})^2 + 0.5(t_{\omega})^2]^{1/2}$. Hence, the 401 nm beam has longer pulse duration, corresponding to the induced delay and a certain percentage ($\sim$50%) of the fundamental pulse duration. The latter is because the energy of the fundamental radiation is in general not high enough in the leading and trailing parts of pulse to effectively generate the second-order harmonic. The value of $t_{2\omega}$ at the output of the 0.3 mm long BBO crystal was estimated to be 72 fs. One can see from these calculations that the 802 and 401 nm pulses were only partially overlapped while entering the semiconductor plasma. This partial overlap decreased the ratio between the overlapped second harmonic and driving pulses and diminished the influence of the 401 nm wave on the output spectrum of generating harmonics.

The second harmonic conversion efficiencies and the temporal overlaps of the driving and second harmonic waves, as well as the pulse durations of 401 nm radiation in the cases of different BBO crystals are summarized below. In the case of abovementioned lengths of crystals allowing the 0.4%, 5%, 9%, 11%, and 13% second harmonic conversion efficiencies, the ratios of overlapped pulses inside the plasma were 0.004, 0.03, 0.04, 0.01, and 0.007 respectively. One can see the prevalence of using the 0.3 and 0.5 mm long BBO, since other crystals did not allow the observation of efficient odd and even harmonics generation due to insignificant overlap inside the semiconductor plasma plume. Moreover, as it was mentioned, the duration of second harmonic pulse increases in the case of longer crystals (65, 72, 105, 140, and 195 fs correspondingly; compare with the 64 fs driving pulse). The decrease of the intensity of second harmonic also diminishes the role of this radiation in the variation of HHG spectra.

The important peculiarity of the two-color pump HHG in the semiconductor plasma is the observation of additional resonantly enhanced harmonics near some strong ionic transitions, analogously to the case of single-color pump. This process is clearly seen in the Figure 5.4 where the 13th and 14th harmonics of the 401 nm pump (corresponding to the 26th and 28th harmonics of 802 nm radiation) were comparable or stronger than the 12th harmonic of this radiation. One can assume the influence of the same group of ionic transitions, which led to the enhancement of 27th harmonic of 802 nm radiation in the case of single color pump (Figure 5.2b), on the efficiency of 26th and 28th harmonics in the case of tellurium plasma.

5.1.4 Quasi-Phase-Matching

The modulation of extended plasma allows the creation of the QPM conditions between the waves of driving and harmonic pulses [12]. To achieve these conditions, one has to modify the shape of extended medium and use a bunch of plasma jets of the sizes equal to the coherence length of harmonics. The coherence length significantly depends on the dispersion of plasma, which can be modified by varying the fluence of heating pulse. Moreover, once the HHG in the extended plasma becomes suppressed due to the influence of a large amount of electrons, the QPM for some groups of enhanced harmonics in the XUV range allows the definition of the electron density in the semiconductor plasma.

The principles of the QPM in the plasma plumes are related with the formation of the medium with modulated density of particles (atoms, ions, and electrons), similar to the formation of the multijet gases, which have previously been used for demonstration of the QPM concept [13]. The multislit mask was used to form multijet plasma. The slit sizes of this mask were 0.3 mm, and the distance between the slits was 0.3 mm. This multislit mask was installed inside the vacuum chamber between the focusing cylindrical lens and target in such a manner, that it allowed the division of the extended 5 mm long plasma onto the eight 0.3 mm long plasma jets.

The thin curve in Figure 5.5 shows the spectrum of enhanced group of harmonics centered in the spectral range of 30 nm (27th harmonic) in the case of eight-jet As plasma. Here the harmonic spectrum from the extended 5 mm long plasma produced on the same arsenic target is also shown at the optimal fluence of heating pulse on the semiconductor surface (thick curve). The optimal fluence in the case of arsenic extended plasma was ~1 J cm^{-2}. The experiments with extended and perforated plasmas were carried out at

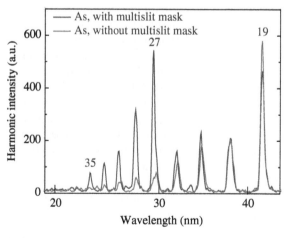

Figure 5.5 Harmonic spectra obtained from the ablation of As using different shapes of plasma formation. Thick curve: extended 5 mm long plasma. Thin curve: eight 0.3 mm long plasma jets. The plasma was produced using the 4 mJ, 370 ps pulses. The intensity of driving pulses was 8×10^{14} W cm^{-2}. *Source:* Ganeev et al. 2015 [5]. Reproduced with permission from AIP Publishing LLC.

similar conditions of ablation. The multislit mask was moved in and out of the path of heating radiation. The whole length of multijet plasma was two times shorter than the one of extended nonperforated plasma. One can see that, even at smaller length of plasma, the yield of the QPM-enhanced harmonics was a few times stronger. Particularly, the 6× enhancement factor for the 27th harmonic was obtained from the arsenic multijet plasma compared with the same harmonic generated from the 5 mm long ablation.

The question may arise as to why the 27th harmonic was enhanced stronger than other orders, while not showing a symmetry of the group of enhanced harmonics (Figure 5.5, thin curve), as was reported earlier in the case of metal plasmas [12]. One can assume that the enhancement of the 27th harmonic was stronger than that of the 25th harmonic due to the joint influence of both QPM and resonance processes. Also note that no resonance enhancement of the 27th harmonic was seen in the case of extended plasma (Figure 5.5, thick curve), contrary to the studies reported in Ref. [14]. One of the reasons of these contradictions could be related with different wavelengths of the fundamental radiation used in Ref. [14], where the resonance enhancement of the 27th harmonic from As plasma was observed ($\lambda_{27H} = 29.37$ nm, $\lambda_{1H} = 793$ nm), and discussed experiments ($\lambda_{27H} = 29.70$ nm, $\lambda_{1H} = 802$ nm). This difference in harmonic wavelengths could cause the detuning from the strong ionic transition, which can be responsible for the enhancement.

The signature of QPM is a quadratic growth of harmonic intensity with the growth of the number (n) of the coherent zones contributing to the overall yield. To prove the role of QPM in the observed peculiarities of harmonic spectra it would be straightforward to investigate the intensity (I_H) of harmonics as a function of the number of plasma jets. This experiment can also confirm the involvement of the coherent accumulation of harmonic yield along the whole length of divided semiconductor plasma, as well as may indicate some incoherency in the contribution of a large amount of zones.

The number of heating zones on the semiconductor surface was shielded step-by-step to form different number of plasma jets (from single jet to eight jets). The anticipated featureless, gradually decreased harmonic spectrum from the single 0.3 mm long plasma jet was similar to that observed in the case of 5 mm long plasma (Figure 5.5 thick curve). With the addition of each next jet, the spectral envelope was changed, with the 27th harmonic intensity in the case of eight-jet configuration becoming 22 times stronger compared with the case of single jet plasma. One can expect the n^2 growth of harmonic yield for the n-jet configuration compared with the single jet once the phase mismatch becomes fully suppressed and the absorption plays an insignificant role, which gives the expected growth factor of 64 in the case of eight-jet semiconductor plasma. The experimentally observed enhancement factor may decrease from the ideal value of n^2 at the conditions when the absorption processes are turned on, or in the case of unequal properties of the jets, which can arise from the heterogeneous excitation of extended target. The broadening of the envelope of QPM-enhanced harmonics with the growth of the number of contributing zones was also observed, which also points out a decrease of selectivity induced by the abovementioned reasons.

5.1.5 Properties of Semiconductor Plasmas

In the following, we analyze the density of the plasmas produced on the semiconductor surfaces by the pulses of different duration. The molecular dynamics simulations for some of the semiconductor targets were carried out using the code ITAP IMD [15]. One can take the advantage of short heating pulses (370 ps), which allowed a direct simulation of the ablation process at the fluence of $1 \, \mathrm{J \, cm^{-2}}$. The optimal delay between the beginning of ablation and the interaction of the plasma particles with the driving pulse is of the order of a few tens of nanoseconds, which cannot be easily simulated directly. In addition, little is known about the heating of already ablated particles. Therefore, for a correct treatment of the delay, which is an important parameter, the following calculation model was used. After leaving the surface of the sample, all particles were considered as noninteracting with the field and were removed from further simulations. For all particles, which left the surface after a given simulation time step, those particles were chosen, which had the kinetic energies sufficient to reach the interaction area ($\sim 100 \, \mu m$ above the surface) exactly after the supposed delay between heating and driving pulses (35 ns).

In order to calculate the relative concentration of neutrals and ions, the single atom ionization probabilities were simulated by time-dependent density functional theory using the Octopus package [16, 17]. For 370 ps pulses, a series of calculations over 50 fs with constant intensities was performed. These calculations allowed the definition of the ionization probability over the envelope of the ablating pulse to yield a time-dependent ionization probability. Then, at every time step of the molecular dynamics simulation, the number of particles with velocities sufficient to reach the interaction region after a given delay was multiplied by this ionization probability. This yielded the concentration of ionized particles in the interaction region. The most important parameter for definition of plasma concentration turned out to be the fluence rather than the intensity of heating radiation. The plasma and electron densities at the distance of $100 \, \mu m$ above the arsenic target, 35 ns from the beginning of ablation using $1 \, \mathrm{J \, cm^{-2}}$ fluence, were calculated to be 3×10^{17} and $2 \times 10^{16} \, \mathrm{cm^{-3}}$ respectively. The details of a three-dimensional molecular dynamical simulation of laser ablation of graphite using the molecular dynamics code ITAP IMD have previously been reported in Ref. [18].

At given HHG conditions (driving pulse is not ultrashort, plasma is not overdense, no resonant absorption is expected), the plasma density influences the HHG yield mainly in two ways. It increases the number of particles participating in HHG, and thus harmonic intensity grows with the increase of the density of emitters. It also increases the number of the free electrons, which influence HHG efficiency in a rather complicated way.

One can see the 15-fold prevalence of the plasma density over the electron density at the conditions corresponding to the maximal HHG. The ratio of plasma and electron densities could be either larger or smaller than 15, depending on the characteristics of ablated material and the conditions at which the highest harmonic yield could be achieved.

The semiconductor targets were ablated in such a manner that the formed plasma did not contain a significant amount of the electrons, which deteriorate the phase-matching conditions. At these conditions, the plasma becomes mostly populated with the neutrals, which are responsible for lower order harmonic generation, rather than the singly charged ions, which are responsible for HHG. At strong excitation of targets, the electron density becomes too large. In that case, the dispersion of plasma will be fully defined by the electrons rather than the neutrals and ions. The phase relations between the driving and harmonic waves at these conditions become far from optimal. Thus the "optimal" ablation is related with the appropriately prepared plasma suitable for the harmonic generation along the whole length of nonlinear medium. Thus the used medium could be qualified as a low-ionized, low-excited plasma containing small amount of free electrons.

5.1.6 Harmonic Cutoffs

In the case of As plasma, the observed high-order harmonics were assumed to be originated predominantly from the singly charged ions. This conclusion was drawn from the three-step model of HHG, taking into account the ionization potential of singly charged arsenic ions and barrier suppression intensity of those particles. The highest possible photon energy of harmonic spectrum is defined by the formula $E_c \approx I_p + 3.17U_p$, where I_p is the ionization potential of the particle emitting harmonics and U_p is the ponderomotive energy of accelerated electron ($U_p = e^2E^2/4m\omega^2 = 9.33 \times 10^{-14}I_L\lambda^2$, where e and m are the electron charge and mass, and E, ω, I_L, and λ are the field's amplitude, frequency, intensity, and wavelength, respectively). To achieve the highest possible cutoff energy, one has to increase the ponderomotive energy or choose the emitter possessing higher ionization potential. The ponderomotive energy of electrons at a fixed wavelength strongly depends on the laser intensity. However, if the nonlinear medium is ionized before the peak of laser pulse, the cutoff energy is defined by the saturation intensity ($I_{BSI} < I_L$), which is related with the barrier-suppression ionization.

To calculate the cutoff energy at the conditions of discussed experiments, the I_{BSI} of low-charged semiconductor ions was estimated using the expression for the barrier-suppression ionization $I_{BSI} = c(I_p)^4/128\pi e^6(Z_i)^2 =$

$4 \times 10^9 \times (I_p)^4/(Z_i)^2$, where c is the light velocity and Z_i is the degree of ionization. The I_{BSI} of the singly charged ions, with the ionization potentials of 16.3, 15.9, 18.6, 21.2, 15.3, 16.5 eV for the Si, Ge, As, Se, Te, and Sb singly charged ions, were calculated to be $\sim 7 \times 10^{13}$, 6×10^{13}, 1.2×10^{14}, 2×10^{14}, 6×10^{13}, and 7×10^{13} W cm^{-2}, respectively. One can see that I_{BSI} of all samples were lower than the laser intensity used in present studies (5×10^{14}–1×10^{15} W cm^{-2}). The harmonic cutoffs were calculated to correspond to the 19th, 17th, 27th, 37th, 17th, and 19th orders, assuming the involvement of singly charged ions in the HHG. These results do not match with the experimentally observed cutoffs (27th, 35th, 35rd, 51st, 31st, 35th harmonics for the Si, Ge, As, Se, Te, and Sb plasmas, respectively).

This discrepancy could be explained by the assumption of the partial involvement of the doubly charged ions as the emitters of high-order harmonics. The involvement of higher charged ions leads to the extension of harmonic cutoff, though the yield of these "additional" harmonics becomes smaller than the yield of lower orders. To prove the involvement of doubly charged ions in the HHG, one has to analyze the plasma characteristics. The spectroscopic analysis of plasma has revealed a presence of the emission lines originated from the doubly charged particles of some of the studied species.

5.2 27th Harmonic Enhancement by Controlling the Chirp of the Driving Laser Pulse During High-Order Harmonic Generation in GaAs and Te Plasmas

The HHG studies with some metal plasmas demonstrated the capability of the generation of an almost monochromatic harmonic radiation through interaction of laser with the ablated plasma. Such an approach paved the way for efficient single harmonic enhancement in XUV range using different plasma sources. In those studies, the harmonic wavelength was tuned by changing the wavelength of the fundamental laser beam by changing the oscillator spectrum. However, changing the oscillator spectrum is not practical because the adjustment of the oscillator spectrum cannot be directly transferred to the final laser spectrum due to gain narrowing and gain saturation processes. Moreover, it is also necessary to re-adjust the stretcher and compressor, which is a cumbersome process. A much simpler approach to tune the harmonic wavelength without modifying the driving laser spectrum is by controlling the chirp of the fundamental radiation [19–21]. In this section, we discuss the spectral tuning of the high-order harmonics produced during the propagation of femtosecond duration laser radiation through a low-excited GaAs [22] and Te [23] plasmas by means of the chirp control of the driving laser radiation.

The pump laser used in these studies was a chirped-pulse amplification Ti:sapphire laser system (Thales Lasers SA), operating at a 10 Hz pulse

Figure 5.6 Schematic of the experimental setup on high-order harmonic generation from GaAs plasma. VC, vacuum chamber; T, target; S, slit; G, grating; L, lenses; MCP, microchannel plate; CCD, charge-coupled device; FP, femtosecond pulse; PP, picosecond pulse. *Source:* Ganeev et al. 2006 [22]. Reproduced with permission from Optical Society of America.

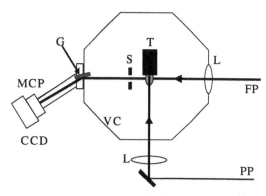

repetition rate. A portion of the uncompressed radiation (pulse energy $E = 30$ mJ, pulse duration $\tau = 300$ ps, central wavelength $\lambda = 793$ nm) was split from the main beam by a beam splitter and used as a heating pulse, as shown in Figure 5.6. This heating pulse was focused by a spherical lens (normal incidence) on the GaAs target kept in a vacuum chamber, which produced an ablation plume predominantly consisting of neutrals and singly charged ions. The focal spot diameter of the heating pulse beam at the GaAs and Te surfaces was adjusted to be ~600 μm. The intensity of the picosecond pulse on the target surface varied between 5×10^9 and 9×10^{10} W cm^{-2}. After some variable delay (20–50 ns), a femtosecond main pulse ($E = 90$ mJ, $\tau = 50$ fs, $\lambda = 793$ nm, 19 nm spectral width) was focused on the area of the GaAs and Te plasmas from a direction parallel to the target surface, using a spherical lens with 500 mm focal length. The maximum intensity of the main femtosecond beam at the focal spot was 4×10^{17} W cm^{-2}. Since this intensity considerably exceeded the barrier suppression intensity of singly charged ions, the position of the laser focus was adjusted by placing it either before the laser plume or after it to optimize the high harmonics output. Intensity of the driving laser pulse at the preformed plasma varied between 2×10^{14} and 8×10^{15} W cm^{-2}. The high-order harmonics were analyzed by a XUVS.

5.2.1 Optimization of HHG in GaAs Plasma

The use of solid target ablation for HHG has some obvious advantages over gas jets such as simplicity, absence of differential pumping, rep-rate operation, and so on. In addition to these, the special advantages of this approach over the conventional gas–jet technique include its capability of generating the plasma with higher density, longer length, and easily variable conditions. This technique also gives new degrees of freedom that can be used for the optimization of HHG. The possibility of the use of any element in the periodic table that can be formed as a solid target may reveal many interesting possibilities for the generation of coherent XUV radiation. The optimization of the plasma conditions can lead to further growth of HHG efficiency.

A majority of the HHG studies from the laser plumes were, so far, focused on the single-atom species, though the HHG from diatomic molecules could offer some advantages compared to the single atoms due to their prolonged structures and delocalized π electrons. While single atoms with their low ionization potentials are likely to see ionization saturation clamp the HHG to low orders, this is not obvious for multi-electron and molecular species where anomalously high ionization saturation intensities have been widely reported in the literature. Further, these diatomic molecules, being excited and evaporated from the solid surface, could possess some properties of nanosized structures with enhanced nonlinear optical response due to local field enhancement. Finally, such structures may possess appropriate electron transitions, which could be used for the resonance enhancement of nonlinear optical response.

The high-order harmonics up to the 43rd order (at $\lambda = 18.4$ nm) were observed in discussed experiment and showed a plateau-like pattern, with the harmonics in the range of 17th–31st orders appearing at nearly equal intensity. Various characteristics of HHG were systematically studied in order to maximize the yield of harmonics from GaAs plasma. The optimal laser plasma was created by the weak focusing of the heating pulse. The laser plasma prior to the interaction with femtosecond pulse dominantly consisted of neutrals and a small amount (~10%) of singly charged ions. The latter was estimated from the spectral analysis of the plasma plume in visible–ultraviolet (UV) range using fiberoptics spectrometer (see Figure 5.7) and in XUV range using XUVS. During the interaction of this plasma with femtosecond pulses, an increase in the intensity of spectral lines corresponding to the singly charged ions and appearance of spectral lines characteristics of the multiply charged ions were observed. This indicated growth of the concentration of the singly charged ions as well as generation of multiply charged ions in the plasma plume. The harmonic generation proved to be most effective when the plume consisted of neutral molecules of GaAs and singly charged ions. The influence of the time delay between the heating and main pulse on the harmonic yield was analyzed.

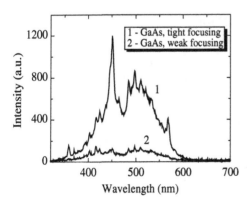

Figure 5.7 Spectral measurements of the GaAs plasma produced at the tight and weak focusing conditions of heating pulse radiation. *Source:* Ganeev et al. 2006 [22]. Reproduced with permission from Optical Society of America.

The harmonic output considerably increased when the delay exceeded 20 ns and then remained approximately constant up to the maximum used delay (57 ns). A decrease in the conversion efficiency was observed with an increase in the heating pulse intensity ($I_{pp} > 2 \times 10^{10}$ W cm^{-2}). This is attributed to the generation of multiply charged ions at higher heating pulse intensities and the ionization-induced defocusing of the main pump beam due to the generation of a large amount of free electrons in the GaAs plume.

The position of the laser focus was adjusted by placing it before the plasma plume or after it to optimize the high harmonics output (Figure 5.8). The change in the position of the laser focus changes the peak intensity of the laser pulse in the plasma plume and thus it changes the temporal profile of the laser beam interacting with the plasma. Figure 5.8 shows the dependence of the 21st harmonic intensity on the focal position of the driving laser radiation. It is seen that more intense harmonics are produced when the laser beam is focused after the plasma plume. Such a peculiarity has been observed previously in the laser–gas jet experiments as well [24], and was attributed to the influence of free electrons in the nonlinear medium leading to the self-defocusing of the driving pulse. The laser intensity for the maximum intensity of harmonics was estimated to be 6×10^{14} W cm^{-2}. A typical high-order harmonic spectrum for this condition for chirp-free laser pulse of 50 fs duration is shown in Figure 5.9.

The appearance of a plateau is an indication that the process involved in HHG follows the three-step mechanism proposed in the early stages of high harmonic studies. It predicts the cutoff energy of harmonic radiation to be given by $E_c \approx I_p + 3.2U_p$, where I_p is the ionization potential and U_p is the ponderomotive potential that corresponds to the energy of free electron in the field of electromagnetic wave ($U_p = 9.33 \times 10^{-14} I_{fp} \lambda^2$). One can examine the generation of harmonics in terms of the interaction of the driving radiation with the GaAs molecules and ions. The observed results cannot be explained if

Figure 5.8 Dependence of 21st harmonic intensity on the focal position of driving laser radiation. *Source:* Ganeev et al. 2006 [22]. Reproduced with permission from Optical Society of America.

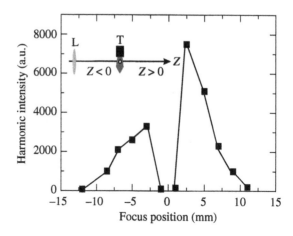

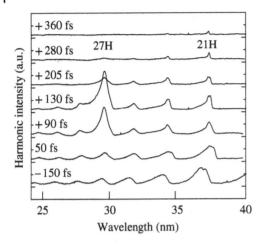

Figure 5.9 Harmonic spectra from GaAs plume as a function of pulse chirp and width. Each curve is shifted vertically to avoid overlap for visual clarity. *Source:* Ganeev et al. 2006 [22]. Reproduced with permission from Optical Society of America.

the plasma plume is assumed to consist only of neutrals. The atomic ionization potential for GaAs is rather low ($I_p = 5.70$ eV) and the barrier suppression intensity for GaAs molecules is estimated to be 3×10^{13} W cm^{-2}. Above this intensity all the GaAs molecules would be ionized. Thus the neutral GaAs molecules can only generate harmonics with a maximum order of 7, which is much smaller than the experimentally observed value of 43. Next, taking the second ionization potential of GaAs to be ~ 18 eV and using the optimal laser intensity of $I_{fp} = 6 \times 10^{14}$ W cm^{-2}, the maximum harmonic energy comes to $E_c \approx 132.6$ eV. The corresponding harmonic order is 83, which is larger than the observed value of maximum harmonic order of 43.

The above discrepancy, which was also observed previously in gas–jet experiments, may be associated with the self-defocusing of femtosecond laser pulse and the phase mismatch between harmonic and driving pulses. The nonlinear refractive index (γ) of the plasma produced under similar conditions was defined to be -2×10^{-18} cm^2 W^{-1} [25]. At a peak laser intensity of 10^{15} W cm^{-2}, the nonlinear addition to the refractive index ($\Delta n = \gamma I$) of the plasma at 793 nm will be -2×10^{-3}. This can produce a strong refractive index gradient resulting in self-defocusing of the femtosecond laser beam at high intensities. Next, at the given density of plasma with $\sim 10\%$ ionization, the coherence length for the 27th harmonic is estimated to be about ~ 0.6 mm. Thus any increase in ionization beyond the optimal plasma condition would severely degrade the phase matching and lower the generation of high harmonics.

5.2.2 Variation of the Chirp of Femtosecond Pulses

We now discuss the effect of introducing chirp in the laser pulse on the harmonic spectrum. The chirp of the main laser pulse was varied by the adjustment of the separation between the gratings in the pulse compressor. A reduction in the grating separation from the chirp-free condition generates positively chirped pulses, and an increase provides negatively chirped pulses. The variations of laser chirp resulted in a considerable variation of the harmonic distribution from laser plasma. In the chirp-free case and for negative chirp, a featureless plateau-like shape of high-order harmonics with a smooth decrease of harmonic intensity was observed. On the other hand, for positive chirp, the harmonic peaks shifted to longer wavelengths. Moreover, in the case of positive chirp, a strong enhancement of 27th harmonic ($\lambda = 29$ nm) intensity compared to that of the neighboring ones was observed (see Figure 5.9, the curves for 90 and 130 fs positively chirped pulses). The intensity of the 27th harmonic was about five times higher than the intensities of nearest harmonics.

The above observation can be explained by the wavelength change in the leading edge of the laser pulse as the pulse is chirped. The initial lower intensity portion of the pulse creates harmonics. As the pulse intensity reaches its peak, the condition for HHG gets spoilt (as in Figure 5.8, when the beam is better focused). By varying the chirp of the laser pulse, one varies the spectral components present in the leading edge of the pulse. This facilitates tuning of the harmonic wavelengths, which allows resonance-induced enhancement of a particular harmonic through its coincidence with some transition in atoms/ions of the plasma plume.

To establish the origin of the resonance-induced enhancement of the 27th harmonic, one has to study the plasma emission characteristics of GaAs plume. Figure 5.10 shows the plasma spectrum of GaAs at the conditions of the tight focusing of heating pulse, out of the optimal conditions of harmonic generation. It is seen that there are no specific emission lines at the wavelength locations of the harmonics in the tight focused plasma. The origin of 29 nm radiation enhancement was further examined by inserting a quarter-wave plate on the

Figure 5.10 Comparison between (a) plasma spectrum and (b) harmonic spectrum of GaAs. It is seen from (c) that the harmonics fully disappear when the femtosecond beam is made circularly polarized. Each curve is shifted vertically to avoid overlap for visual clarity. *Source:* Ganeev et al. 2006 [22]. Reproduced with permission from Optical Society of America.

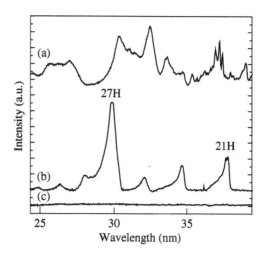

path of the femtosecond beam to make it circularly polarized. No harmonics were observed in the case of circular polarization, as it was expected assuming the nonlinear optical origin of harmonic radiation (Figure 5.10).

The GaAs plasma plume has shown the generation of high-order harmonics, and especially a significant enhancement of the 27th harmonic. A question arises as to which of the two constituent atoms, namely Ga and As, is responsible for this enhancement. Hence HHG studies were carried out on GaP and As plumes. In the case of GaP plume, no any specific peculiarity at the location of the 27th harmonic i.e. at 29 nm was obtained (Figure 5.11). At the same time, the As plasma showed a high intensity peak at the latter position. This clearly established that the 27th harmonic intensity enhancement takes place due to the As ions/atoms (see also Subsection 5.1.2). The spectral measurements of As plasma (Figure 5.11) also did not show any line radiation at the wavelength location of the 27th harmonic.

Intensity enhancement of some harmonic orders has been reported earlier in laser–gas jet interaction. For instance, roles of resonances and re-collision in atoms have been discussed in terms of strong-field atomic phenomena. However, harmonic intensity enhancement was predicted over a broad range of harmonics. Further, using an optimized laser pulse shape, the enhanced 27th harmonic in Ar more than an order of magnitude was reported in Ref. [26]. Generation of arbitrary shaped spectra of HHG by adaptive control of the pump laser pulse in laser–gas jet experiments was also demonstrated [27]. However, in both the above studies, harmonic intensity enhancement occurred for some neighboring harmonics as well, in contrast to discussed studies using the preformed plasmas from solid targets.

Next, a strong dependence of the harmonic radiation spectrum on the chirp of the laser pulse has also been observed in laser–jet experiments [20]. It was shown that, for identical pulse durations, distinct harmonic peaks can be

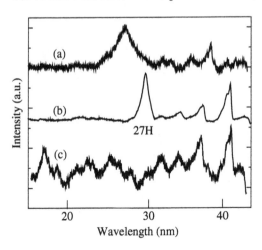

Figure 5.11 (a) Plasma spectrum of ablated arsenic plume, (b) harmonic spectrum from As plasma, and (c) harmonic spectrum from GaP plasma. Each curve is shifted vertically to avoid overlap for visual clarity. *Source:* Ganeev et al. 2006 [22]. Reproduced with permission from Optical Society of America.

observed for positively chirped excitation pulses, while for negatively chirped pulses the harmonic peaks become irregular. This behavior was explained by the simulations, which combine the chirp of the laser with the intrinsic phase shift of the harmonics.

In discussed studies, the case of the HHG at low density of particles in the plasma was considered, where one expects collective phase-matching effects to be minimized compared to the single-atom effects. It was observed that, both for positively and negatively chirped pump pulses, the individual harmonic peaks are well defined and discrete. In contrast, previous reports on laser–jet HHG experiments have shown a distinct difference between the HHG spectra driven by negatively and positively chirped pulses, which were related with both the influence of dynamically induced negative chirp and the positive chirp induced by strong laser radiation and ionized medium due to the self-phase modulation (SPM) of the laser pulse [21].

Meanwhile, no significant difference between the sharpness of harmonics in different parts of plateau region was observed in the case of chirp-free laser pulses propagating through laser plasma. One cannot expect the influence of SPM on the spectral distribution of harmonics since used experimental conditions (low-density plasma, moderate laser intensities) restricted the possibility of the influence of strongly ionized medium on the phase characteristics of generated harmonics. The ionized medium, with higher electron density in the center than in the outer region, acts as a negative lens, leading to defocusing of the laser beam in a plasma and hence to a reduction in the effective harmonic generation volume. In addition, the rapidly ionizing high-density medium modifies the temporal structure of the femtosecond laser pulse due to the SPM. The conditions were maintained when no significant ionization of the plasma by the driving laser pulse takes place, by keeping the laser intensity in the vicinity of the plume below the barrier suppression intensity for singly charged GaAs ions.

5.2.3 Observation of Single-Harmonic Enhancement Due to Quasi-Resonance with the Tellurium Ion Transition at 29.44 nm

In this subsection, we analyze the enhancement of a single harmonic at the wavelength of 29.44 nm by using a tellurium laser ablation plume [23]. The details of experimental setup are described in the previous subsection. Figure 5.12 shows the spectra of the HHG from laser-ablation tellurium plume at the wavelength of (i) 20–45 nm and (ii) 13–20 nm. Only odd harmonics were observed in those studies, while a strong emission at 29.44 nm was obtained (Figure 5.12a). The conversion efficiency of this harmonic (H27) was measured to be 10^{-6}, from which one can determine the pulse energy of the 27th harmonic (12 nJ) at the pump laser energy of 12 mJ. The high-order harmonics up to the 51st order at the wavelength of 15.58 nm were obtained,

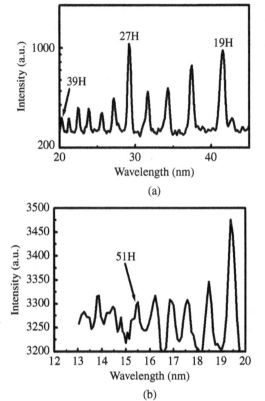

Figure 5.12 HHG spectra from the tellurium and silver laser-ablation plumes were obtained at the wavelength of (a) 13–35 and (b) 13–20 nm. A strong 27th harmonic at the wavelength of 29.44 nm was obtained. The spectra (a) and (b) were accumulated using 10 and 100 shots, respectively. Highest harmonic order obtained in these studies from the Te plume was the 51st one ($\lambda = 15.59$ nm). *Source:* Suzuki et al. 2007 [23]. Reproduced with permission from Optical Society of America.

although the intensity of the 51st harmonic was too low (Figure 5.12b). The conversion efficiency of the 51st harmonic was measured to be 10^{-8}.

To investigate the mechanism of the 27th harmonic enhancement for the laser-ablation tellurium plume, the central wavelength of pump laser pulse was detuned away from 795 nm. Figure 5.13 shows the HHG spectra from the laser-ablation tellurium plume measured by using two different laser wavelengths (795 and 788 nm). One can see that, by changing the pump laser wavelength from 795 to 788 nm, the enhancement of the intensity for the 27th harmonic was decreased compared to the neighboring harmonics, whereas the intensities of neighboring harmonics remained almost the same at the pumping laser wavelengths of 795 and 788 nm.

A Te II ion has been shown to possess a strong $4d^{10}5s^25p^3\,{}^2D_{5/2} \rightarrow 4d^95s^25p^4({}^3P)^2F_{7/2}$ transition at the wavelength of 30.86 nm [8]. The gf-value of this transition has been reported to be 1.832, and this value 2–10 times exceeds those of other transitions of Te II in this spectral range. One can attribute the enhancement of the 27th harmonic by using the 795 nm radiation to the quasi-resonance conditions with this transition driven by AC-Stark shift. By changing the pump laser central wavelength from 795 to 788 nm, the wavelength of the 27th harmonic has been changed from 29.44 to 29.18 nm.

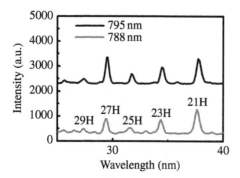

Figure 5.13 HHG spectra from tellurium laser-ablation plume for pump laser with central wavelength of 795 and 788 nm. The black and gray curves are the pumping laser wavelength of 795 and 788 nm, respectively. The enhancement of 27th harmonic radiation decreased compared to the neighboring harmonics as the central wavelength of laser radiation became shorter. *Source:* Suzuki et al. 2007 [23]. Reproduced with permission from Optical Society of America.

Therefore, the wavelength of the 27th harmonic pumped by 788 nm radiation becomes farther away from the $4d^{10}5s^25p^3\,^2D_{5/2} \rightarrow 4d^95s^25p^4(^3P)^2F_{7/2}$ transition (30.86 nm). As a result, the resonance enhancement for the 27th harmonic generated by 788 nm radiation becomes weaker compared to the case of 795 nm pump. The strong resonance enhancement was not observed at another wavelength region because the *gf*-value at this transition of tellurium was low.

Figure 5.14 shows the enhancement factor as a function of the wavelength difference between the harmonic and the strong oscillator strength by using indium, tin, antimony, and tellurium. In discussed studies, the intensity of the 27th harmonic produced from the laser-ablation tellurium plume only few times exceeded those of the neighboring harmonics. In previous research, the reported intensities of the 13th harmonic for indium, the 17th harmonic for tin, and the 21st harmonic for antimony were 20–200 times higher than those of neighboring harmonics. For indium and tin, the wavelength difference between the enhanced harmonic and the ionic transition possessing by the

Figure 5.14 The enhancement factor of the harmonic as a function of the wavelength difference between the harmonic and the strong oscillator strength transition. The black solid circles is tellurium, the gray solid squares is tin, and the black triangles and squares are antimony and indium, respectively. *Source:* Suzuki et al. 2007 [23]. Reproduced with permission from Optical Society of America.

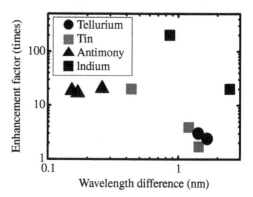

high oscillator strength were 0.86 and 0.49 nm, respectively. In discussed work, however, the difference between the wavelengths of the 27th harmonic and corresponding Te II transition was 1.42 nm, and comparing with the abovementioned studies, this difference is mostly away. For these wavelength differences of 1.4 nm by using tin, the enhancement factor was estimated to be two as can be seen in Figure 5.14, and this enhancement factor by using tin with the 778 nm wavelength pumping is almost similar to the tellurium case. To enhance the enhancement factor, the wavelength of the enhanced harmonic should therefore be close to the wavelength of the strong oscillator strength transition. As a consequence, it may be possible to increase the enhancement factor for the 27th harmonic by using the 810 nm pumping laser pulse because the wavelength of the 27th harmonic is estimated to be 30 nm. The wavelength of the used laser system could not be tuned toward the wavelength of 810 nm. Another reason of a low enhancement of the 27th harmonic generating in tellurium plasma could be related with the difference in oscillator strength of the transitions involved in the cases of different plumes.

5.3 Resonance Enhanced Twenty-First Harmonic Generation in the Laser-Ablation Antimony Plume at 37.67 nm

In this section, a strong enhancement of the single harmonic at the wavelength of 37.67 nm using the antimony laser-ablation plume is discussed [9]. Figure 5.15 shows the typical HHG spectra from the antimony and silver laser-ablation plumes at the wavelength range of 10–30 nm. The XUV spectrum after propagation of the femtosecond radiation through the antimony plasma has shown high-order harmonics up to the 55th order at the cutoff wavelength of 14.45 nm. The conversion efficiency of the 55th harmonic was measured to be 2×10^{-7}, while the harmonic efficiency in the range of 15th to 27th harmonics was close to 1.2×10^{-6}. The details of the absolute conversion efficiency calibration of the spectrometer were described in Ref. [28]. Using the silver laser-ablation plasma, the same cutoff (H55) was observed, though the second ionization potentials of these materials (16.53 eV for Sb and 21.49 eV for Ag) were different.

Figure 5.16 shows the HHG spectrum from the antimony laser-ablation plume at the wavelengths of 33–60 nm. A strong 21st harmonic at the wavelength of 37.67 nm was obtained. The intensity of the 21st harmonic was 20 times higher than those of the 23rd and 19th harmonics. The conversion efficiency of the 21st harmonic was measured to be 2.5×10^{-5}, and thus the pulse energy of the 37.67 nm radiation of 0.3 μJ was obtained from the pump laser energy of 12 mJ. By changing the pump laser polarization from the linear

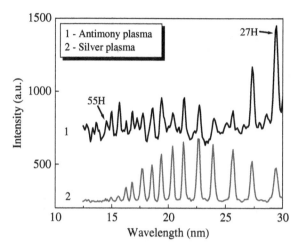

Figure 5.15 HHG spectra from the laser-ablation antimony and silver plumes irradiated by femtosecond laser pulse. The black and gray lines are the harmonics from the antimony and silver plumes, respectively. The spectrum of the HHG from antimony at the wavelength range of 10–30 nm was accumulated using 100 shots. The HHG from the silver plume was accumulated using 10 shots. The curves are shifted vertically to avoid overlap for visual clarity. *Source:* Suzuki et al. 2007 [9]. Reproduced with permission from Optical Society of America.

Figure 5.16 HHG spectrum at the wavelength range of 33–65 nm from the laser-ablation antimony plume. This spectrum was accumulated during 10 shots. The intensity of 21st harmonic was measured to be 20 times higher than those of the 23rd and 19th harmonics. *Source:* Suzuki et al. 2007 [9]. Reproduced with permission from Optical Society of America.

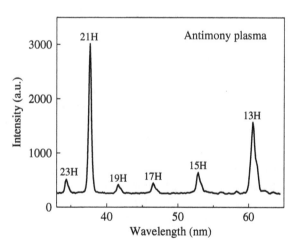

polarization to the circular one using the quarter-wave plate, the 37.67 nm radiation was completely disappeared. This tendency is consistent with that of HHG, which allows concluding that the strong emission at the wavelength of 37.67 nm is generated through the HHG.

To investigate the mechanism of the 21st harmonic enhancement in Sb plume, the central wavelength of the laser pulse was tuned in the range from

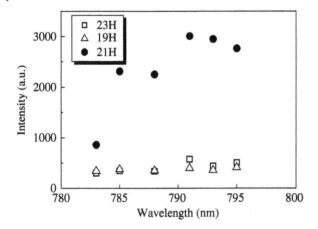

Figure 5.17 Intensities of the 21st (solid circles), 23rd (open squares), and 19th (open triangles) harmonics as the functions of the pump laser wavelength. *Source:* Suzuki et al. 2007 [9]. Reproduced with permission from Optical Society of America.

795 to 783 nm. Figure 5.17 shows the intensities of the 19th, 21st, and 23rd harmonics as the functions of the pump laser wavelength. By changing the laser wavelength from the longer wavelength side (795 nm) to the shorter one, the intensity of the 21st harmonic was initially gradually increased and then abruptly decreased. The highest intensity of the 21st harmonics was observed at 791 nm. At the same time, the intensities of the 19th and 23rd harmonics remained almost the same at the wavelength range of 795–783 nm. In the past work, the strong Sb II transitions of $4d^{10}5s^22p^3P_2$-$4d^95s^25p^3(^2D)^3D_3$ and $4d^{10}5s^22p^1D_2$-$4d^95s^25p^3(^2D)^3F_3$ at the wavelengths of 37.82 and 37.55 nm, respectively, have been reported and analyzed [6]. The *gf* values of $4d^{10}5s^22p^3P_2$-$4d^95s^25p^3(^2D)^3D_3$ and $4d^{10}5s^22p^1D_2$-$4d^95s^25p^3(^2D)^3F_3$ transitions have been calculated to be 1.36 and 1.63, respectively, which was six to seven times higher than those of the neighbor transitions. The enhancement of the 21st harmonic of the 791 nm radiation was highest in our studies, though the 21st harmonic ($\lambda = 37.67$ nm) was slightly away from the $4d^{10}5s^22p^3P_2$-$4d^95s^25p^3(^2D)\ ^3D_3$ transition ($\lambda = 37.82$ nm). Probably, in this case, the enhancement of 21st harmonic was due to the resonance with the $4d^{10}5s^22p^3P_2$-$4d^95s^25p^3(^2D)\ ^3D_3$ transition driven by the AC-Stark shift. By changing the pump laser wavelength from 791 to 788 nm, the intensity of the 21st harmonic gradually decreased because this harmonic drifts away from the $4d^{10}5s^22p^3P_2$-$4d^95s^25p^3(^2D)\ ^3D_3$ transition. However, the enhancement of the 21st harmonic of the 785 nm radiation was higher than that of 788 nm radiation. The reason of this intensity enhancement was attributed to the resonance with the $4d^{10}5s^22p^1D_2$-$4d^95s^25p^3(^2D)\ ^3F_3$ transition driven by the AC-Stark shift. Further decrease of the 21st harmonic wavelength led to

the mismatching with the above transitions. As a result, the intensity of the 21st harmonic pumped by the 783 nm radiation was considerably decreased.

In discussed study, the enhancement in intensity of the single harmonic belonging to the mid-plateau region was smaller compared with the enhancement in the intensity of the 13th harmonic generated from indium plume reported in the previous studies [29]. The latter harmonic belongs to the beginning of the plateau distribution in the case of harmonic generation from indium plasma. The enhancement in intensity of the 13th harmonic (×200) generated from In plasma considerably exceeded that for the 21st (×20) harmonic generated from the Sb plasma. The reason of such a difference is perhaps related with the difference in oscillator strengths of the transitions involved.

References

1 Akiyama, Y., Midorikawa, K., Matsunawa, Y. et al. (1992). *Phys. Rev. Lett.* 69: 2176.

2 Ganeev, R.A. (2012). *High-Order Harmonic Generation in Laser Plasma Plumes*. London: Imperial College Press.

3 De Nalda, R., López-Arias, M., Sanz, M. et al. (2011). *Phys. Chem. Chem. Phys.* 13: 10755.

4 Ganeev, R.A., Suzuki, M., and Kuroda, H. (2014). *Phys. Plasmas* 21: 053503.

5 Ganeev, R.A., Suzuki, M., Yoneya, S., and Kuroda, H. (2015). *J. Appl. Phys.* 117: 023114.

6 D'Arcy, R., Costello, J.T., McGuinness, C., and O'Sullivan, G. (1999). *J. Phys. B: At. Mol. Opt. Phys.* 32: 4859.

7 Gaynor, L., Murphy, N., Kilbane, D. et al. (2005). *J. Phys. B: At. Mol. Opt. Phys.* 38: 2895.

8 Murphy, N., Costello, J.T., Kennedy, E.T. et al. (1999). *J. Phys. B: At. Mol. Opt. Phys.* 32: 3905.

9 Suzuki, M., Baba, M., Kuroda, H. et al. (2007). *Opt. Express* 15: 1161.

10 Tauheed, A. and Hala, A. (2012). *Phys. Scr.* 85: 025304.

11 Esteves, D.A., Bilodeau, R.C., Sterling, N.C. et al. (2011). *Phys. Rev. A* 84: 013406.

12 Ganeev, R.A., Suzuki, M., and Kuroda, H. (2014). *J. Phys. B: At. Mol. Opt. Phys.* 47: 105401.

13 Seres, A., Yakovlev, V.S., Seres, E. et al. (2007). *Nat. Phys.* 3: 878.

14 Ganeev, R.A., Naik, P.A., Singhal, H. et al. (2007). *Opt. Lett.* 32: 65.

15 Stadler, A., Mikulla, R., and Trebin, H.-R. (1997). *Int. J. Mod. Phys.* 8: 1131.

16 Marques, A.L., Castro, A., Bertsch, G.F., and Rubio, A. (2003). *Comput. Phys. Commun.* 151: 60.

17 Castro, A., Appel, H., Oliveira, M. et al. (2006). *Phys. Status Solidi B* 243: 2465.

18 Ganeev, R.A., Hutchison, C., Witting, T. et al. (2012). *J. Phys. B: At. Mol. Opt. Phys.* 45: 165402.

19 Chang, Z., Rundquist, A., Wang, H. et al. (1998). *Phys. Rev. A* 58: R30.

20 Kim, H.T., Lee, D.G., Hong, K.-H. et al. (2003). *Phys. Rev. A* 67: 051801.

21 Kim, H.T., Kim, I.J., Lee, D.G. et al. (2004). *Phys. Rev. A* 69: 031805.

22 Ganeev, R.A., Singhal, H., Naik, P.A. et al. (2006). *J. Opt. Soc. Am. B* 23: 2535.

23 Suzuki, M., Baba, M., Ganeev, R.A. et al. (2007). *J. Opt. Soc. Am. B* 24: 2686.

24 Bellini, M., Corsi, C., and Gambino, M.C. (2002). *Laser Part. Beams* 20: 277.

25 Ganeev, R.A., Suzuki, M., Baba, M., and Kuroda, H. (2006). *J. Opt. Soc. Am. B* 23: 1332.

26 Bartels, R., Baskus, S., Zeek, E. et al. (2000). *Nature* 406: 164.

27 Pfeifer, T., Walter, D., Winterfeldt, C. et al. (2005). *Appl. Phys. B* 80: 277.

28 Ganeev, R.A., Baba, M., Suzuki, M., and Kuroda, H. (2005). *Phys. Lett. A* 339: 103.

29 Ganeev, R.A., Suzuki, M., Baba, M. et al. (2006). *Opt. Lett.* 31: 1699.

6

Resonance Processes at Different Conditions of Harmonic Generation in Laser-Produced Plasmas

6.1 Application of Picosecond Pulses for HHG

Small intensity of picosecond pulses significantly decreases the chances to observe resonance-induced processes during harmonic generation. However, in some cases, this process occurs even at these unfavorable conditions. Frequency multiplication of laser radiation in gases for a long time was one of the most efficient methods for obtaining short-wavelength coherent radiation. This is facilitated by three most important factors for using gases as nonlinear media: their transparency in the extreme ultraviolet (XUV) region, the possibility of using anomalous dispersion to achieve phase matching for low-order harmonics, and the high nonlinear susceptibilities under resonant and quasiresonant conditions.

The possibility of analyzing these properties using picosecond pulses has been demonstrated more than 40 years ago [1, 2], when low-order frequency multiplication was carried out in a two-component medium consisting of alkali metal vapor in a rare gas. However, extended media are required for the efficient conversion of fairly high-power radiation [3], and it is technically difficult to make a long cell with a homogeneous distribution of alkali metal vapor. On the other hand, rare gases possess higher breakdown and multi-photon ionization thresholds than do metal vapors, and so the most suitable nonlinear media from this point of view are rare gases.

The important issue here is the use of resonance conditions, which has been demonstrated in the case of low-order harmonic generation in gases using picosecond pulses. Particularly, 3rd harmonic generation in xenon, from the 3rd harmonic of a Nd:YAG laser, is quasiresonant. The lines of two allowed transitions are closely situated, having mismatches of 0.8 and 1.0 nm and oscillator strengths of 0.0968 and 0.395, respectively [4]. Consequently, the nonlinear susceptibilities are relatively high for the rare gas and are only weakly dependent on the radiation frequency, and so the experimental and theoretical results do not agree everywhere [5].

Resonance Enhancement in Laser-Produced Plasmas: Concepts and Applications,
First Edition. Rashid A. Ganeev.
© 2018 John Wiley & Sons, Inc. Published 2018 by John Wiley & Sons, Inc.

All those pioneering studies were carried out to analyze lowest order harmonics using relatively long (picosecond) pulses. In the following, we discuss the application of picosecond pulses for high-order harmonic generation (HHG) in the plasmas demonstrating resonance enhancement of single higher order harmonic.

6.1.1 High-Order Harmonic Generation of Picosecond Laser Radiation in Carbon-Containing Plasmas

A search for new plasma media and definition of the best experimental conditions, such as pulse duration of driving laser field, for efficient HHG in different spectral ranges is a way for further enhancement of harmonic yield, particularly by resonance method. The carbon-containing materials such as graphite, fullerenes, and carbon nanotubes have proven that, at optimal ablation conditions, they can be considered as attractive plasma media for the harmonic generation in the 40–80 nm spectral range using the femtosecond laser pulses [6–8]. Propagation effects during the HHG of subpicosecond KrF laser ($\lambda = 248$ nm, $t = 700$ fs) in carbon plasma were analyzed in Ref. [9], and the harmonics up to the 13th order were reported. The studies [6, 7] have shown some peculiarities distinguishing the carbon plasma from other species used for frequency conversion of the femtosecond pulse propagating through the plasma plume. This may be attributed to the specific properties of carbon atoms and ions, or creation of specific conditions of plasma formation.

Application of picosecond pulses for plasma HHG has some positive preferences as compared with femtosecond pulses. This is related with higher energy density of harmonics at the same level of HHG efficiency. Also, for this purpose, relatively low-cost laser setups can be used. The motivation of the following discussed studies of the HHG in carbon-containing plasmas using relatively long laser pulses [10] was based on the following reasons: (i) stronger energy conversion efficiency for lower order harmonics, (ii) application of picosecond pulses, and (iii) availability of strong transitions of ionic and neutral carbon in the vicinity of the 7th harmonic of Nd:YAG laser radiation. The goal of those studies was a search of the conditions for the generation of energetic coherent picosecond pulses in the range of 80–220 nm using various carbon-containing plasma plumes, as well as resonance enhancement of single harmonic in the vicinity of strong ionic transitions of carbon.

6.1.1.1 Experimental Arrangements and Results

The passive-mode-locked Nd:YAG laser ($\lambda = 1064$ nm) has generated a train of pulses of 38 ps duration and 1.5 Hz pulse repetition rate. The two-stage amplification of a single pulse was followed by splitting of this pulse into two parts, one (heating pulse) with the energy of 5 mJ, which was used for plasma formation on the target surface, and another (probe pulse) with the energy of up to 28 mJ, which was used, after some delay, for frequency conversion

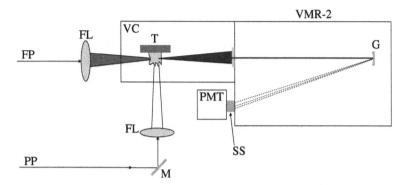

Figure 6.1 Experimental setup for the HHG in laser plasma using the picoseconds pulses. FP, fundamental probe picosecond pulse; PP, heating picosecond pulse; M, mirror; FL, focusing lenses; VC, vacuum chamber; T, target; VMR-2, vacuum monochromator; G, grating; SS, sodium salicylate; PMT, photomultiplier tube. *Source:* Ganeev et al. 2012 [10]. Reproduced with permission from The Optical Society of America.

in the prepared plasma (Figure 6.1). The heating pulse was focused inside the vacuum chamber containing various targets. The size of the ablated area was 0.5 mm. The plasma plume size at a distance of 50–100 µm above the target surface was considered the same, taking into account the divergence of expanding plasma (30°). The ablated area was $\sim 2 \times 10^{-3}$ cm², and the intensity of heating pulse on the surface was $\sim 6 \times 10^{10}$ W cm⁻². The same sizes of plasma were maintained during those experiments. The maximum intensity of probe pulse at the focus was 4×10^{13} W cm⁻². The delay between these two pulses during most of the experiments was maintained at 25 ns, which was optimal for efficient harmonic generation in carbon-containing plasmas. The probe and harmonic pulses were separated in the vacuum monochromator, and the harmonic radiation in the range of 50–300 nm was detected using the sodium salicylate scintillator and photomultiplier tube (PMT).

The measurements of conversion efficiency of the HHG were carried out using the following procedure. At the first step, the 4th harmonic signal was measured using a "monochromator + sodium salicylate + PMT" detection system at known energy of the 4th harmonic of 1064 nm radiation generating in the nonlinear crystals. This allowed the calibration of monochromator at the wavelength of 266 nm. Since the quantum yield of sodium salicylate is the same in a broad spectral range between 40 and 350 nm, the calibration of registration system at 266 nm allowed calculating the conversion efficiency along the abovementioned spectral range. The monochromator allowed observing the harmonics down to $\lambda = 50$ nm.

The targets were made of various carbon-containing materials (graphite, glassy carbon, pyrographite, boron carbide, silicon carbide, and pencil lead). The bulk samples of these materials cut with the sizes of $5 \times 5 \times 2$ mm³ were used in those studies. A three-coordinate manipulator made it possible to move the target along the *z*-axis and thus control the zone of the interaction of the probe radiation with the plasma relative to the target plane.

In the following, we discuss some special features of HHG in carbon-containing plasmas as the media for efficient lower order harmonic generation in the vacuum ultraviolet (VUV) range using the 38 ps, 1064 nm probe pulses. At soft ablation conditions, the carbon-containing plasma mainly consists of excited neutral atoms and singly ionized carbon, which was confirmed by the spectral measurements in the near ultraviolet and visible ranges (Figure 6.2).

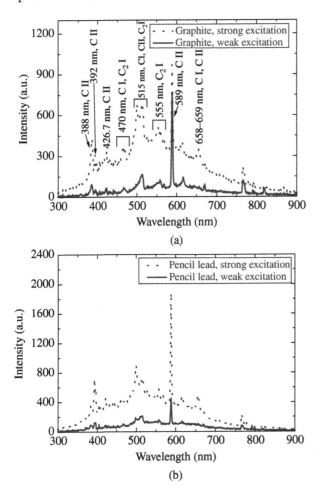

(a)

(b)

Figure 6.2 Plasma emission spectra from the (a) graphite, (b) pencil lead, and (c) glassy carbon at weak (solid lines, $I = 7 \times 10^{10}$ W cm^{-2}) and strong (dotted lines, $I = 1.5 \times 10^{11}$ W cm^{-2}) excitation of target surfaces. *Source:* Ganeev et al. 2012 [10]. Reproduced with permission from The Optical Society of America.

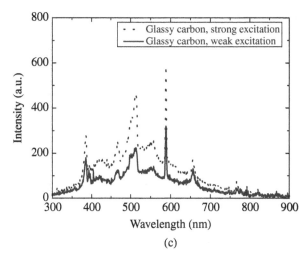

Figure 6.2 *(Continued)*

Figure 6.2a shows the HHG spectrum obtained in graphite plasma plume. The plateau-like shape of the highest harmonics (above the 13th order) does not exactly represent the approximately equal intensity harmonics. Other carbon-containing plumes showed the same properties (see, for example, the harmonic spectra from pencil lead and glassy carbon plasmas, Figure 6.2b,c). It should be noted that the harmonics disappeared after the overexcitation of target surfaces and generation of strong C III and C IV lines in the plasma spectra at high intensities of heating pulse. The maximal conversion efficiency from graphite plasma in the range of plateau (15th–21st harmonics) was measured to be $\sim 0.7 \times 10^{-6}$.

The important peculiarity of all these studies was the generation of the efficient 7th harmonic (Figure 6.3a–c). Its efficiency considerably (three to seven times) exceeded that of the lower (5th) order harmonic, contrary to expected dependence defined by a perturbative theory of lower order harmonics [11]. The enhanced 7th harmonic was a common feature of this class of carbon-containing plasma plumes, excluding the boron carbide plasma (Figure 6.4). In particular, the maximum conversion efficiencies of the 5th and 7th harmonics in the glassy carbon plasma were 10^{-5} and 6×10^{-5}, respectively. The possible reasons of such behavior of low-order harmonics generating from carbon plasma are discussed in the Section 6.1.1.2. Note that, in the case of various metal plasmas (Mn, Cu, etc.), this peculiarity (i.e. enhanced 7th harmonic) was not observed. The example of harmonic spectrum from metal (Mn) plasma is presented on Figure 6.3d, which shows a monotonic decrease of each next order of harmonics over all spectral range up to the 11th order,

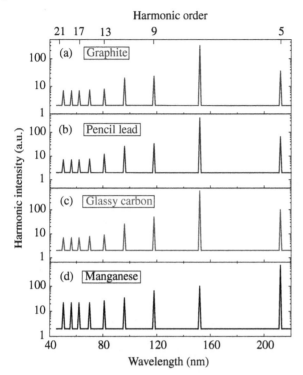

Figure 6.3 Harmonic spectra from the (a) graphite, (b) pencil lead, (c) glassy carbon, and (d) manganese plasmas. *Source*: Ganeev et al. 2012 [10]. Reproduced with permission from The Optical Society of America.

above which the harmonics showed a plateau-like behavior described by the nonperturbative three-step model of HHG [12–14].

Various properties of the 7th and neighboring harmonics were analyzed. Figure 6.5a shows the dependences of the 5th and 7th harmonics on the angle of rotation of the half-wave plate, which caused variation of the polarization of driving radiation from linear (at 0°) to circular (at 45°). Small deviation from linear polarization led to a considerable decrease of the 5th and 7th harmonic intensities. The application of circularly polarized laser pulses led to the complete disappearance of harmonic emission, as it should be assuming the origin of HHG [15].

The delay between the heating and driving pulses is crucial for the optimization of HHG. A typical dependence of 7th harmonic intensity on the delay between pulses is presented in Figure 6.5b in the case of pyrographite plasma. At the initial stages of plasma formation and spreading out of the target surface, the concentration of particles (neutrals and singly charged ions) is insufficient, since the particles possessing the velocities in the range of 5×10^3 m s^{-1} cannot reach the optical axis of the propagation of probe beam ($\sim$50–100 µm above the target surface). The increase in delay allows the appearance of plasma particles along the path of converting pulse, which leads to the growth of HHG

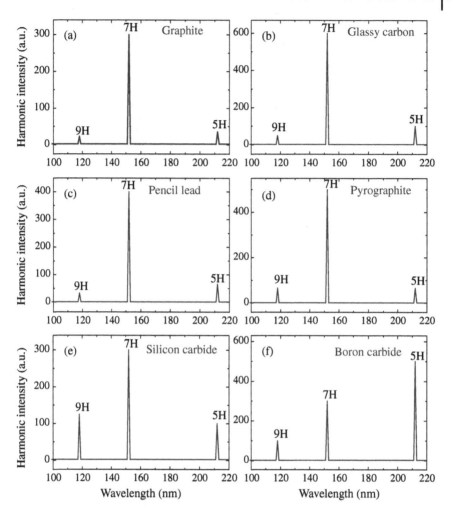

Figure 6.4 Distributions of the 5th, 7th, and 9th harmonic intensities in the cases of different carbon-containing plasmas. (a) graphite, (b) glassy carbon, (c) pyrographite, (d) pencil lead, (e) silicon carbide, and (f) boron carbide. *Source:* Ganeev et al. 2012 [10]. Reproduced with permission from The Optical Society of America.

efficiency. Further increase in delay leads to saturation of the HHG at ~25 ns and gradual decrease of conversion efficiency at longer delays (>40 ns).

There was a similarity in harmonic-delay dependence for all the samples under investigation. This behavior is caused by the prevailing influence of carbon atoms and ions, as the main emitters of harmonics. While maintaining the same velocity (~5×10^5 cm s^{-1}, in accordance with the estimates based on the thermodynamic approach [16] and a three-dimensional molecular

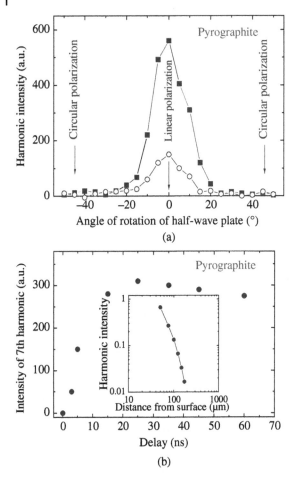

Figure 6.5 (a) Polarization dependence of 5th (circles) and 7th (squares) harmonic intensities and different angles of rotation of the half-wave plate. (b) The dependence of 7th harmonic intensity on the delay between the heating and probe pulses. Inset: Harmonic intensity as a function of the distance between the target surface and probe beam axis for the 7th harmonic. *Source*: Ganeev et al. 2012 [10]. Reproduced with permission from The Optical Society of America.

dynamical simulation of laser ablation of graphite using the molecular dynamics code ITAP IMD of the spreading of the carbon particles from the target after ablation [17]), one can calculate the time of their arrival in the area of probe picosecond pulse propagation, which was ~100 μm above the target surface, as 20 ns, which is close to the experimental observations. One can note that the delay dependence was similar for any harmonics in the generating spectra from the graphite plasma. Boron carbide showed the same features, since in that case the main contributors to harmonic yield were the singly charged ions and neutrals as well. This target had the advantage to withstand the laser ablation for a longer period compared with other targets under study; however, the nonlinear optical properties of B_4C-ablated plume were approximately the same as for graphite and other samples, excluding the absence of single harmonic enhancement in the former case.

Harmonic intensity considerably depended on the distance between the optical axis of driving beam and target surface (see the inset in Figure 6.5b), due to the change of plasma concentration above the target. The log–log dependence of harmonic intensity (I_h) on the distance between the optical axis of driving beam and target surface (x) for the 7th harmonic generation in pyrographite plasma was corresponded to $I_h \sim x^{-3}$.

The important parameter for optimization of harmonic emission is the fluence of the heating pulse on the target surface. The dependences of HHG efficiency on the energy of heating pulse were measured, while maintaining the same size of ablation beam on the targets ($\sim$500 μm). Figure 6.6 shows these dependences for different harmonics from the pyrographite target. The common feature of these studies was an abrupt decrease of harmonics at irradiation of targets using stronger heating pulses. The reason of these observations is related with the overexcitation of the target, which leads to the appearance of the abundance of free electrons in the plasma plume. The latter causes a phase mismatch between the waves of the driving field and harmonics. This effect is especially important for lower order harmonics.

No calculations were carried out regarding the plasma density at the optimal conditions of the excitation of carbon-containing targets using the 38 ps pulses. Meanwhile, the calculations at similar conditions ("optimal" carbon plasma, 25 ns delay, 150 μm distance from the target, 8 ps heating pulse, intensity of the

Figure 6.6 Dependences of harmonic intensity on the heating pulse energy for the 7th (upper curve), 9th (middle curve), and 11th (bottom curve) harmonics generating in pyrographite plasma. *Source*: Ganeev et al. 2012 [10]. Reproduced with permission from The Optical Society of America.

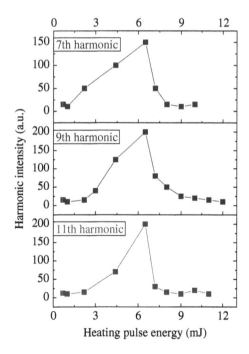

heating pulse on the surface 2×10^{10} W cm^{-2}) have been reported in Ref. [17]. The concentration of carbon plasma at the abovementioned conditions of target ablation by an 8 ps pulse allowing efficient harmonic generation was found to be 2.6×10^{17} cm^{-3}.

6.1.1.2 Discussion

It has been shown previously that efficient harmonic generation is observed only in the case when the visible and UV plasma emission is originated dominantly from neutral and singly ionized carbon species [18]. The present studies have also confirmed this peculiarity (Figure 6.2). The broad features near 470, 515, and 555 nm (Figure 6.2) could be assigned to the bands of excited C_2 molecules. These bands have previously been observed in the ablation of graphite at several wavelengths [19–21]. Other lines in the spectra presented in Figure 6.2 are attributed to the neutral and singly charged carbon. The main goal of these spectral studies was to analyze what the consistency of carbon plasma was at the conditions of efficient HHG. In the meantime, the analysis of optical spectra in the visible and UV ranges does not inform about the presence of highly ionized species, which can be revealed by collecting the plasma emission in the VUV range. It is worth noting that the discussed observations of harmonic spectra [10] did not show a presence of plasma emission from high-charged particles in the studied spectral range (50–250 nm), which confirms that these experiments were carried out at a soft ablation regime of carbon-containing targets.

The observed difference between the HHG from the plasmas produced on boron carbide and all other carbon-containing targets can be attributed to different influence of the neutral and ionic transitions of B_4C in the vicinity of 152 nm on the harmonic response compared with other species. The tuning of the frequencies of those transitions in the case of B_4C compared with carbon can drastically change the conditions of resonance enhancement for the 7th harmonic. The difference in the nonlinear optical properties of boron carbide and other carbon-containing plasmas also demonstrates that the former molecule probably was presented in the plume mostly without the disintegration during laser ablation; otherwise the response from singly charged carbon ions could cause the enhancement of 7th harmonic like in other cases. It means that plasma HHG can be served as a precise tool for the nonlinear spectroscopy analysis of the atomic physics and structure of materials.

Carbon possesses some strong neutral and ionic transitions in the vicinity of the 7th harmonic of Nd:YAG laser radiation. Those are the $2s^2 2p^2 \, {}^1D-2s^2 2p3d$ $^1D^0$ (C I, $\lambda = 148.2$ nm), $2s^2 2p^2 \, {}^3P-2s2p^3 \, {}^3D^0$ (C I, $\lambda = 156.0$ nm), and $1s^2 2s^1 \, {}^2S-$ $1s^2 2p \, {}^2P^0$ (C IV, $\lambda = 154.8$ nm) transitions [22]. One can note that, at used experimental conditions, the excitation of triply charged carbon ions is almost impossible, since the targets were excited using the moderate fluencies. Probably, the transitions from neutrals ($\lambda = 148.2$ and 156.0 nm) are responsible for the observed enhancement of 7th harmonic ($\lambda = 152$ nm). Another option for the explanation of the enhanced nature of 7th harmonic with regard to the 5th order could be the absorption of the latter harmonic in the carbon-containing plasma. However, previous studies of carbon plasma did not show strong absorption lines in the vicinity of 5th harmonic ($\lambda = 213$ nm) [22]. Thus the most probable reason of observed single harmonic property in the vicinity of 152 nm is the resonance-induced enhancement of the 7th harmonic. This enhancement can be explained by the micro- and macroprocesses influencing HHG.

The enhancement of harmonics depends on the medium density and bandwidth of laser radiation. The variations of the plasma components and fundamental wavelength result in the tuning of a selected harmonic frequency in the plateau region. The free electrons and ac Stark shift change the phase mismatch and the optimal laser frequency at which the efficient selection of a single harmonic is achieved. So, the intensity enhancement of harmonic due to propagation effects can be even stronger than that due to resonance processes. This approach can be applied to the discussed results, which demonstrated the enhancement of single harmonic in the vicinity of strong transitions of the neutral carbon.

Below, we analyze the calculations of the contributions of the resonant nonlinear susceptibility enhancement and the radiation propagation effect during nonlinear wave interaction in a plasma medium within the framework of perturbation theory, since in discussed experiments the maximum amplitude of a laser field (1.7×10^{10} V m^{-1}) was ~30 times less than the intra-atomic field of unexcited carbon atoms (4.7×10^{11} V m^{-1}). The nonlinear medium basically consisted of the atoms and singly charged ions of carbon. The carbon atoms can be presented in singlet and triplet states. According to the estimations, which have been carried out using the equations presented in Refs. [11, 23] for the observed transitions from unexcited triplet $2s^2 2p^2\, ^3P$ and singlet $2s^2 2p^2\, ^1D$ states of carbon atoms and from unexcited carbon ions, the ratios of the absolute values of 7th and 5th order nonlinearities $r = |\chi^{(7)}/\chi^{(5)}|$ do not exceed ~7×10^{-12} esu. In calculations, the values of transition frequencies and oscillator strengths were taken from the Ref. [22]. Those estimations have shown that, due to closeness to the resonance conditions, the value of resonant 7th-order nonlinearity is approximately six times higher than the nonresonant one. One can note that, for carbon atoms, the frequency shift from seven-photon resonance is about ~1600 cm^{-1}.

For estimations of the variations of harmonic signal the relation defined for the case when the diffraction length of laser beam R_d is longer than the nonlinear medium length L was used (see for example [11, 16]). At laser intensity of the order of 10^{13} W cm^{-2} and absence of the generalized phase mismatch influence, the ratio of the intensities of 7th and 5th harmonics $\gamma = I_7/I_5$ does not exceed 5×10^{-2} at harmonic generation from the atomic and ionic ground states. At the peak, laser intensity used (4×10^{13} W cm^{-2}) this relation should be increased by 16 times. So, at this level of laser intensity used in the experiment, the 5th harmonic should be stronger than the 7th harmonic, once one assumes the resonant enhancement of the 7th-order nonlinearity alone. Contrary to these estimates, in the experiment, the 7th harmonic was three to seven times stronger than the 5th one at rather broad range of the variations of laser intensity.

Let us consider the influence of the phase conditions on the HHG and analyze the propagation effect. The phase mismatches for the 5th and 7th harmonics are positive once one considers the generation of harmonics from the carbon atoms in singlet ground state and also from the carbon ions in ground state. The phase mismatch of the electronic gas is also positive. In that case, the total phase mismatch Δk should be positive. In the meantime, during generation of 7th harmonic in the media containing carbon atoms in triplet ground state, the value of phase mismatch is negative. If the medium mostly consists of atomic carbon in triplet state, the addition of ~5% of the electronic gas is necessary for the compensation of phase mismatch. In that case the positive value of Δk is increased for the compensation of phase mismatch. The calculations show that, at strong phase mismatch, the dependence of the harmonic intensity on the medium length demonstrates the oscillating features, and the conditions of stronger 7th harmonic compared with the 5th one are possible. In that case the relative increase of 7th harmonic will be caused by a suppression of 5th harmonic intensity. However, at a medium length of 500 μm, the relative increase of the 7th harmonic can be achieved at the concentration of the carbon atoms in the triplet ground states exceeding 1.2×10^{19} cm^{-3}.

The estimations, which were carried out for the HHG from the excited state $2s^2 2p3s\ ^3P^0$ of carbon atoms, have shown that the presence of a small part of excited particles (~10^{-3}) is capable to considerably change the total linear phase mismatch and nonlinear susceptibilities responsible for harmonic generation and self-action effects. The analysis of this excited state is of special interest, since in that case a rather strong single-photon resonance with the laser radiation can be observed (e.g. transition $2s^2 2p3s\ ^3P^0 - 2s^2 2p3p\ ^3D$, frequency shift ~37 cm^{-1}).

Regarding the main finding of resonance enhancement of single harmonic in the carbon plasma, one can consider it not as just a clear evidence of considerable enhancement of single harmonic in the spectral range where no such phenomenon was reported so far, but rather as a tool for the spectroscopic studies of various atomic and ionic transitions possessing different oscillator strengths.

6.1.2 Resonance Enhancement of the 11th Harmonic of 1064 nm Picosecond Radiation Generating in the Lead Plasma

A few plasmas [24–30] have proven that, at appropriate experimental conditions, they can be considered as the excellent media for the enhanced single harmonic generation in the 80–40 nm spectral range using the ultrashort (femtosecond) laser pulses. Some harmonics (particularly the 13th harmonic from the In plasma and the 17th harmonic from the Sn plasma in the case of 800 nm femtosecond radiation) demonstrated the considerable enhancement attributed to the influence of the ionic transitions. The mechanisms of such effect were discussed in some recent publications [31–34].

In this subsection, we analyze the results of studies of the HHG in a lead plasma using the relatively long (picosecond) pulses. The enhancement of the 11th harmonic generating in the Pb plasma in the case of 1064 nm picosecond probe pulses resembles the abovementioned peculiarities of the resonance-induced enhancement of harmonics using the 800 nm femtosecond radiation. We compare the enhanced harmonic in the cases of laser ablation of pure Pb and its alloy. We also analyze the variation of resonance-induced enhancement of the 11th harmonic of Nd:YAG laser from this plasma in the presence of different gases [35]. Pb, Sn:Pb alloy, and Sn were used as the targets for those experiments. A three-coordinate manipulator made it possible to change the targets and to control the zone of the interaction of the probe radiation with the plasma relative to the target surface. Various gases were inserted inside the vacuum chamber containing plasma plumes to analyze the influence of the dispersion properties of these gases on the variations of harmonic spectra. The chamber allowed the insertion of the gases and was connected to the vacuum monochromator, so the gas was also inserted in this monochromator. The absorption of the resonance-enhanced 11th harmonic of 1064 nm radiation at the used densities of gases was insignificant.

In the following, we discuss some specific features of lead and alloy plasmas as the media of HHG using the picosecond laser pulses. The goal of those studies was to define whether the components of Pb-containing plasma influence the resonance-enhanced harmonic generation efficiency. Changing a composition of plasma from pure lead was accomplished using an alloy of this material, which consisted of tin and lead (at a volume ratio of ~5 : 3). Plasma harmonic spectra from pure lead and Pb:Sn alloy were also compared with those from pure tin. To analyze the influence of propagation effect, the variations of harmonic spectra from lead plasma was studied while adding different gases in the vacuum chamber thus changing the dispersion inside the plasma plume.

6.1.2.1 Analysis of Resonantly Enhanced 11th Harmonic

Figure 6.7 shows the HHG spectra obtained in the Sn, Pb, and Sn:Pb alloy plasma plumes. The harmonics from tin plasma (Figure 6.7, upper panel) showed a gradual decrease of each next harmonic's intensity, which is a common feature in the case of most of the plasma harmonics experiments. Contrary to that, a lead plasma demonstrated the enhanced 11th harmonic (Figure 6.7, middle panel), which was stronger than the lower orders. This peculiarity of lead harmonics was maintained at different conditions of

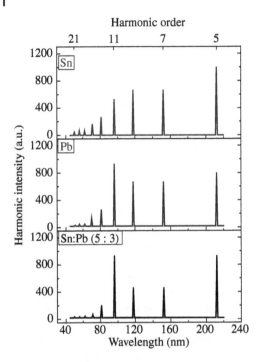

Figure 6.7 Harmonic spectra from the Sn, Pb, and Sn:Pb (5 : 3) alloy plasmas. *Source*: Boltaev et al. 2014 [35]. Reproduced with permission from The Optical Society of America.

experiments by varying the confocal parameter, plasma length, and so on. The concentration of plasma ($\sim 2 \times 10^{17}$ cm^{-3}) was insufficient for the absorption of the lower order harmonics to create the conditions of stronger 11th harmonic compared with the lower ones.

The harmonics from the mixture of lead and tin plasmas comprised both harmonic spectra, while maintaining the enhanced 11th harmonic (Figure 6.7, bottom panel). It is seen from the results of plasma mixture that the role of non-resonant plasma components (i.e. the tin in the case of ablating Sn:Pb alloy) was insignificant and did not lead to the considerable variation of the envelope of harmonic spectra compared with the pure Pb plasma plume. There are no specific reasons in different enhancement factors of the 11th harmonic in the pure Pb and Sn:Pb plasmas. The ratio I_{11H}/I_{5H} from shot to shot was close to 1, or slightly larger, while in both plasmas stronger 11th harmonic with regard to the 9th and 7th orders was observed. This observation points out the insignificant influence of the dispersion properties of additional plasma component on the relation between the phases of 11th harmonic and 1064 nm radiation.

The comparative studies of harmonic generation efficiency were carried out by moving the translating stage up and down, thus allowing to measure the HHG conversion efficiency (in arbitrary units) in different plasmas without changing the conditions of experiments (delay, intensity of the pump and probe pulses, distance from the target to the axis of propagation of the probe radiation, confocal parameter of the probe radiation, etc.). These studies allowed comparing the efficiencies of harmonics from various metal plasmas during one set of measurements just by changing the targets at similar conditions. The meaning of "arbitrary units" refers to the voltage signal measured from the photomultiplier registering the luminescence from the sodium salicylate. No absolute measurements of HHG conversion efficiency were carried out for each set of studies, since it was unpractical due to necessity in the changes of the conditions of experiments. However, in some cases, the absolute values of conversion efficiency were measured by using the technique described in Ref. [36]. The conversion efficiency of the HHG was defined using the following procedure. At the first step, the 4th harmonic signal was measured using a "monochromator + sodium salicylate + PMT" detection system using the calibrated energy of the 4th harmonic of 1064 nm radiation generating in the nonlinear crystals. This allowed the calibration of monochromator and registration system at the wavelength of 266 nm. Since the quantum yield of sodium salicylate has the same value along a broad range between 30 and 350 nm, the calibration of registration system at 266 nm allowed the calculation of the conversion efficiency along the whole spectral range. The conversion efficiency of the 11th harmonic from the Pb plasma was measured to be 3×10^{-6}.

The enhanced harmonic from lead plasma was analyzed at different experimental conditions. The delay between pump and probe pulses is crucial for optimization of the HHG. A typical dependence of the 11th harmonic intensity on the delay between pulses is presented in Figure 6.8a. At the initial stages of plasma formation, the concentration of particles (neutrals and singly charged ions) in the interaction area is insufficient, since the particles possessing the velocities of $\sim 5 \times 10^5$ cm s^{-1} do not reach the optical axis of probe beam propagation (~ 100 μm above the target surface). The increase in delay allowed the

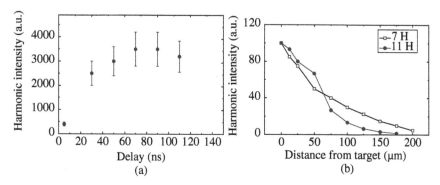

Figure 6.8 (a) The dependence of the 11th harmonic intensity on the delay between the pump and probe pulses. (b) Harmonic intensity as a function of the distance between Pb target surface and probe beam axis for the 7th (squares) and 11th (circles) harmonics. *Source:* Boltaev et al. 2014 [35]. Reproduced with permission from The Optical Society of America.

appearance of plasma particles along the path of probe pulse, which caused the growth of HHG efficiency. Further increase in delay led to saturation of the HHG at ~70 ns and gradual decrease of conversion efficiency at longer delays (>110 ns). The influence of the distance between the target and the optical axis of propagation of the probe radiation at fixed delay was also analyzed (Figure 6.8b). This distance was varied by a manipulator, which controlled the position of the target relative to the waist of the probe picosecond radiation.

The common feature of these studies was an abrupt decrease of harmonics at the irradiation of targets above the optimal level. The term "optimal level" refers to the conditions of plasma ablation when the components of ablation do not cause the growth of the impeding processes, which restrict, or even diminish, or in some cases entirely cancel the harmonic generation. Those include (i) the excess of free electron concentration leading to the strong variation of the dispersion properties of nonlinear medium and to the phase mismatch between the interacting waves, (ii) the excess in absorption of the XUV radiation, and (iii) the intense emission from the plasma.

The origin of these processes is related with the formation of highly ionized plasma, which leads to the appearance of the abundance of free electrons. The latter cause a phase mismatch between the waves of probe and harmonic fields. This effect is especially important for the lower order harmonics. One can note that a decrease of harmonic efficiency in highly ionized plasma has been reported for higher order harmonics as well [37–41], though this effect was less abrupt than that observed in present studies. Figure 6.9a shows the influence of the energy of pump pulse on the intensity of the 11th harmonic generating in the lead plasma. Harmonic intensity increased up to the pump pulse energy of 3 mJ. Further growth of pump energy led to a decrease of harmonic intensity.

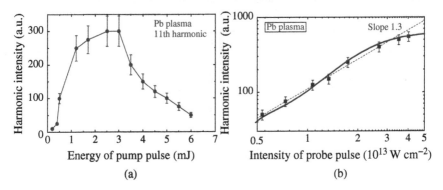

(a)   (b)

Figure 6.9 (a) The dependence of the harmonic intensity on the pump pulse energy for the 11th harmonic generating in a lead plasma. (b) Comparison of experimental (squares, dashed curve) and calculated (solid curve) dependences of the 11th harmonic yield on the probe pulse intensity. *Source:* Boltaev et al. 2014 [35]. Reproduced with permission from The Optical Society of America.

Lead plasma is not a first plasma medium where the enhancement of some specific harmonic was reported. It is just a sample showing such feature while using the 1064 nm radiation. Mn, In, Te, As, Sn, and Cr were among the plasma plumes where the enhancements of single harmonic of 800 nm radiation of Ti:sapphire laser were reported [42]. Notice that no enhancement of single harmonic from these metal plasmas was observed using the Nd:YAG laser radiation, and vice versa the 800 nm femtosecond lasers were not able to produce the enhanced single harmonic from the lead plasma.

The analysis of the influence of the probe intensity on the 11th harmonic yield from the lead plasma was carried out as well (Figure 6.9b). The slope of this dependence was close to 1.3 (dashed line). As the frequency shift from the 11th photon resonance between ground and exited states $6s^26p$ $^2P_{1/2}$–$6s^28d$ $^2D_{3/2}$ (96.72 nm) of Pb II was only 4.5 cm^{-1}, the analysis of harmonic yield was carried out using the resonant approach [11, 23, 43]. In this approach, the squared module of the off-diagonal elements of the density matrix defines the resonant harmonic intensity. In the adiabatic limit the off-diagonal elements of the density matrix for the resonant transition can be written as

$$\sigma \sim [u_q E^q + uE_q \exp(i\Delta kz)]\{[1 + \Delta(E)]^2 + v|E|^{2q}\}^{-1/2} \qquad (6.1)$$

where u_q is the normalized composite matrix element for q-photon transition, u is the normalized dipole matrix element of the resonant transition, Δk is the phase mismatch, and E and E_q are the amplitudes of laser and resonant harmonic fields, respectively. The value of $v \sim u_q^2$ describes the saturation of the resonant transition. In discussed studies, the laser field did not exceed the intra-atomic field of singly charged ions. The calculated I_{11H} (I_{probe}) dependence using the fittings of AC Stark shift, relaxation time, and generalized matrix elements is presented as the solid line in Figure 6.9b.

No data on the oscillator strengths of the 6p–8d transitions of Pb II could be found in literature. Notice that the influence of resonances in the enhancement of single harmonic is defined by both the oscillator strength of nearby transition and the detuning between the wavelengths of harmonic and transition. There is no need for exact coincidence of the resonance transition and harmonic. The optimal detuning defines how strong the nonlinear optical response could be expected. Indeed, the detuning of 11th harmonic from 96.72 nm transition may be the reason for the insignificant enhancement of this harmonic, as was observed in discussed studies. Probably, another reason of a weak enhancement is a small oscillator strength of this transition. One can note that all previous observations of resonantly induced harmonic enhancement were obtained at different detuning of harmonics with regard to the exact positions of ionic transitions. Moreover, the exact coincidence of these wavelengths may cause some restrictions in the harmonic enhancement due to the influence of self-absorption.

6.1.2.2 Variation of Resonance Enhancement by Insertion of Gases

The role of resonances on the high-order nonlinear optical response of medium could be monitored by two methods: tuning of the driving radiation wavelength and changing the phase relations of the interacting waves. The former option, being often used in the case of resonance-induced harmonic studies using the broadband Ti:sapphire lasers allowing both direct tuning of radiation wavelength and manipulation of harmonic wavelength through the chirp variation, cannot be applied in the case of Nd:YAG laser. It is impossible to tune the wavelength of this laser. The only clear proof for the involvement of the resonance on the yield of 11th harmonic could be the variation of the phase mismatch between the two waves (1064 and 96 nm) in the vicinity of this ionic transition. In the following, we analyze experiments allowing such a variation of relative phases of waves by adding different gases in the plasma. This approach in the studies of the resonance enhancement is not new. It was used in the early stage of low-order harmonic and parametric generation in gases, which showed the attractiveness of the manipulation of the phase relations in the vicinity of resonances by introducing the variable ratios of positive and negative dispersions using different gases [11].

The insertion of gases in the vacuum chamber containing targets led to variations of harmonic spectra compared with the case of pure lead plasma. Four noble gases (He, Ar, Kr, Xe) possessing different dispersions, as well as different absorption, in the region of interest (50–250 nm) were used in those studies. Figure 6.10a,b shows the Pb harmonic spectra in the presence of light (He) and heavy (Xe) gases. The former gas possesses weak absorption until 65 nm and thus its influence on the variations of harmonic spectra could be attributed mostly to the optical dispersion properties of the gas changing the phase-matching conditions for resonant and nonresonant harmonics. The helium pressure up to 13.3 kPa was used above which the optical breakdown of the gas (without ignition of the metal plasma) was observed. One can see the abrupt decrease of the 11th harmonic with the increase of He pressure. Meanwhile, some harmonics, in particular the 9th one, showed less decrease, and the 5th harmonic became even stronger compared with the case of the plasma formation at vacuum conditions. Approximately same features were observed in the case of insertion of the xenon gas, though a decrease of the 11th harmonic was less pronounced.

The plasma harmonics from graphite ablation allowed observing a strong 7th harmonic of Nd:YAG laser radiation (152 nm) (see Section 6.1.2.1). The enhancement of this harmonic was attributed to the closeness with the ionic transition of carbon and the improvement of the phase-matching conditions for this particular case. Figure 6.10c shows the spectra of carbon harmonics at different pressures of xenon. One can see that a decrease of strong 7th harmonic with the growth of gas pressure (while the neighboring harmonics did not follow this abrupt change of intensity) resembles the one shown in the case of the 11th harmonic generating in the lead plasma (Figure 6.10a). The comparable variations of Pb and C harmonic spectra may lead to the following conclusions. The enhancement of harmonics at plasma conditions (i.e. without the insertion of the gases) in both these cases is originated from better matching of the phases of probe and harmonic waves for some specific orders (11th and 7th harmonics from the lead and carbon plasmas, respectively). The addition of the medium possessing positive dispersion worsens the phase matching between those waves and correspondingly decreases the intensities of the 7th

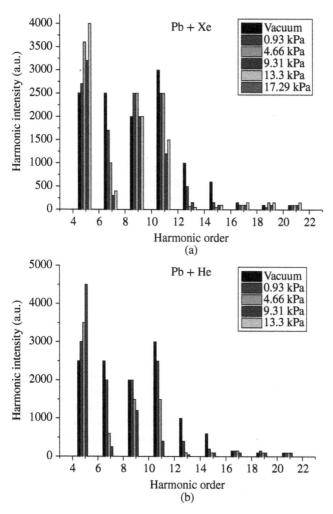

Figure 6.10 Variations of plasma harmonic spectra at different pressures of gases. (a) Pb plasma, He gas, (b) Pb plasma, Xe gas, and (c) carbon plasma, Xe gas. *Source*: Boltaev et al. 2014 [35]. Reproduced with permission from The Optical Society of America.

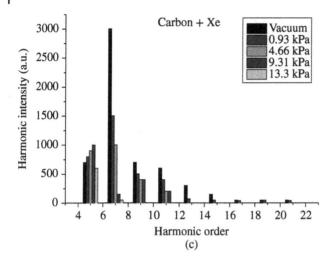

Figure 6.10 (*Continued*)

and 11th harmonics. The role of the gas on other harmonics is also defined by the positive addition to the refractive index of the plasma–gas mixture. However, in that case, the dispersion properties of gas play less decisive role compared with the resonance-enhanced harmonics.

The role of gas absorption on the observed properties of resonant harmonic was analyzed as well. At most unfavorable conditions (2 m long optical path inside the monochromator, He gas pressure 7 kPa) the absorption starts at 23 eV, which corresponds to the 23 : 1.17 eV = 19th harmonic, while the transmittance for lower order harmonics is equal to 1 (1.17 eV corresponds to the energy of 1064 nm probe photons). No significant absorption of the 11th harmonic in Ar and Ne was observed at used experimental conditions. The absorption in the cases of Ne and Ar starts at 19 and 14.5 eV, which should lead to the suppression of 17th and 13th harmonics, but not the 11th one ($E_{11H} = 12.9$ eV). The insertion of Xe led to 30% absorption of the 11th harmonic, since the transmission in this gas started to decrease at ~12 eV (for Kr this value is 13.8 eV and this gas does not absorbs the 11th harmonic).

Those studies have shown that the insertion of gas leads to the change of propagation effect on the harmonic generation. Probably, the combination of microprocesses (i.e. closeness of harmonic wavelength and ionic transition, absorption processes) and macroprocesses (propagation effect comprising the joint influence of Gouy phase, dispersion of plasma close to the resonant transitions, and dispersion of neutral gas) cause the observed variations of harmonic spectra from the plasma plume at different gas pressures. Notice that the concentration of lead particles in the area of interaction with the probe pulse (~2×10^{17} cm^{-3}) becomes comparable with one of the gases

$((1-10) \times 10^{17} \text{ cm}^{-3})$ inside the plasma volume. One can remind that, at the early stages of gas harmonic studies, the inclusion of additional gaseous component allowed the enhancement (or decrease) of generating harmonics in the gas mixtures in the VUV range due to achievement of the phase-matching conditions between the interacting waves [44–46].

6.2 Size-Related Resonance Processes Influencing Harmonic Generation in Plasmas

6.2.1 Resonance-Enhanced Harmonic Generation in Nanoparticle-Containing Plasmas

Nanoparticle-containing media subject to intense laser pulses show strong low-order nonlinear optical response (e.g. nonlinear refraction and nonlinear absorption), as well as can emit the short-wavelength coherent radiation through the low-order harmonic generation. The excellent nonlinear optical properties of nanoparticles may cause the improvement of the high-order harmonic efficiency as well. Initially, studies of nanoparticle-induced harmonic generation were limited to exotic clusters of noble gases. The physical origin of this process in the gas clusters was mostly related to standard atomic harmonic generation, modified by the fact that the atoms in nanoparticles are disposed close to each other [47–55]. The increase of HHG conversion efficiency in those studies was attributed to the growth of the number of emitters. Particularly, the cross-section of recombination of the accelerated electron with the parent particle in the case of nanoparticles is higher compared with the atoms. The mechanisms of the HHG from gas nanoparticles have been analyzed in a few studies [56, 57]. Among them the ionization and recombination to the same ion, to the neighboring ions, and to the whole nanoparticle have been compared. The experiments with gas nanoparticles have revealed some difficulties in disentangling the harmonics produced by different species (i.e. monomers and nanoparticles of different sizes).

Similar studies of the HHG in laser-produced plasmas consisted of nanoparticles or monomers showed that, at equal experimental conditions, the former species provide considerably stronger harmonic yield, thus pointing out the advantages in application of the ablated nanoparticles for harmonic generation in the longer wavelength range of XUV [58]. It is possible to produce the laser plasmas containing different nanoparticles due to availability of various metal-based nanoparticles. The enhancement of harmonics in the nanoparticle-containing plasmas compared with monomer-containing plasmas has been studied in various laboratories [6, 7, 59–62].

In the meantime, additional option could be a resonance enhancement of single harmonic in the nanoparticle-containing plasmas. There are no fundamental restrictions in the resonance enhancement independently on the sizes of harmonic emitters. To achieve the resonance-induced growth of harmonics from large particles one has to find the conditions of optimal plasma formation during ablation of the nanoparticle-containing targets. In that case one can simultaneously observe two mechanisms of the growth of harmonic yield (i.e. nanoparticle-induced and resonance-induced enhancement of harmonics).

In this subsection, the results of studies of three nanoparticle-containing plasmas (In_2O_3, Sn, and Mn_2O_3) allowing both the efficient generation of lower order harmonics and the resonance enhancement of single (13th, 17th, and 33rd respectively) harmonics are discussed [63]. The comparison of the harmonic spectra obtained using the plasmas produced on the bulk and nanoparticle materials of the same origin showed that the enhancement of single harmonics occurs in the case of strong excitation of the nanoparticle-containing targets.

6.2.1.1 Experimental Arrangements

The 802 nm, 370 ps, 10 Hz repetition rate uncompressed pulses of Ti:sapphire laser were used to ablate targets (Figure 6.11). These heating pulses were focused using the 200 mm focal length cylindrical lens inside the vacuum chamber contained an ablating material to produce the extended plasma plume. The intensity of heating pulses (I_{hp}) on a plain target surface was varied in the range of 1×10^9–5×10^9 W cm^{-2}, which corresponded to the fluencies (F) of 0.37–1.85 J cm^{-2}. The 64 fs driving pulses from the same laser were used, 45 ns from the beginning of ablation, for harmonic generation in the plasma plumes. The driving pulses were focused using the 400 mm focal length spherical lens onto the extended plasma from the orthogonal direction, at a

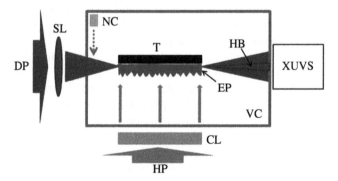

Figure 6.11 Experimental setup for plasma harmonic generation. DP, driving pulse; HP, heating pulse; SL, spherical lens; CL, cylindrical lens; VC, vacuum chamber; T, target; EP, extended plasma; NC, nonlinear crystal (BBO); HB, harmonic beam; XUVS, extreme ultraviolet spectrometer. *Source:* Ganeev et al. 2015 [63]. Reproduced. with permission from © IOP Publishing.

distance of ~100 μm above the target surface. The intensity of 802 nm driving pulses in the focus area was varied up to 7×10^{14} W cm^{-2}. The harmonic emission was analyzed by an XUV spectrometer containing a gold-coated cylindrical mirror and a 1200 grooves/mm flat field grating (FFG) with variable line spacing. The spectrum was recorded by a microchannel plate (MCP) detector with the phosphor screen, which was imaged onto a charge-coupled device (CCD) camera. The translation of MCP along the focusing plane of FFG allowed the observation of harmonics in different regions of XUV.

In_2O_3, Mn_2O_3, and Sn nanoparticles were studied as the ablating targets. The powders of these nanoparticles were glued on the 5 mm long glass plates and then installed in the vacuum chamber. The harmonic spectra from the ablation of 5 mm long In, Mn, and Sn bulk targets were also discussed.

The sizes of Mn_2O_3 nanoparticles were in the range of 40–60 nm, the Sn nanoparticles had the sizes of <100 nm, and In_2O_3 nanoparticles were in the range of 20–70 nm. The purities of nanoparticles were 98+%, 99.7%, and 99.99% respectively. The morphology of nanoparticles before and after ablation was analyzed using the transmission electron microscopy (TEM). The TEM images of nanoparticles before their ablation are shown in Figure 6.12a. The original nanoparticles were dissolved in methanol and rinsed. The drop of

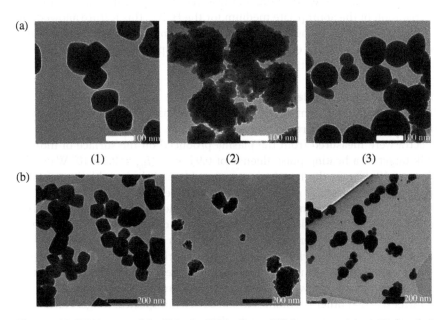

Figure 6.12 TEM images of the (1) In_2O_3, (2) Mn_2O_3, and (3) Sn nanoparticles (a) before their ablation and (b) after deposition during ablation. White lines on the top panels correspond to 100 nm. Black lines on the bottom panels correspond to 200 nm. *Source:* Ganeev et al. 2015 [63]. Reproduced with permission from © IOP Publishing.

this suspension was dried and analyzed by TEM. The sizes of nanoparticles were in the range of 40–130 nm. The tin nanoparticles were presented in the sphere-like shape, while the oxide nanoparticles showed both irregular structures (Mn_2O_3) and elliptical or squared shapes (In_2O_3).

The analysis of the ablated material, which was deposited on nearby glass plates, showed the nanoparticles similar to the original ones (Figure 6.12b). This observation confirmed the presence of nanoparticle species in the plasma plume at the used fluencies of heating pulse. In most cases, both the original and deposited nanoparticles were individually isolated. The morphology of the debris deposited on the glass plates during ablation of bulk targets was also analyzed. No nanoparticles were observed in that case at the heating pulse fluencies up to 1.5 J cm^{-2}. This observation points out the absence of nanoparticles and presence of monomers in the plasma plumes at the conditions corresponding to the harmonic generation from the ablation of bulk metals.

The nanoparticle powders were prepared as targets by mixing with cyanoacrylate-based glue and coating the surface of glass substrate. Thus the material directly surrounding the nanoparticles was a polymer (epoxy glue), which has a considerably lower ablation threshold than the metallic bulk materials. Therefore, the nanoparticle-carrying polymer begins to ablate at relatively low intensities, resulting in the lower laser intensity required for the preparation of the appropriate medium for the HHG. This feature allowed for easier formation of the optimum plasma conditions, which also resulted in a better HHG conversion efficiency from the nanoparticle-containing plumes compared to those from the bulk target.

6.2.1.2 In$_2$O$_3$ Nanoparticles

In the following, the comparative studies of the harmonics generated in the monomer- and nanoparticle-containing plasmas are discussed. The HHG spectra were measured from the plasma produced on the surface of indium bulk target at a heating pulse fluence of 0.9 J cm^{-2} ($I_{hp} \approx 2.5 \times 10^9$ W cm^{-2}). Figure 6.13a shows the harmonic spectra obtained during ablation of bulk indium and measured at different acquisition times (30, 10, and 1 s). One can see a significant enhancement of the 13th harmonic over other harmonics. The upper panel of Figure 6.13a shows the spectrum of saturated 13th harmonic collected during 300 laser shots. These conditions of registration were chosen to show the spectral distribution of weak neighboring harmonics. The analysis of this emission during unsaturated regime of registration showed that the 13th harmonic was a few ten times stronger than the 21st harmonic, which in turn exceeded other harmonics in the 30–75 nm spectral range. Harmonic cutoff in that case was in the region of 40s orders.

The ablation of indium oxide nanoparticles at the fluence $F = 1.3$ J cm^{-2} led to the emission of incoherent radiation from the laser-produced plasma, without the appearance of harmonics during propagation of femtosecond

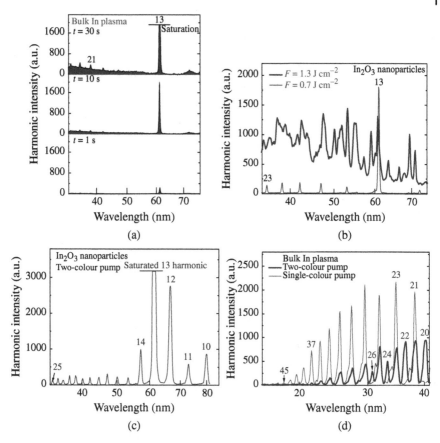

Figure 6.13 Harmonic spectra in the cases of (a) ablation of bulk indium (see text), (b) ablation of indium oxide nanoparticles using the fluencies 0.7 J cm^{-2} (thin curve) and 1.3 J cm^{-2} (thick curve) on the target surface, (c) application of two-color pump of the indium oxide nanoparticle-containing plasma (13th harmonic was collected at saturation level to demonstrate the odd and even harmonic distribution), and (d) shortest wavelength harmonics generated during single- and two-color pumps of the plasma produced on the indium bulk target. The harmonic orders are shown on the graphs. *Source:* Ganeev et al. 2015 [63]. Reproduced with permission from © IOP Publishing.

pulses through such plasma plume (Figure 6.13b, thick curve). The decrease of heating pulse fluence (from 1.3 to 0.7 J cm^{-2}) allowed the formation of plasma conditions when the intense lower order harmonics became visible in the spectrum, which did not contain the plasma emission (thin curve). Strong enhanced 13th harmonic was observed at these conditions, though the enhancement of this harmonic over neighboring ones was not as high as in the case of bulk indium ablation. This nanoparticle-containing plasma allowed harmonic generation up to the 27th order. Further decrease of heating pulse fluence ($F = 0.3$ J cm^{-2}) led to considerable decrease of the 13th harmonic and appearance of the featureless plateau-like spectrum along the whole range of harmonic generation. Polarization measurements of the enhanced 13th harmonic yield from nanoparticle plasma showed the anticipated decrease of its intensity with the change of the polarization of driving radiation from linear to elliptical and circular, which is a characteristic feature of this nonlinear optical process.

Two pulses of different wavelengths (802 and 401 nm) interacting in the nanoparticle-containing plasma were also used to analyze the influence of weak second field on the growth of harmonic yield. The additional motivation of these two-color studies of the HHG from nanoparticle-containing plasmas was related with the analysis of possible enhancement of the nearby even (12th and 14th) harmonics generated close to the resonance-enhanced odd (13th) harmonic. One can assume that the resonances, which were responsible for enhancement of the 13th harmonic, could also modify the harmonic yield of the neighboring even orders. Part of the driving pulse was converted to the second harmonic in the nonlinear crystal (beta barium borate (BBO), type I) placed inside the vacuum chamber on the path of the focused femtosecond beam at a distance of 150 mm from the plasma plume (Figure 6.11). In that case, the orthogonal polarizations of two pumps were used for harmonic generation. Numerous gas and plasma HHG experiments have shown that the use of the orthogonal polarizations of two pumps is a sufficient and simple way to significantly modify the generating spectrum. The 0.5 mm long BBO crystal was used, which allowed a partial overlapping of two pumps in the plasma medium despite the delay induced by group velocity dispersion between the 802 and 401 nm pulses, while producing a sufficient amount of 401 nm photons. The second-harmonic conversion efficiency in this crystal was 9%. The driving and second-harmonic beams were focused inside the plasma plume.

The application of the two-color pump of nanoparticle-containing plasma at the optimal fluence of heating radiation (0.7 J cm^{-2}) allowed achieving the growth of the 12th harmonic, which was close to the resonance transition of In II responsible for the enhancement of the 13th harmonic (Figure 6.13c). This spectrum is presented at the conditions of saturation of the enhanced 13th harmonic for better visibility of the relative intensities of other odd and even harmonics (particularly, compare the intensities of 12th and other orders). The two-color pump of the plasma produced on the indium bulk target showed similar harmonic spectrum. Figure 6.13d presents the comparative spectra of the shortest wavelength harmonics generated in the cases of single- and two-color pumps. The harmonic cutoffs in these two cases were the 45th and 37th orders, while the even harmonics generated up to the 26th order.

The influence of strong ionic transition $4d^{10}5s$ $^{21}S_0 \rightarrow 4d^95s^25p(^2D)$ $^1P_1$ or some other nearby transitions of In II on the 12th harmonic yield led to enhancement of this radiation compared with the lower order (10th and 11th) harmonics even using the weakly overlapped leading and trailing parts of two pulses in the nanoparticle-containing extended plasma. The origin of enhancement could be attributed to both the single ion response and propagation effect. Note the absence of the 12th harmonic in the emission spectrum comprising only 10th, 14th, 18th, etc. orders in the case of using the single-color 401 nm pump of nanoparticle-containing plasma, which confirms the origin of this radiation in the two-color pump experiments as a result of the interaction of two waves in the plasma plume. In the case of two-color pump scheme, the range of enhanced harmonics was extended toward the shorter wavelength region (31st order), in spite of relatively small ratio of the overlapped driving pulses inside the plasma (0.04). The application of extended plasma diminished the delay between 401 and 802 nm pulses due to positive dispersion of this medium.

These experiments have demonstrated that the resonance enhancement in nanoparticle-contained plasmas occurred only at the heating fluencies allowing the formation of ionized species of nanoparticles, while further growth of target excitation led to the deterioration of HHG conditions. Note that, in the case of bulk indium ablation, the enhanced 13th harmonic was observed in a broad range of the fluencies of heating pulses, thus indicating the appearance of the emitters of this harmonic at both weak and strong excitation of indium target.

6.2.1.3 Mn_2O_3 Nanoparticles

Similar feature was observed in the case of manganese oxide nanoparticles. Figure 6.14 (panel 1) shows the low-order harmonic spectrum generated in the plasma produced on the manganese bulk target at the fluence $F = 0.8\,J\,cm^{-2}$. One can see the plateau-like harmonic distribution. The same low-order harmonic spectrum was observed in the case of Mn_2O_3 nanoparticle-contained plasma at the optimal conditions of excitation of the nanoparticle-covered target ($F = 0.7\,J\,cm^{-2}$, panel 2). The two-color pump allowed generation of odd and even harmonics from this nanoparticle-contained plasma ($F = 0.7\,J\,cm^{-2}$, panel 4). Overexcitation of nanoparticle-contained target using the stronger fluence of heating pulse ($F = 1.3\,J\,cm^{-2}$) led to strong incoherent emission and disappearance of harmonics (panel 3).

The higher order harmonics produced from the laser ablation of Mn bulk target showed the short-wavelength emission with the extended harmonic spectrum and enhanced 33rd order. In the case of the weak ablation of manganese oxide nanoparticles ($F = 0.6\,J\,cm^{-2}$) the harmonics were extended up to the 39th order, without the appearance of the enhanced 33rd harmonic. With the growth of target excitation ($F = 1\,J\,cm^{-2}$), the enhanced 33rd harmonic, though not as strong as in the case of the ablation of Mn bulk target, was observed. Further growth of F led to deterioration of the conditions of harmonic generation.

Those measurements have shown that the intensity of the lower order harmonics generated from the nanoparticle-containing plasma produced at

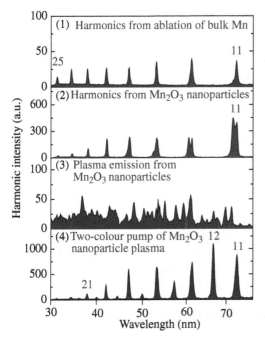

Figure 6.14 Panel 1: lower order harmonic spectrum obtained from the plasma produced on the bulk manganese target ($F = 0.8\,\text{J cm}^{-2}$). Panel 2: harmonic spectrum obtained from the plasma produced on the target contained Mn_2O_3 nanoparticles at optimal conditions of ablation ($F = 0.8\,\text{J cm}^{-2}$). Panel 3: plasma spectrum obtained at overexcitation of the Mn_2O_3 nanoparticle-containing target ($F = 1.3\,\text{J cm}^{-2}$). Panel 4: harmonic spectrum obtained using the two-color pump of the plasma produced on the target contained Mn_2O_3 nanoparticles ($F = 0.7\,\text{J cm}^{-2}$). *Source*: Ganeev et al. 2015 [63]. Reproduced with permission from © IOP Publishing.

$F = 0.8\,\text{J cm}^{-2}$ was six times stronger compared with those generated from the plasma produced on the Mn bulk target (Figure 6.14). In the meantime, the harmonic cutoff from the nanoparticle-containing plasma at these conditions of target excitation was notably lower compared with the one from the plasma produced on the solid target (39th and 107th harmonics respectively). The growth of excitation led to the appearance of similarity in the harmonic spectra from the plasmas produced during nanoparticle and bulk-target ablations. Lower order harmonics from nanoparticle-containing plasma became weaker, while showing the resonance enhancement of the 33rd harmonic followed with the extension of harmonic cutoff up to the 50s orders. This variation of the pattern of harmonic spectrum could be attributed to the appearance of ionized species of Mn_2O_3 nanoparticles. The disintegration of nanoparticles during ablation may cause the appearance of Mn_2O_3 molecules and other oxides of Mn in the plasma volume. Those species could be responsible for the appearance of the harmonic spectra similar to the case of ablation of the bulk manganese target.

The enhancement of a group of harmonics starting from the 33rd order, compared with the nearest lower order harmonics, has been attributed to the influence of the resonance transitions of Mn II in this spectral region (~52 eV). Here, we do not discuss the origin of this enhancement since it has been analyzed in detail in a few previous publications [64, 65], but just suggest that a growth of the 33rd harmonic was attributed to the closeness with the giant 3p → 3d resonances of manganese ions. The discussed studies have shown the prevailing role of neutrals rather than ions as the emitters of lower order harmonics from the Mn_2O_3 nanoparticle-containing plasma, while some small amount of singly charged nanoparticles could be responsible for the generation of weak higher order (H17–H39) harmonics. Stronger excitation of target created larger amount of ionized nanoparticles. At these conditions, the resonance enhancement of the 33rd harmonic became significantly pronounced, compared with the 31st harmonic.

6.2.1.4 Sn Nanoparticles

Tin plasma allowed the enhanced 17th harmonic generation, similarly to previous studies [66, 67]. Figure 6.15a shows the resonant enhancement of

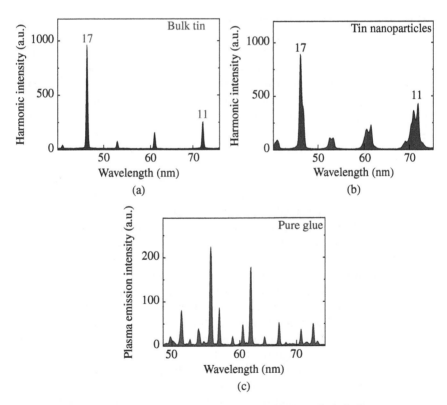

Figure 6.15 Harmonic spectra from the plasmas produced (a) on the Sn bulk target ($F = 0.8\,J\,cm^{-2}$) and (b) on the target contained Sn nanoparticles ($F = 0.6\,J\,cm^{-2}$). (c) Emission spectrum of emission from the pure glue ablation. No harmonics were observed in this plasma at various conditions of ablation. *Source:* Ganeev et al. 2015 [63]. Reproduced with permission from © IOP Publishing.

single harmonic in the tin-contained plasma using the bulk target ablated at a fluence $F = 0.8\,\mathrm{J\,cm^{-2}}$. In the discussed studies, the strong 17th harmonic was obtained with an enhancement factor of about 10× compared with the neighboring harmonic orders. In the case of Sn nanoparticle-contained target, similar enhancement of single harmonic was observed as well, thus pointing out the appearance of ionized clusters at relatively weak excitation of the target ($F = 0.6\,\mathrm{J\,cm^{-2}}$, Figure 6.5b). At lower fluencies, the enhancement of 17th harmonic was significantly suppressed.

The Sn II resonances, which influence the enhancement of single harmonic, correspond to the wavelengths in the region of 47 nm. The observed enhancement of the 17th harmonic of 802 nm radiation ($\lambda = 47.17\,\mathrm{nm}$) was related with the influence of the $4d^{10}5s^25p\ ^2P_{3/2} \rightarrow 4d^95s^25p^2$ transitions of Sn II ions. Some of these transitions possess large oscillator strengths in the photon energy range of 24.9–27.3 eV [68].

The spectral lines of harmonics from Sn nanoparticles were broadened at the shorter wavelength side and showed additional peaks (Figure 6.15b) compared with the narrow lines of the harmonics generated in the monomer-containing plasma produced during ablation of bulk tin (Figure 6.15a). The same feature was observed in the case of the lower order harmonics generated from the manganese oxide clusters (see panel 2 of Figure 6.14). Note that this modification of harmonic spectra was observed only in the case of nanoparticle-containing plasmas, similarly to those analyzed in Ref. [69].

6.2.1.5 Discussion

Several studies of the variations of harmonic bandwidth have been reported in Refs. [70–72]. Those studies were performed under the conditions of the broadening of the laser bandwidth. The laser pulses underwent a significant self-phase modulation due to ionization of the medium. At these conditions, a considerable heterogeneous broadening of the spectral distribution is expected. This effect can be explained by the wavelength change in the leading edge of the laser pulse due to self-phase modulation of the radiation propagating through the laser plasma. The initial lower intensity portion of the pulse creates harmonics. As the pulse intensity reaches its peak, the additional ionization can change the distribution of laser spectrum along the whole pulse. It has been shown in the previous studies of gas and plasma HHG using the chirped pulses that the leading edge of the pulse mainly contributes to the HHG. Laser radiation becomes negatively chirped along the propagation through the extended plasma. The part of harmonics produced with negatively chirped laser pulses becomes blue-shifted because the harmonics produced in the leading edge of the laser pulse come from the blue part of the laser spectrum. This assumption qualitatively explains the blue-sided broadening of lower order harmonics.

Figure 6.12 presents the images of original and deposited nanoparticles. One can see a similarity in the spherical and irregular shapes of those nanoparticles. One can assume from these results that neither the disintegration due to ablation nor aggregation on the substrate showed the changes in the shape of nanoparticles. Note that independent of the shape of nanoparticles, the whole tendency of harmonic distribution from those species remained similar to the bulk target ablation. The glue without nanoparticles was separately ablated, which did not lead to harmonic generation during propagation of femtosecond pulses through the glue-contained plasma. The ablation of pure glue at any fluence led to the emission of incoherent radiation (Figure 6.15c). The glass substrates also did not play any role in HHG.

The density of nanoparticles shown in Figure 6.12a was significantly lower than the one of the glued nanoparticles on the glass surface. The entire ablation of the whole set of material was carried out during a few ten seconds, which corresponded to a few thousand laser shots at 10 Hz repetition rate. The plasma density was notably small ($\sim 10^{17}$ cm^{-3}) to cause a significant absorption of generated harmonics. The dependence of the harmonic yield (I_H) as a function of nanoparticle plasma length (l) was analyzed. It was found that it follows the quadratic rule ($I_H \propto l^2$) up to $l = 3.5$ mm, with some declination of the slope of this dependence at longer lengths. These measurements show that the phase matching was fulfilled along almost all length of interaction without the significant influence of free electrons and absorption.

6.2.2 High-Order Harmonic Generation from Fullerenes

Molecules provide a unique opportunity to study HHG process. Systems exhibiting large polarizability (vis-a-vis collective motion of electrons) can be potentially used to increase high harmonic yield. Moreover, the HHG process can be influenced by molecular orientation [73] and quantum interferences inherent to multi-electron systems [74]. The HHG from molecules itself can be used as a versatile tool to probe and characterize the multi-electron dynamics and retrieve structural information [75]. With this aim, HHG in several diatomic, triatomic [74], and organic molecules was studied [76]. As the molecular system increases in complexity, the theory of HHG, which is based on single active electron approximation by assuming that only the valence electron responds to the incident laser field while the other electrons are frozen, has to be modified to include the multi-electron dynamics.

In this subsection, the experimental observations of high-order harmonics generated in fullerenes produced by laser ablation are discussed and compared with those produced in carbon. In C_{60}, an extended cutoff with harmonics up to the 19th order was observed. The spectrum exhibits an enhancement of the harmonics (11th–15th order) that lie within the plasmon resonance, indicative of the multi-electron dynamics. The harmonic yield in fullerenes is 25 times larger compared to carbon [77].

C_{60} was chosen as a nonlinear medium because (i) it is highly polarizable ~80 Å^3 [78], (ii) it is stable against fragmentation in intense laser fields due to very large number of internal degrees of freedom leads to the fast diffusion of the excitation energy, (iii) it exhibits giant plasmon resonance at ~20 eV [79], (iv) has large photoionization cross-sections [79], and (v) multi-electron dynamics is known to influence ionization and recollision [80] that are central to HHG process. The saturation intensities of different charge states of C_{60} are higher compared to isolated atoms of similar ionization potential [81].

Previous studies on fullerenes have demonstrated generation of 2nd [82], 3rd [83], and 5th [84] harmonics. Experiments on HHG in C_{60} beyond the 5th order do not exist. Theoretical studies on HHG from C_{60} involved extending the three-step model [85] and using dynamical simulations [86]. In the latter, higher order harmonics were shown to be due to multiple excitations and could be easily generated even with a weak laser field. Both studies reveal how HHG can be used to probe the electronic and molecular structure of C_{60}.

Bulk carbon and two types of fullerene samples were used in the discussed studies as targets for laser ablation. In one fullerene sample, the powder (98% C_{60}, 2% C_{70}) was mixed with epoxy and fixed onto glass or silver substrates leading to an inhomogeneous distribution of fullerene nanoparticles. The second sample is a fullerene film on glass substrate. The films are grown by evaporating C_{60} powder (99%) in a resistively heated oven at 600 °C. The effusive beam of C_{60} molecules is deposited onto glass substrate maintained at liquid nitrogen temperature. The growth conditions, which depend on the rate of evaporation (oven temperature) and deposition time, determine the thickness and the quality of the film. The thickness of the film used in the experiment was about few microns.

For laser ablation, a heating pulse that was split from the uncompressed Ti:sapphire laser (pulse duration $t = 210$ ps, wavelength $\lambda = 800$ nm, pulse repetition rate 10 Hz) was focused on the target placed in a vacuum chamber by using a plano-convex lens (focal length $f = 150$ mm, see inset in Figure 6.16). After a delay (variable between 6 and 74 ns), the femtosecond pulse (pulse energy 8–20 mJ, $t = 35$ fs, $\lambda = 800$ nm) was focused on the plasma from the orthogonal direction using a plano-convex lens (focal length $f = 680$ mm).

Figure 6.16 shows the harmonic spectra obtained from ablation of bulk carbon target, C_{60} powder fixed in epoxy on silver, and C_{60} film. HHG produced in the ablation plume of bulk carbon targets exhibit plateau-like harmonic spectrum up to the 25th order. The harmonic spectra from targets containing C_{60} powder in epoxy and C_{60} film were significantly different in comparison with the bulk carbon target under identical experimental conditions. (i) Harmonics lying in the spectral range of surface plasmon resonance in C_{60} (20 eV, $\lambda = 62$ nm) were enhanced. (ii) The harmonic yields were larger

Figure 6.16 Harmonic spectra obtained in the plasma plumes produced from (bottom panel) bulk carbon target, (middle panel) C_{60} powder-rich epoxy, and (upper panel) C_{60} film. The dashed curve in the top panel corresponds to the photoionization cross-sections near plasmon resonance. Inset shows the experimental set-up of HHG in fullerenes. MP: main pulse; PP: picosecond pulse; DL: delay line; C: grating compressor; FL: focusing lenses; T: target; XUVS: extreme ultraviolet spectrometer; G: gold-coated grating; MCP: micro-channel plate; CCD: charge-coupled device. *Source:* Ganeev et al. 2009 [77]. Reproduced with permission from American Physical Society.

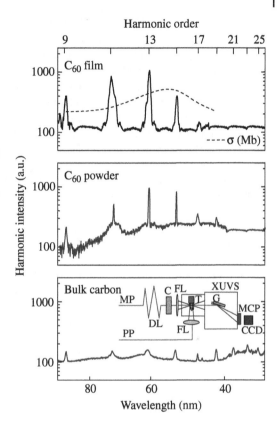

by a factor of 20–25 for 13th harmonic. (iii) The harmonic cutoff in C_{60} was lower (19th order) than carbon but extended beyond the value (11th order) as predicted by the three-step model. (iv) The 11th and 13th order harmonics in C_{60} were more intense than the 9th harmonic. Although the sensitivity of used detection system decreases for longer wavelengths at around 70 nm, in most cases, where various bulk targets and atoms were used, considerably stronger 9th harmonic was obtained.

Varying the experimental conditions such as temporal delay between the heating pulse and femtosecond pulse (6, 24, and 74 ns) and their intensities did not considerably change harmonic yield in C_{60}. The delay is not important in C_{60} targets due to lower ablation thresholds. The used heating pulse intensity was $2 \times 10^9 \, \mathrm{W \, cm^{-2}}$, 10 times lower than bulk carbon target. Increasing the intensity of the femtosecond pulse did not lead to an extension of the cutoff in fullerenes, which is a sign of saturation of the HHG in this medium. Moreover, at relatively high laser intensities, a decrease in harmonic output was observed. At such intensities, multiple ionization of C_{60} occurs leading to high free electron density causing phase mismatch. Similarly, an optimal heating pulse intensity exists above which harmonics in C_{60} became weaker due to fragmentation of fullerenes, increase of the free electron concentration, phase mismatch, and self-defocusing. By calibrating the detection system using techniques previously reported [36], the efficiency of the 11th–15th harmonics (between 50 and 70 nm) from fullerene plume was estimated to be $3 \times 10^{-6} – 3 \times 10^{-5}$.

The structural integrity of the fullerenes ablated off the surface should be intact until the driving pulse arrives. So, the heating pulse intensity is a very sensitive parameter and was kept between 2×10^9 and 8×10^9 W cm^{-2}. At lower intensities the concentration of clusters in the ablation plume is low, while at higher intensities one can expect fragmentation. The temperature at the surface after the absorption of a 1 mJ heating pulse was estimated to be in the range of 600–700 °C, which was above the evaporation threshold of fullerenes (~300 °C) but below the temperature of fragmentation (~1000 °C). This estimation is valid for both types of fullerene targets.

C_{60} films produced slightly more intense and stable harmonics with low shot-to-shot variation compared to the powder–epoxy mixture. This is due to the homogeneous distribution of particles in the film. In both types of fullerene targets the density of the ablation plume decreases for successive laser shots due to evaporation of C_{60} from the ablation area. As a result the harmonic intensity decreases, as shown in Figure 6.17 for C_{60} film. After about 10 laser

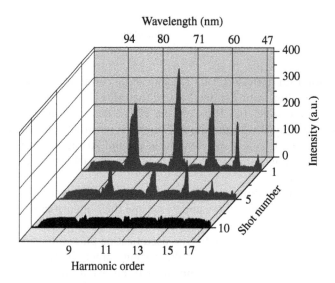

Figure 6.17 Variation of harmonic spectra observed at the consecutive shots on the same spot of fullerene film. *Source*: Ganeev et al. 2009 [77]. Reproduced with permission from American Physical Society.

pulses at the same target position, harmonic generation almost disappeared, unless target was moved to a fresh spot on the fullerene film.

The influence of fullerene density on HHG efficiency can be understood in terms of three length parameters. For optimum HHG, the length of the nonlinear medium should be (a) larger than the coherence length, which is defined by the phase mismatch between the fundamental and harmonic fields and depends on density, and (b) smaller than the absorption length of the medium, which inversely depends on the density and ionization cross-sections. Photoionization cross-sections of C_{60} are well known experimentally and theoretically. It exhibits a giant plasmon resonance at ~20 eV [79] (around the 11th, 13th, and 15th harmonics).

We now discuss two important observations, namely (i) extension of harmonic cutoff and (ii) enhancement of harmonics in the vicinity of plasmon resonance. In the three-step model for HHG, the cutoff harmonic is given by $3.17 U_p + I_p$ (where $I_p = 7.6$ eV is the ionization potential of C_{60} and the U_p is the ponderomotive energy). The intensity of ~10^{14} W cm^{-2} was used during discussed studies, which is above the saturation intensity of the first two charge states of C_{60} [87]. The saturation intensity of C_{60}^+ is 5×10^{13} W cm^{-2} in close agreement with the theoretical value of 4×10^{13} W cm^{-2} [81]. Accordingly, if the HHG is from neutral C_{60} the cutoff should be at 11th harmonic. In contrast, the harmonics up to 19th order were achieved.

Higher cutoff could be due to multiphoton excitation of surface plasmon (20 eV) by the incident laser field (1.55 eV). If ionization starts from a plasmon state and the electron returns to the ground state upon recombination, the plasmon energy is converted into photon energy extending the cutoff [88]. Recombination into orbitals, other than the highest occupied molecular orbital of C_{60}, with higher ionization potentials [89] can also result in extension of cutoff.

High harmonics in C_{60} or in any complex multi-electron system will have two contributions – the usual harmonics generation process and the physical mechanisms that lead to enhancement of harmonics (9th–15th in C_{60}) around the frequencies at which the system displays collective electron oscillations (20 eV in C_{60} with a full width at half maximum of ~10 eV). Plasmon excitation under two different scenarios can lead to enhancement of high harmonics. (i) The recolliding electron excites the plasmon upon recombination, which then decays by emitting photons. This leads to enhancement of the harmonics in the vicinity of plasmon resonance [89]. Such a mechanism would be wavelength-dependent. It was shown that at longer wavelengths the HHG spectrum resembles that of atomic systems. (ii) The laser field directly excites the surface plasmon through multiphoton process, which then decays back by emitting high-energy photons. Similar bound–bound transitions were theoretically shown to exist [85] in C_{60}. Such a mechanism will be independent of recollision process and can be revealed by ellipticity measurements.

References

1 Miles, R.B. and Harris, S.E. (1973). *IEEE J. Quantum Electron.* QE-9: 470.

2 Puell, H., Spanner, K., Falkenstein, W. et al. (1976). *Phys. Rev. A* 14: 2240.

3 Kung, A.H., Young, J.F., and Harris, S.E. (1973). *Appl. Phys. Lett.* 22: 301.

4 Ganeev, R.A., Kulagin, I.A., Usmanov, T., and Khudaiberganov, S.T. (1982). *Sov. J. Quantum Electron.* 12: 1637.

5 Zych, L.J. and Young, J.F. (1978). *IEEE J. Quantum Electron.* QE-14: 147.

6 Elouga Bom, L.B., Pertot, Y., Bhardwaj, V.R., and Ozaki, T. (2011). *Opt. Express* 19: 3077.

7 Pertot, Y., Elouga Bom, L.B., Bhardwaj, V.R., and Ozaki, T. (2011). *Appl. Phys. Lett.* 98: 101104.

8 Theobald, W., Wülker, C., Schäfer, F.R., and Chichkov, B.N. (1995). *Opt. Commun.* 120: 177.

9 Ganeev, R.A. (2009). *Phys. Usp.* 52: 55.

10 Ganeev, R.A., Boltaev, G.S., Satlikov, N.K. et al. (2012). *J. Opt. Soc. Am. B* 29: 3286.

11 Reintjes, J.F. (1984). *Nonlinear Optical Parametric Processes in Liquids and Gases.* Orlando: Academic Press.

12 Krause, J.L., Schafer, K.J., and Kulander, K.C. (1992). *Phys. Rev. Lett.* 68: 3535.

13 Corkum, P.B. (1993). *Phys. Rev. Lett.* 71: 1994.

14 Lewenstein, M., Balcou, P., Ivanov, M.Y. et al. (1994). *Phys. Rev. A* 49: 2117.

15 Manakov, N.L. and Ovsyannikov, V.D. (1980). *Sov. Phys. JETP* 52: 895.

16 Ganeev, R.A., Kulagin, I.A., Suzuki, M. et al. (2005). *Opt. Commun.* 249: 569.

17 Ganeev, R.A., Witting, T., Hutchison, C. et al. (2012). *Phys. Rev. A* 85: 015807.

18 Ganeev, R.A. (2012). *J. Mod. Opt.* 59: 409.

19 Anselment, M., Smith, R.S., Daykin, E., and Dimauro, L.F. (1987). *Chem. Phys. Lett.* 134: 444.

20 Rohlfing, E.A. (1988). *J. Chem. Phys.* 89: 6103.

21 Acquaviva, S. and De Giorgi, M.L. (2002). *Appl. Surf. Sci.* 197–198: 21.

22 NIST (2013). Atomic spectra database (Ver. 4.1.0). http://physics.nist.gov/asd (accessed 08 March 2018).

23 Kulagin, I.A. and Usmanov, T. (2000). *Quant. Electron.* 30: 520.

24 Ganeev, R.A., Suzuki, M., Ozaki, T. et al. (2006). *Opt. Lett.* 31: 1699.

25 Ganeev, R.A., Singhal, H., Naik, P.A. et al. (2006). *J. Opt. Soc. Am. B* 23: 2535.

26 Suzuki, M., Baba, M., Ganeev, R. et al. (2006). *Opt. Lett.* 31: 3306.

27 Suzuki, M., Baba, M., Kuroda, H. et al. (2007). *Opt. Express* 15: 1161.

28 Ganeev, R.A., Strelkov, V.V., Hutchison, C. et al. (2012). *Phys. Rev. A* 85: 023832.

29 Haessler, S., Strelkov, V., Elouga Bom, L.B. et al. (2013). *New J. Phys.* 15: 013051.

30 Ganeev, R.A., Witting, T., Hutchison, C. et al. (2013). *Phys. Rev. A* 88: 033838.

31 Strelkov, V. (2010). *Phys. Rev. Lett.* 104: 123901.

32 Milošević, D.B. (2010). *Phys. Rev. A* 81: 023802.

33 Frolov, M.V., Manakov, N.L., and Starace, A.F. (2010). *Phys. Rev. A* 82: 023424.

34 Tudorovskaya, M. and Lein, M. (2011). *Phys. Rev. A* 84: 013430.

35 Boltaev, G.S., Ganeev, R.A., Kulagin, I.A., and Usmanov, T. (2014). *J. Opt. Soc. Am. B* 31: 436.

36 Ganeev, R.A., Baba, M., Suzuki, M., and Kuroda, H. (2005). *Phys. Lett. A* 339: 103.

37 Akiyama, Y., Midorikawa, K., Matsunawa, Y. et al. (1992). *Phys. Rev. Lett.* 69: 2176.

38 Kubodera, S., Nagata, Y., Akiyama, Y. et al. (1993). *Phys. Rev. A* 48: 4576.

39 Wahlström, C.-G., Borgström, S., Larsson, J., and Pettersson, S.-G. (1995). *Phys. Rev. A* 51: 585.

40 Ganeev, R.A., Redkorechev, V.I., and Usmanov, T. (1997). *Opt. Commun.* 135: 251.

41 Krushelnick, K., Tighe, W., and Suckewer, S. (1997). *J. Opt. Soc. Am. B* 14: 1687.

42 Ganeev, R.A. (2009). *Open Spectrosc. J.* 3: 1.

43 Anikin, V.I., Gora, V.D., Drabovich, K.N., and Dubovik, A.N. (1976). *Sov. J. Quantum Electron.* 6: 174.

44 Bjorklund, G.C. (1975). *IEEE J. Quantum Electron.* 11: 287.

45 Hilbig, R., Lago, A., and Wallenstein, R. (1984). *Opt. Commun.* 49: 297.

46 Ganeev, R.A., Gorbushin, V.V., Kulagin, I.A., and Usmanov, T. (1986). *Appl. Phys. B* 41: 69.

47 Donnelly, T.D., Ditmire, T., Neuman, K. et al. (1996). *Phys. Rev. Lett.* 76: 2472.

48 Tisch, J.W.G., Ditmire, T., Fraser, D.J. et al. (1997). *J. Phys. B* 30: L709.

49 Hu, S.X. and Xu, Z.Z. (1997). *Appl. Phys. Lett.* 71: 2605.

50 Tajima, T., Kishimoto, Y., and Downer, M.C. (1999). *Phys. Plasmas* 6: 3759.

51 Véniard, V., Taïeb, R., and Maquet, A. (2001). *Phys. Rev. A* 65: 013202.

52 Vozzi, C., Nisoli, M., Caumes, J.-P. et al. (2005). *Appl. Phys. Lett.* 86: 111121.

53 Fomichev, S.V., Zaretsky, D.F., Bauer, D., and Becker, W. (2005). *Phys. Rev. A* 71: 013201.

54 Pai, C.-H., Kuo, C.C., Lin, M.-W. et al. (2006). *Opt. Lett.* 31: 984.

55 Kundu, M., Popruzhenko, S.V., and Bauer, D. (2007). *Phys. Rev. A* 76: 033201.

56 Moreno, P., Plaja, L., and Roso, L. (1994). *Europhys. Lett.* 28: 629.

57 Ruf, H., Handschin, C., Cireasa, R. et al. (2013). *Phys. Rev. Lett.* 110: 083902.

58 Ganeev, R.A. (2014). *J. Opt. Soc. Am. B* 31: 2221.

59 Elouga Bom, L.B., Ganeev, R.A., Abdul-Hadi, J. et al. (2009). *Appl. Phys. Lett.* 94: 111108.

60 Singhal, H., Ganeev, R.A., Naik, P.A. et al. (2010). *J. Phys. B: At. Mol. Opt. Phys.* 43: 025603.

61 Ganeev, R.A., Hutchison, C., Castillejo, M. et al. (2013). *Phys. Rev. A* 88: 033803.

62 Singhal, H., Naik, P.A., Kumar, M. et al. (2014). *J. Appl. Phys.* 115: 033104.

63 Ganeev, R.A., Suzuki, M., Yoneya, S., and Kuroda, H. (2015). *J. Phys. B* 48: 165401.

64 Ganeev, R.A., Elouga Bom, L.B., Kieffer, J.-C., and Ozaki, T. (2007). *Phys. Rev. A* 76: 023831.

65 Ganeev, R.A., Witting, T., Hutchison, C. et al. (2012). *Opt. Express* 20: 25239.

66 Ganeev, R.A., Elouga Bom, L.B., Kieffer, J.-C., and Ozaki, T. (2007). *Phys. Rev. A* 75: 063806.

67 Ganeev, R.A., Chakera, J.A., Naik, P.A. et al. (2011). *J. Opt. Soc. Am. B* 28: 1055.

68 Duffy, G., van Kampen, P., and Dunne, P. (2001). *J. Phys. B* 34: 3171.

69 Ganeev, R.A., Singhal, H., Naik, P.A. et al. (2009). *J. Appl. Phys.* 106: 103103.

70 Tosa, V., Kim, H.T., Kim, I.J., and Nam, C.H. (2005). *Phys. Rev. A* 71 (063): 808.

71 Froud, C.A., Rogers, E.T., Hanna, D.C. et al. (2006). *Opt. Lett.* 31: 374.

72 Ganeev, R.A., Singhal, H., Naik, P.A. et al. (2009). *J. Opt. Soc. Am. B* 26: 2143.

73 Lein, M., Hay, N., Velotta, R. et al. (2002). *Phys. Rev. A* 66: 023805.

74 Kanai, T., Minemoto, S., and Sakai, H. (2005). *Nature* 435: 470.

75 Itatani, J., Levesque, J., Zeidler, D. et al. (2001). *Nature* 432: 867.

76 Hay, N., Castillejo, M., De Nalda, R. et al. (2000). *Phys. Rev. A* 61: 053810.

77 Ganeev, R.A., Elouga Bom, L.B., Abdul-Hadi, J. et al. (2009). *Phys. Rev. Lett.* 102: 013903.

78 Ballard, A., Bonin, K., and Louderback, J. (2000). *J. Chem. Phys.* 113: 5732.

79 Hertel, I.V., Steger, H., de Vries, J. et al. (1992). *Phys. Rev. Lett.* 68: 784.

80 Bhardwaj, V.R., Rayner, D.M., and Corkum, P.B. (2004). *Phys. Rev. Lett.* 93: 043001.

81 Jaroń-Becker, A., Becker, A., and Faisal, F.H.M. (2006). *Phys. Rev. Lett.* 96: 143006.

82 Hoshi, H., Manaka, T., Ishikawa, K., and Takezoe, H. (1997). *Jap. J. Appl. Phys.* 36: 6403.

83 Banfi, G.P., Fortusini, D., Bellini, M., and Milani, P. (1997). *Phys. Rev. B* 56: R10075.

84 Kafafi, Z.H., Lindle, J.R., Pong, R.G.S. et al. (1992). *Chem. Phys. Lett.* 188: 492.

85 Zhang, G.P. (2005). *Phys. Rev. Lett.* 95: 047401.

86 Ciappina, M.F., Becker, A., and Jaroń-Becker, A. (2007). *Phys. Rev. A* 76: 063406.

87 Bhardwaj, V.R., Rayner, D.M., and Corkum, P.B. (2003). *Phys. Rev. Lett.* 91: 203004.

88 Zanghellini, J., Jungreuthmayer, C., and Brabec, T. (2006). *J. Phys. B* 39: 709.

89 Ruggenthaler, M., Popruzhenko, S.V., and Bauer, D. (2008). *Phys. Rev. A* 78: 033413.

7

Comparison of the Resonance-, Nanoparticle-, and Quasi-Phase-Matching-Induced Processes Leading to the Growth of High-Order Harmonic Yield

7.1 Introduction

Generation of the high-order harmonics of ultrashort laser pulses by various means (during specular reflection from the surfaces [1], as well as during propagation through the gases [2] and laser-produced plasmas (LPPs) [3]) has been proven to be an effective method for the formation of coherent short-wavelength sources. The main obstacle in application of these sources is their low fluence, which was caused by small conversion efficiency of the high-order harmonic generation (HHG). The history of HHG studies includes various approaches, such as the application of nanoparticles (NPs) and clusters as the harmonic emitters [4, 5], the two-color pump (TCP) of gases and plasmas [6, 7], the application of extended gas-filled fibers [8], the use of feedback loops using various adaptive algorithms [9], the resonance enhancement of single harmonic in the plateau-like region [10], and the quasi-phase-matching (QPM) of harmonics by different means [11–13], which were considered among the most reliable methods for the enhancement of harmonic yield. Meanwhile, the maximal reported conversion efficiencies in gases and plasmas (10^{-5} to 10^{-4}) [14, 15] reached the plateau, and further growth of this parameter of HHG seems unrealistic.

Currently, most of the efforts in harmonic generation, at least in the case of gaseous media, were shifted toward the analysis of attosecond pulses generation and their applications, as well as the studies of the molecular orientation of gaseous emitters through the studies of harmonic spectra. Nevertheless, the attempts in the growth of harmonic yield by different means still remain on agenda. Both abovementioned "old" approaches and "new" proposals in the analysis of harmonic yield enhancement are under consideration in the laboratories dealing with this method of generation of the coherent extreme ultraviolet (XUV) radiation.

The modification of harmonic spectrum may allow various applications in the spectroscopy of harmonic emitters. The analysis of these spectra is crucial for the studies of the orientational properties of the molecules existing in the

Resonance Enhancement in Laser-Produced Plasmas: Concepts and Applications,
First Edition. Rashid A. Ganeev.
© 2018 John Wiley & Sons, Inc. Published 2018 by John Wiley & Sons, Inc.

gaseous and plasma-like states [16]. The shift of driving laser pulses toward longer wavelength region offers some perspectives in the extension of harmonic cutoff due to the $E_{cutoff} \propto \lambda^2$ rule [17]. The analysis of the nonlinear optical response of complex molecules under the action of near infrared (NIR) field allows definition of the specific properties of these species [18].

The above consideration leads to the anticipation in the advantageous applications of a few methods of harmonic amendment in a single set of experiments. The particular interest here is related with the comparison of the collective mechanisms of accumulation of the nonlinear optical response of medium (so-called macroprocesses) and the mechanisms related with the individual properties of single emitters (so-called microprocesses). Those mechanisms include the fulfillment of the conditions of QPM of the driving and harmonic waves along the whole medium and the conditions of the coincidence of the individual harmonic in the plateau region and the ionic transition possessing strong oscillator strength. Both these processes could be further amended through the implementation of TCP technique in the NIR range.

Although the QPM processes were demonstrated in the gaseous media, no significant resonance-induced enhancement of single harmonic was reported using gas HHG approach. On the contrary, LPP have proven to be the effective media for these two (macro- and micro) processes. On one hand, QPM in LPP has recently been reported in the case of silver plasma [19], though other ablated materials have also proven to be suitable for these purposes [20]. On the other hand, resonance enhancement of single harmonic has been frequently observed during plasma HHG. In, Sn, Cr, Mn, and a few other, predominantly semiconductor, targets have shown the attractiveness for ablation and further generation of enhanced single harmonic during propagation of the ultrashort pulses through the LPP.

The plasmas containing small molecules of silver produced during ablation of the bulk silver and silver NPs at the conditions suitable for efficient harmonic generation of the ultrashort pulses propagating through the laser-produced plasmas can also lead to the growth of harmonic yield. The time-of-flight mass-spectroscopy (ToFMS) studies of plasmas confirmed the presence of these species in the plasmas. It is suggested that the harmonics generated during the ablation of NP and bulk targets could be enhanced due to the appearance of these small species (Ag_2, Ag_3) at the moment of propagation of the ultrashort laser pulses through such plasmas. Similar observation has been reported in the case of carbon NPs-containing plasmas, when the clusters containing 5–25 atoms demonstrate strong growth of harmonics. Comparison of this process with the one related with resonance enhancement of harmonics allows achieving best conditions for coherent XUV radiation generation.

In this chapter, we show and compare three methods of harmonic enhancement. We discuss the advances in using collective processes of enhancement together with the growth of single harmonic induced by the individual properties of emitters for the enhancement of harmonic yield by different means. We analyze the indium plasma to show the growth of a group of harmonics around the maximally enhanced 30s orders. The tuning of maximally enhanced harmonic was carried out using the variation of the conditions of plasma formation. Simultaneously, the wavelength of driving pulse was adjusted to generate the enhanced H21 of the TCP (1310 to 655 nm) using optical parametric amplifier (OPA), which was matched with the abovementioned transition of In II. We compare these principally different processes of harmonic enhancement and show the advantages of the joint application of as much as possible methods of HHG amendment (QPM-induced enhancement, resonance-induced enhancement, and TCP-induced enhancement) in a single experiment.

7.2 Quasi-Phase-Matched High-Order Harmonic Generation in Laser-Produced Plasmas

The phase mismatch (PMM) problem suppresses HHG of ultrashort laser pulses in both gaseous and plasma media. The PMM is related with the dephasing between the propagated harmonic field and the laser-induced polarization ($\Delta k = qk_1 - k_q$) caused predominantly by the dispersion of medium enhanced by the presence of free electrons. The difference in the velocities of these waves leads to the conditions when, at some distance from the beginning of the medium, the phase shift becomes close to π. After propagation of this distance (coherence length, $L_{coh} = \pi/\Delta k$) the constructive accumulation of harmonic photons reverts to destructive when the newly generated XUV photons being in reverse phase compensate for the earlier generated photons, thus decreasing the harmonic yield. The QPM allows diminishing this restriction of harmonic conversion efficiency by different means. Particularly, the QPM has been demonstrated by using the multiple gas puffs [11, 21, 22] and multijet plasmas [13]. However, in the latter case what remained unclear was the separate role in the QPM mechanism of the two groups of free electrons: those generated during target ablation versus those coming from tunnel ionization in the driving field.

The PMM induced by plasma dispersion for the qth harmonic is defined by the relation

$$\Delta k_{disp} = qN_e e^2 \lambda / 4\pi m_e \varepsilon_0 c^2 \tag{7.1}$$

where λ is the wavelength of driving radiation; N_e, m_e, and e are the density, the mass, and the charge of the electron; c is the light velocity; and ε_0 is the vacuum permittivity [23]. The coherence length of this harmonic is

$$L_{coh} = \pi/\Delta k \approx \pi/\Delta k_{disp} \approx 4\pi^2 m_e \varepsilon_0 c^2 / qN_e e^2 \lambda \tag{7.2}$$

From this expression the coherence length (in mm) at the conditions of using the 800 nm driving laser could be presented as

$$L_{coh} \approx 1.4 \times 10^{18}/(N_e \times q_{qpm}) \tag{7.3}$$

where q_{qpm} is the harmonic order in the QPM region showing the highest enhancement and N_e is the electron density in the plasma jets measured in cubic centimeters. This simple formula allows defining the electron density by knowing the coherence length (which is in fact the size of a single plasma jet at the conditions of the QPM in the multijet structure) and the maximally enhanced harmonic order. The decrease of heating pulse fluence on the surface of the ablating target should lead to a decrease of electron density (due to less ablation- and tunnel-induced electrons) in the plasma plume followed by the shift of q_{qpm} toward the shorter-wavelength region. Similarly, one can anticipate that, for the plasma jets of different sizes, the maximally enhanced harmonics will also be tuned along the XUV spectrum.

In this section, we analyze the experimental and theoretical studies of the QPM-enhanced harmonics generated in the laser-produced plasmas [24]. Particularly, we discuss the influence of driving and heating pulse energies on the enhancement and broadening of the groups of QPM harmonics in the case of modulated plasma and show the tuning of those enhanced harmonics by tilting the multislit mask (MSM) placed in front of the ablating Ag target. We also discuss the results of calculations of the QPM conditions in the laser-produced plasmas and analyze two cases: (i) joint influence of the electrons appeared during ablation and tunnel ionization on the QPM of high-order harmonics inside the extended Ag plasma and (ii) the case when influence of the ablation electrons becomes insignificant. The calculations show that the first scenario is closer to the experimental results.

7.2.1 Experimental Arrangements

The uncompressed radiation of the Ti:sapphire laser operated at 10 Hz repetition rate was used as a heating pulse (central wavelength $\lambda = 802$ nm, pulse duration 370 ps, pulse energy up to $E_{hp} = 4$ mJ) for extended plasma formation. The heating pulse was focused using a 200 mm focal-length cylindrical lens inside the vacuum chamber containing an ablating target to create the extended plasma plume above the target surface (Figure 7.1a). The focusing of the heating pulse on the target surface produced the extended nonperforated plasma (Figure 7.1b, upper panel). The intensity of the heating pulses on a plain target surface was varied up to 5×10^9 W cm^{-2}. The compressed driving pulse from the same laser with the energy of up to $E_{dp} = 5$ mJ and 75 fs pulse duration was used, after 45 ns from the beginning of ablation, for the harmonic generation in the plasma plume. The driving pulse was focused using a 400 mm focal-length spherical lens onto the prepared plasma from the

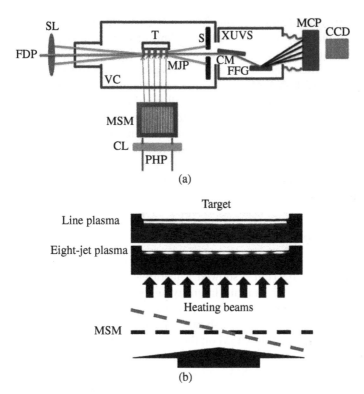

Figure 7.1 (a) Experimental scheme for the harmonic generation in the multijet plasmas. FDP, femtosecond driving pulse; PHP, picosecond heating pulse; SL, spherical lens; CL, cylindrical lens; MSM, multislit mask; VC, vacuum chamber; T, target; MJP, multijet plasma; S, slit; XUVS, extreme ultraviolet spectrometer; CM, cylindrical mirror; FFG, flat field grating; MCP, microchannel plate detector; CCD, charge-coupled device camera. (b) Images of the extended imperforated (upper panel) and eight-jet (bottom panel) plasma formations produced on the surface of the silver target. The number of jets was increased and the sizes of jets were decreased by tilting the MSM. *Source*: Ganeev et al. 2015 [24]. Reproduced with permission from American Physical Society.

perpendicular direction, at a distance of ~100 μm above the target surface. The confocal parameter of the focused driving beam was 18 mm. The intensity of the driving pulse in the focal area was varied up to 9×10^{14} W cm^{-2}. The harmonic emission was analyzed by an extreme ultraviolet spectrometer (XUVS) containing a gold cylindrical mirror and a 1200 grooves/mm flat field grating with variable line spacing. The spectrum was recorded on a microchannel plate detector with a phosphor screen, which was imaged onto a charge-coupled device (CCD) camera.

Silver was used as the target for ablation. The size of the target where the ablation occurred was 5 mm. To create multijet plasmas, an MSM was used.

The size of the slits was 0.3 mm with the distance between them also 0.3 mm. The MSM was installed between the focusing cylindrical lens and target such to divide the continuous 5 mm plasma in eight 0.3 mm long plasma jets with ~0.3 mm separation. The image of these jets was captured from the top of the vacuum chamber and is shown in the bottom panel of Figure 7.1b. It was possible to increase the number of plasma jets by tilting the MSM, as shown in Figure 7.1b. Tilting the mask at 45° allowed the formation of 11 jets. The energy of the heating pulse decreased after propagation through the MSM; however, the fluence of this radiation on the target surface remained unchanged, since the size of the ablated area was also decreased. It means that the electron and plasma densities in the cases of extended homogeneous and multijet plasmas were almost equal.

7.2.2 Experimental Observations of QPM

The harmonic spectra from the 5 mm extended silver plasma at a fluence of heating radiation of ~0.9 J cm^{-2} showed a featureless plateau-like shape with a gradual decrease of harmonic intensity in the shorter wavelength region (Figure 7.2, upper panel). The harmonic cutoff was in the region of the 50s orders. Other panels of Figure 7.2 show four spectra of the harmonics obtained using the insertion of the MSM on the path of the heating beam. The MSM was tilted at different angles (0°, 15°, 30°, and 45°) with regard to the heating beam. The tilting of the MSM led to a decrease of the sizes of single jets and to the growth of the number of jets. A characteristic QPM-induced growth of the groups of harmonics in different spectral ranges was observed in all these cases. The maximum enhancements were achieved for the H37, H41, H43, and H51, respectively. The maximally enhanced harmonic orders approximately corresponded to the rule when the product $L_{coh} \times q_{qpm}$ (where L_{coh} corresponded to the sizes of single jets) was maintained constant at a fixed electron density (since $L_{coh} \times q_{qpm} \approx 1.4 \times 10^{18}/N_e$). A significant enhancement of the groups of harmonics compared to the extended imperforated plasma was due to involvement of the multijet plasmas in the harmonic generation allowing the achievement of QPM.

To analyze the role of driving pulse intensity on the variations of QPM conditions different energies of these pulses were applied. The aim of these studies was to define whether the increase of the cross-section of tunneling ionization caused by the growth of driving pulse intensity leads to the variation of the QPM conditions due to the growing number of free electrons. More discussion on that matter will be presented in the following subsections. Here we discuss the observation, which pointed out the insignificance of the tunneled electrons that additionally appeared in the variations of QPM at the conditions when the driving pulse energy was changed from 1.85 to 5 mJ, which corresponded to the intensities ~3 × 10^{14} and ~9 × 10^{14} W cm^{-2}, respectively.

Figure 7.2 Harmonic spectra generated in different multijet silver plasmas produced by tilting the MSM. Upper panel shows the harmonic spectrum generated in the 5 mm Ag plasma. *Source*: Ganeev et al. 2015 [24]. Reproduced with permission from American Physical Society.

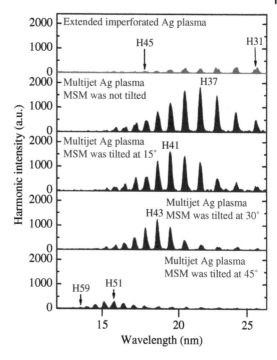

Figure 7.3 shows two harmonic spectra obtained in the eight-jet Ag plasma using the 1.85 and 5 mJ driving pulses. The q_{qpm} in these two cases were approximately the same (H37 and H39). The difference between these two spectra is mostly related with the broadening of the envelope of enhanced harmonics in the latter case, which points out the decrease of selectivity of a small group of harmonics due to less favorable conditions of QPM for different parts of the driving beam. Another observation is related with the difference in the divergences of the harmonics generated in the multijet and imperforated plasmas. In the case of the 1.85 mJ pulse, the sizes of these harmonics were approximately equal to each other (see the inset in Figure 7.3a showing the spatial distributions of H37 in the far field). The growth of the driving pulse energy from 1.85 to 5 mJ led to a significant spatial modulation of the harmonics generated in the extended plasma, while the sizes of QPM-enhanced harmonics generated in the multijet plasma showed some insignificant broadening compared with the previous case (inset in Figure 7.3b).

Thus the growth of the tunneling ionization cross-section leading to the appearance of additional free electrons in the multijet medium did not spoil the conditions of QPM. Figure 7.4 (empty circles) shows the near similarity of the q_{qpm} along a broad range of variations of the driving pulse energy. A considerably different behavior of QPM was observed once the heating pulse energy was changed, which led to variation of the electron density of

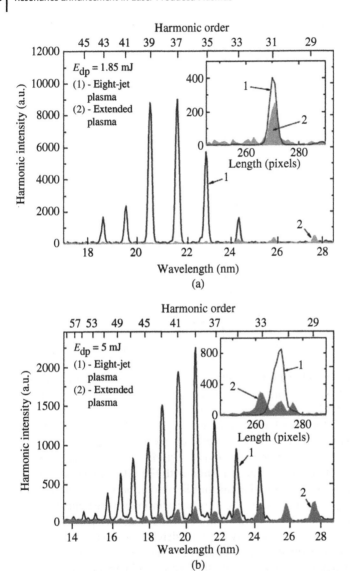

Figure 7.3 Variation of the spectral and spatial shapes of the harmonics generated in the extended and multijet silver plasmas using the (a) 1.85 mJ and (b) 5 mJ driving pulses. Insets in (a) and (b) show the far-field images of the 37th harmonic in the cases of (1) multijet and (2) extended imperforated plasmas. One can see the equal divergences of the harmonics generated in the extended and perforated plasmas in the case of the 1.85 mJ driving pulse, while in the case of the 5 mJ pulse, the harmonic beam from the extended plasma became more divergent and strongly modulated compared with the one generated from the eight-jet plasma. *Source*: Ganeev et al. 2015 [24]. Reproduced with permission from American Physical Society.

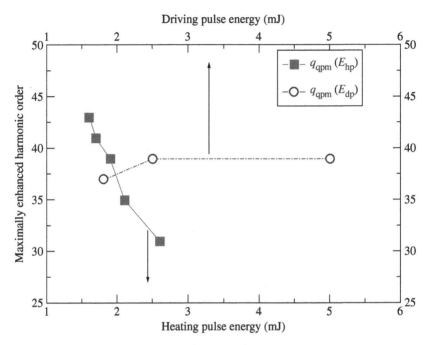

Figure 7.4 Variations of maximally enhanced harmonic order at different energies of heating (filled squares) and driving (empty circles) pulses. One can see that the threefold growth of driving pulse energy (from 1.85 to 5 mJ) did not significantly change the q_{qpm}. In the meantime, the q_{qpm} was strongly dependent on the variation of the heating pulse energy. Solid and dotted lines are inserted for better viewing of the influence of heating and driving pulse energies on the QPM conditions. *Source*: Ganeev et al. 2015 [24]. Reproduced with permission from American Physical Society.

plasma (Figure 7.4, filled squares). The comparative studies of the harmonic spectra generated from the eight-jet silver plasma at different fluencies of the heating pulse (370 ps) on the target surface were accomplished by changing the energy of heating radiation using calibrated filters. The energy of the heating pulse was varied to analyze the influence of this parameter on the dynamics and shape of the envelope of harmonic distribution using the same multijet plasma. The group of harmonics gradually tuned toward the lower orders (i.e. longer-wavelength region) with the growth of the heating pulse energy. A decrease of N_e at weaker excitation of the target should lead to the optimization of the QPM for higher q_{qpm} to keep the product $q_{qpm} N_e$ unchanged at the fixed spatial characteristics of plasma jets, which was demonstrated in Figure 7.4. Thus the additional electrons appearing during stronger ablation of targets significantly influence the QPM conditions. However, whether these additional electrons were attributed to those produced during either ablation or tunnel ionization of additional silver particles remains unclear. In the following subsection, we address this issue by analyzing the spatial distribution of harmonics in the extended plasma in two separate cases.

7.2.3   Modeling HHG in Plasma Plumes

A 3D nonadiabatic numerical model [25] was adapted and used to simulate the propagation of the driving pulse in the plasma medium and the process of HHG. The code follows the physical process and proceeds in three main steps.

1) The propagation of the driving laser pulse is governed by the wave equation:

$$\nabla^2 E_1(r,z,t) - \frac{1}{c^2}\frac{\partial^2 E_1(r,z,t)}{\partial t^2} = \frac{\omega^2}{c^2}(1 - \eta_{eff}^2)E_1(r,z,t) \tag{7.4}$$

where η_{eff} is the effective refractive index of the medium expressed by:

$$\eta_{eff}(n_a, n_e, r, z, t) = \eta_2(n_0)I(r,z,t) - \frac{\omega_p^2(n_e, r, z, t)}{2\omega^2} \tag{7.5}$$

where the first term accounts for the intensity-dependent optical Kerr-effect, while the second is the plasma term which depends on the total electron density n_e. In the experimental configuration used here one has to take into account multiple ionization of the Ag. The time-dependent populations of the species existing in the interaction region (neutrals, Ag^+, Ag^{2+}, Ag^{3+}) were calculated by solving the system of rate equations:

$$dn_0(t)/dt = -n_0(t)w_0(t),$$
$$dn_i(t)/dt = n_{i-1}(t)\,w_{i-1}(t) - n_i(t)\,w_i(t), i = 1, 2$$
$$dn_3(t)/dt = n_3(t)w_3(t) \tag{7.6}$$

The ionization rate and respectively the density of the i-times ionized species were denoted by w_i and n_i. For the intensities used in HHG experiments one can reasonably limit the calculation to triple ionized Ag. The w_i ionization rates are calculated with the Ammosov–Delone–Krainov (ADK) model [26] by using the ionization potentials for Ag, Ag^+, and Ag^{2+}. The total number of electrons is estimated as:

$$N_e(t) = \sum in_i(t). \tag{7.7}$$

2) In a second step the polarization $P_{nl}(r,z,t)$ of a single atom in interaction with the propagated laser pulse $E_1(r,z,t)$ was calculated by solving the Lewenstein integral [27]. This uses the strong-field approximation and the single active electron approximation. In this framework, in calculations it was always needed to choose the "active species," either Ag^+ or Ag^{2+}, so that in the Lewenstein integral the amplitude of the initial and final state can be that of the Ag^+ (or Ag^{2+}). Despite the fact that this seems to be a drawback, yet it allows to distinguish between (and compare) the contributions to the HHG coming from Ag^+ and Ag^{2+}.

3) The harmonic field is calculated as the coherent sum of the elementary dipole emissions. The propagation of the harmonic field is described by the wave equation:

$$\nabla^2 E_h(r,z,t) - \frac{1}{c^2}\frac{\partial^2 E_h(r,z,t)}{\partial t^2} = \mu_0 \frac{\partial P_{nl}(r,z,t)}{\partial t^2} \tag{7.8}$$

where the source term contains the nonlinear polarization of the medium calculated in step 2.

In the end one can obtain the harmonic field in every (r,z) point in the interaction region and in time/frequency. This allows following the build-up of the harmonic field in the medium and the evolution of the phase-matching conditions for selected harmonic orders.

Since neutral Ag has low ionization potential, the medium is completely ionized to Ag^+ mainly by the heating pulse and by the low-intensity optical cycles in the leading edge of the driving pulse. It is therefore reasonable to assume in modeling that the species responsible for harmonic generation could be Ag^+ or Ag^{2+}. Indeed, it was calculated that at high driving pulse intensities also the Ag^{2+} may further ionize and give a contribution to the HHG process. However, as seen in Figure 7.5, for a peak intensity of 3.5×10^{14} W cm^{-2} Ag^+ completely ionizes in the leading part of the 94 fs pulse, while the fractional population

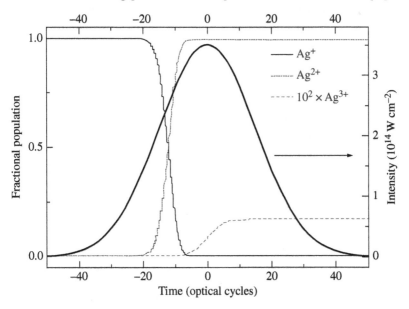

Figure 7.5 Time evolution of pulse envelope and fractional populations for a medium initially composed of Ag^+ ions. In this case harmonics are produced mainly by Ag^+ in the leading part of the driving pulse. *Source:* Ganeev et al. 2015 [24]. Reproduced with permission from American Physical Society.

of Ag^{3+} reaches a value of only 0.2×10^{-3} by the end of the laser pulse. It is therefore reasonable to assume that at these laser intensities the main HHG signal comes from Ag^+ in the presence of electrons produced by ablation and by femtosecond pulse ionization.

The problem of finding the real time–space distribution of ions and electrons after ablation in the multijet geometry is not a trivial one and will be discussed in Section 7.2.4. We intend here to emphasize solely the effects of the presence or absence in the HHG process of the electrons produced previously by ablation. We present in the following the results obtained in modeling two extreme cases: (i) At the time when the driving pulse interacts with the plasma no ablation electrons are present; in this case one can consider HHG in Ag^+ ions and only electrons produced by further ionization are present in the medium. (ii) Again Ag^+ are the active ions producing harmonics but all the electrons resulted from the ablation are present at the time when the driving pulse arrives, and their concentration and spatial distribution is the same as the concentration and spatial distribution of Ag^+ ions along the interaction region; here this distribution is uniform. Probably the physical reality is somewhere between these extreme cases: in a multijet configuration, electrons originating from ablation will have a broader distribution than the active ions that will follow more closely the step-like configuration given by the mask.

The simulations were performed in an extended plasma medium; therefore, the results presented below allow a qualitative interpretation instead of direct comparison with the experimental data. In Figure 7.6, the build-up of two harmonic orders along the plasma length (z-axis) and radius of beam waist (r axis) in the absence and in the presence of ablation electrons are compared, respectively. The other generation conditions are identical, namely $I_{dp} = 5 \times 10^{14} \, W \, cm^{-2}$, $p = 1 \, Torr$ (i.e. $3.3 \times 10^{16} \, cm^{-3}$), Ag^+ as active species. Harmonic order H25, which is a plateau harmonic, is continuously constructed along the cell and is the strongest at the exit of the interaction region. Without ablated electrons H25 is built on-axis, while with the electrons present the constructive build-up region is moved off-axis. In the case of the cutoff harmonic order H45, the presence of ablated electrons causes more fundamental change in the way how the harmonic is constructed. While without preexistent electrons H45 constructs all along the cell with long coherence length, the presence of ablation electrons causes the shortening of the coherence length according to Eq. (7.3) and characteristic Maker fringes [28] appear.

Figure 7.7 has the same structure as Figure 7.6, but the pressure was changed to 2 Torr ($6.6 \times 10^{16} \, cm^{-3}$). When no ablation electrons are considered, the way how both H25 and H45 are constructed is similar to the 1 Torr case. However, considering that all the electrons from ablation are present in the cell, the scenario for harmonic build-up changes. According to Eq. (7.3), $L_{coh} \sim 1/N_e$, so twice as much electrons result in half coherence length, therefore already for H25 one can start seeing the Maker fringes. The effect is more visible for H45,

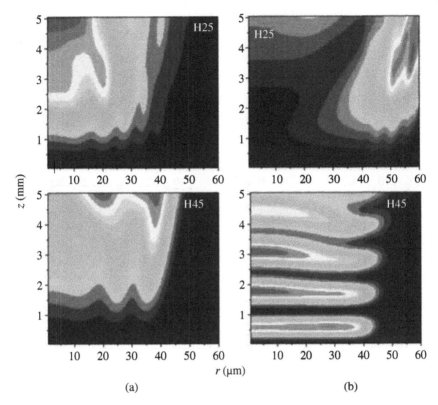

Figure 7.6 Spatial (r,z) maps of harmonic orders H25 (plateau) and H45 (cutoff). Parameters: $I_{dp} = 5 \times 10^{14}$ W cm^{-2}, $p = 1$ Torr (3.3×10^{16} cm^{-3}), Ag^{+}. (a) No ablated electrons are present and (b) all ablated electrons are included. *Source*: Ganeev et al. 2015 [24]. Reproduced with permission from American Physical Society.

where the coherence length shortening almost exactly follows the rule from Eq. (7.3). In addition, one can see that the intensity of H45 seriously drops in the second half of the plasma medium. This can be explained as follows: the plasma medium is placed after the focus of the driving pulse, therefore the pulse intensity will decrease due to geometrical defocusing. In addition, pulse intensity further reduces also due to the strong plasma defocusing as propagation proceeds. H45 becomes a harmonic in the cutoff and less dipole is emitted in the second half of propagation.

In Figure 7.8, the influence of the driving pulse intensity on the QPM conditions is presented. When the presence of ablated electrons is not taken into account, the harmonic field for H35 builds up constructively all along the plasma length, similarly to the cases presented in Figures 7.6 and 7.7. When all the ablated electrons are present during HHG, the Maker fringes appear,

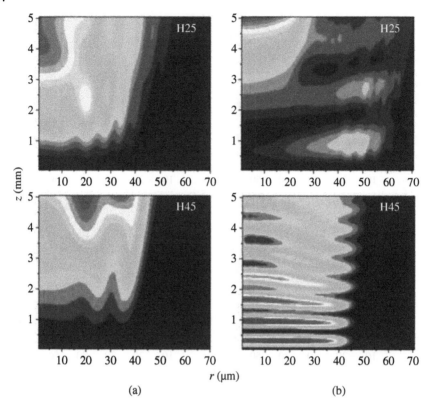

Figure 7.7 Spatial (r,z) maps of harmonic orders H25 (plateau) and H45 (cutoff). Parameters: $I_{dp} = 5 \times 10^{14}$ W cm^{-2}, $p = 2$ Torr (6.6×10^{16} cm^{-3}), Ag$^+$. (a) No ablated electrons are present and (b) all ablated electrons are included. *Source:* Ganeev et al. 2015 [24]. Reproduced with permission from American Physical Society.

as expected. The conditions of QPM are less favorable at higher intensity. The reason is the appearance of additional electrons during tunnel ionization (just similarly to gas HHG). The increased number of tunneled electrons decreases the coherence length. One can see five fringes when $I_{dp} = 3 \times 10^{14}$ W cm^{-2} and six fringes when $I_{dp} = 5 \times 10^{14}$ W cm^{-2}. The radial extension of the generation volume is larger for higher intensity, which comes as a compensation and might increase the total harmonic yield when measuring the radially integrated power spectrum. The overall influence of the driving pulse intensity on the fringe-structure is not significant, which is in qualitative agreement with the experimental observation that the driving pulse intensity has little influence on the QPM conditions as well as on the maximally enhanced harmonic order.

Efficient QPM induced by a perpendicularly propagating terahertz pulse in the case of a very similar Maker fringes harmonic structure has been suggested

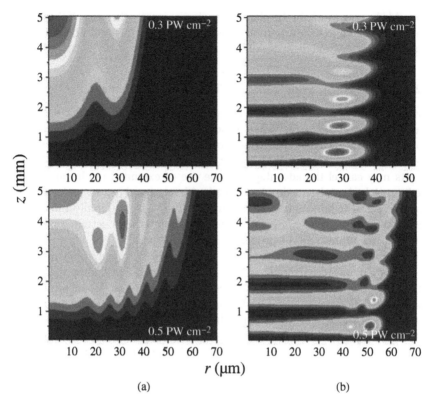

Figure 7.8 Spatial (*r,z*) maps of harmonic H35. Parameters: $I_{dp} = 3 \times 10^{14}$ W cm^{-2} (two upper panels) and 5×10^{14} W cm^{-2} (two bottom panels), $p = 2$ Torr (6.6×10^{16} cm^{-3}), Ag$^+$. (a) No ablated electrons are present and (b) all ablated electrons are included. *Source*: Ganeev et al. 2015 [24]. Reproduced with permission from American Physical Society.

in Ref. [29]. In that study, the wavelength of the terahertz pulse was related to the coherence length, in analogy with the MSM applied in the present experimental arrangement. The above results suggest a possible mechanism for the QPM observed in experiment for multijet configurations. Indeed, the presence of ablated electrons produces a periodic phase matching with an almost constant coherence length. The presence of only ionization electrons, without ablation electrons does not induce a periodic variation of harmonic intensity. This can be clearly seen in Figures 7.6 and 7.7 for H45, but also for lower orders. If an ablated region containing Ag$^+$ ions extends for a distance equal to the coherence length of H45, for example, one will have H45 and neighboring harmonics enhanced. The next nonablated region will not contain ions but will restore the phase-matching condition in the presence of ablated electrons. We emphasize that the presence of ablated electrons is not only favorable but necessary in the phase-restoring process; otherwise, if the pulses would propagate in vacuum, their phases would vary insignificantly and the phase restoring would be impossible.

7.2.4 Discussion and Comparison of Theory and Experiment

HHG of laser radiation in extended gaseous and plasma media has long been considered as an inefficient process also due to the medium length, which exceeds the coherence length of high-order harmonics. Particularly, the abundance of free electrons appearing in plasma during laser ablation of targets may cancel the advantages of the use of extended nonlinear optical medium. A higher plasma density created with increased heating pulse energy may decrease the harmonic yield because the PMM grows rapidly and causes destructive interference between the dipole generated at a certain z position in the plasma plume and the propagated harmonic field.

The onset of destructive interference can be shifted to higher plasma densities, and enhanced harmonic yields can be obtained by dividing the whole interaction length of plasma d into M sections of thickness d/M and separating them such that dispersion-induced changes in the laser phase and amplitude between two adjacent sections can shift the phase of the atomic dipole oscillations of the qth-order harmonic by π at the input of the next ablated section with respect to the exit of the previous one. As in the case of gas medium, separation of the extended plasma formation into a few small-sized plasma jets may restore the proper phase relationship between the driving and harmonic waves and allow for further enhancement of the yield in different spectral ranges. In order to create QPM conditions, one has to maintain a modulation of the coupling between the driving and harmonic waves, which is periodic along the generation path. At these conditions, the enhancement of harmonics will occur only when the interference between the propagated harmonic signal and the newly generated dipole becomes constructive along the path of propagation through the whole set of jets. That is what was analyzed in the experimental part of the discussed work.

Below we address the following issue directly related to the usefulness of the proposed method. As it has already been highlighted, the electrons in the plasma are composed of those produced during ablation of the target and those appearing during tunnel ionization of plasma components by the driving pulse. The question is, how can one estimate and separate the contribution of either of these groups of electrons on the PMM of HHG in the laser-produced plasma? The discussed work gives indications for resolving this problem.

Distinguishing the role of electrons produced during ablation from those coming from tunnel ionization of plasma in the variation of the wave numbers of interacting driving and harmonic waves along the whole medium is a complicated issue, yet clearly understandable in the frames of the reported experimental results of plasma HHG. In the following, we present two scenarios to emphasize the role of the two groups of electrons in the variation of the QPM conditions in the case of plasma harmonic generation.

7.2.4.1 Scenario 1

If one considers a simple addition of tunneled electrons to the already existing electrons from laser ablation, then the plasma dispersion should significantly increase the PMM of HHG. The barrier suppression intensity of silver atoms is well below 1×10^{14} W cm^{-2}, which means that all neutrals in the plasma will be ionized by the heating pulse and the initial part of the driving pulse. Once we assume that the density of free electrons at these conditions becomes the order of plasma density (in the discussed case $\sim 5 \times 10^{17}$ cm^{-3}, which was estimated using the molecular dynamics code calculations), then the PMM should fully stop the growth of harmonic yield at the initial stages of propagation of the driving pulse through the extended plasma. Particularly for the 30s harmonics (20–25 nm), L_{coh} at this electron density will be equal to ~ 0.1 mm (see Eq. (7.3)). It means that at larger sizes of medium the PMM should stop the growth of harmonic yield in this spectral range.

Once one considers the restrictions of HHG from the point of view of the barrier suppression of singly and doubly charged ions, then even more frustrating predictions could be drawn. The barrier suppression intensities of the Ag$^+$ and Ag^{2+} are equal to 1×10^{14} and 3.7×10^{14} W cm^{-2}, respectively. It means that at $I_{dp} > 7 \times 10^{14}$ W cm^{-2} used in discussed experiments, both singly and doubly charged ions will be further ionized. The electron density at these conditions will be three times the plasma density ($N_e \sim 1.5 \times 10^{18}$ cm^{-3}). One can expect an even worse scenario for the phase matching in the extended plasma once one assumes that all these "transient" electrons cause the growth of plasma dispersion during propagation of the driving pulse. The PMM in that case will be $\Delta k \approx 1000$ cm^{-1}. At these conditions, the harmonic yield becomes limited to the gain achieved during propagation of a very short distance. One can calculate that at this N_e the coherence length of the 30s harmonics will be ~ 0.03 mm. After this distance, the fluctuations of harmonic intensity are expected in the range between 0 and I_H (0.03 mm).

Fortunately, this scenario does not happen. Numerous experiments using extended plasma media (0.5 mm and longer) have shown that the harmonic emission increases in these conditions. Moreover, the influence of the plasma length on the harmonic yield analyzed in Refs. [30, 31] has demonstrated a steady growth of conversion efficiency. Some declination from the anticipated quadratic rule ($I_H \propto d^2$) was observed at the lengths of a few mm. However, the saturation did not cancel the harmonic emission.

7.2.4.2 Scenario 2

The electrons that originated from ablation had (at least partially) gone from the plume due to the higher velocity compared to the heavier ions at the moment when the plasma cloud reached the optical axis of the femtosecond pulse propagation. In this scenario one can disregard this group of electrons. At these conditions, the main role in PMM would be related (as in the case of gas HHG) with the involvement of the tunneled electrons only. The driving laser pulse experiences the dispersion from the free electrons generated in plasma during tunnel ionization. Probably, the plasma concentration at the moment of propagation of the femtosecond pulse was not as high as was assumed from the molecular dynamics code simulations (5×10^{17} cm^{-3}). The incorrectness of former calculations could be related with the long delay between the beginning of ablation and the plasma formation at the optical axis of the driving pulse. Once we assume that it was a few times smaller (8×10^{16} cm^{-3}), the results presented in Figure 7.4 could be explained as follows. The increase of heating pulse energy led to the growth of plasma concentration and correspondingly the concentration of electrons appeared during tunnel ionization (not ablation; those disappear from the plasma due to higher velocity in accordance with our scenario). Correspondingly, the maximally enhanced harmonic order significantly depends on the heating pulse energy. As for variations of driving pulse energy, the cross-section of tunnel ionization of ions does not significantly change at these conditions, since the used intensity has already been larger than the barrier suppression intensity. That is why we did not see significant variation of maximally enhanced harmonic order when varying the driving pulse intensity.

Which of these two scenarios better describes the real experimental situation? We presented here two extreme cases, but most probably a combination of the two scenarios applies. The most reasonable picture is that we have a periodic variation of both electrons and ions, following the periodicity of ablated and nonablated regions. The electron density distribution along the whole interaction region has lower contrast and broader regions of high density (as they spread more easily also along the driving beam axis), while ions have the regions of high density almost equal to the length of the ablated regions and density almost zero in the nonablated regions.

Correlating the experimental results with the qualitative interpretation of the simulations one cannot ignore the electrons produced by ablation, since the enhancement of the harmonics in multiregion configurations was obtained by using masks of dimensions comparable to the coherence lengths seen in the case of combining the influence of the two groups of electrons (see Figures 7.6 and 7.7). Despite the fact that simulations were performed in extended plasma, the results indirectly confirm the efficiency of the experimental multijet geometry in the following way: simulations performed by taking into account the presence of ablated electrons clearly showed the short coherence length for cutoff harmonics. Applying a mask with periodicity close to the coherence length of the selected spectral region creates the conditions for phase restoring in the domains where the PMM dominates, finally resulting in QPM with measurable enhancement of the harmonic yield, as seen in the experimental results.

7.3 Influence of a Few-Atomic Silver Molecules on the High-Order Harmonic Generation in the Laser-Produced Plasmas

7.3.1 Introduction

Harmonic generation using propagation of the ultrashort pulses through the laser-produced plasma has become an attractive method for generation of the coherent XUV radiation. Meanwhile, this method continues to demonstrate the insufficient conversion efficiency toward the short-wavelength region, similar to the HHG in the gaseous media.

The ablation of NP-contained targets offers the opportunity in the growth of the high-order harmonic yield, similar to resonance-induced enhancement. The crucial requirement here is the "mild" ablation of NPs to avoid the disintegration and aggregation of these species. Additionally, the high fluencies of heating pulses for laser ablation may cause the appearance of a large amount of free electrons, which is the main impeding factor preventing the efficient HHG in the LPP due to the growing PMM between the interacting waves. The formation of NP-contained plasmas has been studied using various methods and targets. Although NPs have already been used as the media for HHG, the mechanism of the enhancement of harmonic yield from these media is yet to be clearly understood.

The higher cross-section of recombination of the electron with parent particles in the NPs, compared with the single-particle atomic and ionic species appearing in LPP, is advantageous in allowing the growth of harmonic emitters. On the other hand, the inner particles of large NPs may not participate in the enhancement of the harmonic yield. Only the electrons, which tunnel out from the surface of NP, will be able to complete their trajectory, whereas the electrons tunneling out from the atoms inside of the NP will find it difficult to complete its trajectory due to multiple collisions on the way. Hence, only surface atoms of NPs can provide the electrons for the HHG. Thus, the inner atoms of large NPs do not participate in HHG, but rather absorb the XUV emission [32]. There is also a collective wavefunction of the electrons in the NP, which acts as the source and the place of recombination for electrons, the effect of which was confirmed for HHG too.

Another problem is the relatively small velocity of NPs, which leads to the notable delay of the appearance of these species in the area of ultrashort pulses propagation during HHG experiments compared with the single-particle components of plasma possessing much higher velocity ($\sim 10^2$ and $\sim 10^5$ ms^{-1}, respectively) due to significantly larger masses of the former species [33, 34]. It takes a few tens of microseconds for NPs to reach the area of interaction (0.2–1 mm above the target surface), contrary to a few tens of nanoseconds in the case of single-particle species [35]. In the meantime, multiple studies have shown that the harmonic yield from the LPPs produced on the NP-containing targets enhances compared with the bulk targets of the same origin while using the short delays between the driving and heating pulses (see, for example, Ref. [36]). It remains a puzzle as to which components are responsible for this enhancement.

Although the morphological studies of deposited material have shown that the "mild" ablation of NPs allows preserving them in the plasmas [37], nothing could be said about their disintegration onto the small components during heating, melting, and evaporation. It has been proven that the physical characteristics of those plasmas, such as the presence of NPs, directly affect the harmonic yield. Meanwhile, it became unclear whether those large NPs with the observed sizes of deposited material in the range of a few nanometers to a few tens of nanometers could be involved in the enhancement of harmonic generation. Since the HHG studies were carried out using the small delays between the ignition of LPP and the propagation of driving pulses ($\sim$30 ns), the involvement of the species with the number of particles of $\sim 10^4$ and larger in the HHG has raised serious doubts.

In this section, the studies of the bulk silver and silver NPs ablated at conditions suitable for the efficient HHG of the 806 and 1310 nm pulses propagating through the LPP are discussed [38]. To analyze the morphological properties of NPs, the transmission electron microscopy (TEM) of ablated species was used. The mass spectroscopy analysis of LPP allows defining the presence of small clusters in the area of interaction with the propagating driving pulses. These studies point out the coincidence of the presence of small silver molecules containing a few atoms or ions and the growth of conversion efficiency, which may lead to the assumption about these species as the probable sources of enhanced harmonic yield from the silver-contained plasmas.

7.3.2 Experimental Setup

The experimental setup for harmonic generation comprised the Ti:sapphire laser (806 nm, 10 Hz), OPA, and HHG scheme using the propagation of the amplified near infrared (NIR) signal radiation from OPA (1310 nm) and its second harmonic (H2, 655 nm) through the extended LPP (Figure 7.9). Part of the amplified uncompressed radiation of pump laser (806 nm) with the pulse energy of 4 mJ was separated from the whole beam and used as a heating pulse for homogeneously extended plasma formation using the 200 mm focal length cylindrical lens installed in front of the targets placed in the vacuum chamber. The intensity and fluence of the heating pulses on the target surface were varied up to 4×10^9 W cm^{-2} and 1.5 J cm^{-2}, respectively. The HHG experiments were carried out using the 1 mJ, 70 fs, 1310 nm signal pulses and 0.2 mJ, 83 fs, 655 nm pulses of the second harmonic of 1310 nm radiation, as well as 806 nm + H2 (806 nm) pump. These pulses were focused $\sim$0.2 mm above the target surface.

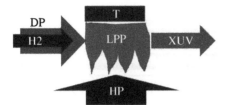

Figure 7.9 Experimental setup for nanopowder and bulk targets ablation for efficient harmonic generation. HP, heating pulse; DP, driving pulse; H2, second harmonic pulse; XUV; T, target; LPP. *Source*: Ganeev 2017 [38]. Copyright 2017. Reproduced with permission from World Scientific Publishing Co Pvt Ltd.

The intensities of focused 1310 and 655 nm pulses inside the LPP were 2×10^{14} and 5×10^{13} W cm^{-2}, respectively. In the case of 806 nm + H2 pump, the intensities of two waves were 6×10^{14} and 1.2×10^{14} W cm^{-2}, respectively. The harmonic emission was analyzed using an XUVS. The details of HHG setup using OPA were described in Ref. [39].

The laser ablation of bulk Ag and Ag NPs by picosecond pulses was studied at the conditions allowing the efficient HHG. Silver NPs (99.95% purity, diameter 20–30 nm) and 5 mm long bulk silver were used as the targets for plasma formation. The NP powders were prepared as targets by mixing with the cyanoacrylate-based glue and coating the surface of 5×5 mm^2 glass substrate. The material directly surrounding the NPs was a polymer. The NP-carrying polymer begins to ablate at relatively low fluencies, resulting in the lower laser intensity required for the formation of the optimal plasma medium for the HHG. This feature allowed for easier formation of the plasma conditions, which resulted in a better conversion efficiency of the lower order harmonics from the NP-contained LPPs compared with the harmonic generation from the ablation of bulk targets of the same origin. The 5 mm long bulk silver was ablated to demonstrate the formation of the NPs and small clusters suitable for the efficient HHG. The ablated material was deposited on nearby glass plates. The TEM images of deposited NPs were analyzed at different fluencies of the heating pulses. The ToFMS of the ablated species was performed to show the appearance of the small clusters originated from the initial NPs and bulk material.

7.3.3 Harmonic Generation and Morphology of Ablated Materials

The propagation of 806 nm and H2 (403 nm) pulses through the plasma produced on the surface of bulk silver using the same fluence (0.6 J cm^{-2}) of heating radiation allowed the generation of the plateau-like spectrum of odd and even harmonics (Figure 7.10a, upper raw image showing H11–H38 harmonics). The use of a driving radiation consisting of the laser field and its second harmonic is

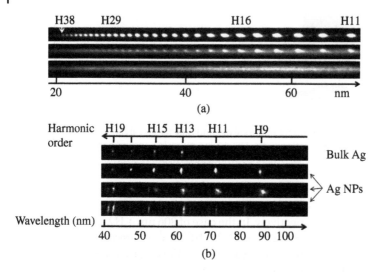

Figure 7.10 (a) Raw images of harmonic spectra using two-color pump (806 + 403 nm) and plasma emission from ablated Ag bulk target at optimal (upper panel) and overexcited (middle panel) ablation. Bottom panel shows the plasma emission without propagation of the driving pulses through the plasma. (b) Raw images of the low-order harmonic spectra from ablated bulk Ag (upper panel) and Ag NPs at optimal ablation (second panel, 0.5 J cm⁻²), stronger ablation (third panel, 0.8 J cm⁻²), and overexcitation (bottom panel, 1.2 J cm⁻², no harmonic emission) of NP target. *Source*: Ganeev 2017 [38]. Copyright 2017. Reproduced with permission from World Scientific Publishing Co Pvt Ltd.

one of the reliable methods for enhancing the harmonic yield. The explanation of HHG enhancement in two-color orthogonally polarized fields was reported in Ref. [6], in terms of the selection of a short quantum path component, which has a denser electron wave packet, and higher ionization rate compared with the single-color pump (SCP) case. Harmonic emission dominated over plasma emission along the whole range of analyzed XUV spectrum (25–65 nm). The growth of heating pulse fluence (second panel, 1 J cm⁻²) led to the appearance of the plasma emission continuum, while the HHG conversion efficiency was decreased. The bottom panel shows the plasma emission spectrum obtained at similar conditions of laser ablation and without the propagation of the driving pulses. The raw images shown in this and other figures are presented for better viewing of the obtained data and direct visual comparison of the variations of harmonic and plasma emission at different conditions of laser ablation.

The use of the LPP produced on the NP target allowed the growth of lower order harmonic yield compared with the above case (Figure 7.10b, two upper raw images corresponding to the bulk Ag and Ag NP ablation, respectively). These experiments were carried out using the SCP (806 nm). The growth of the fluence of heating pulses on the NP target led to the appearance of plasma emission lines alongside the decrease of harmonic emission (third raw image).

Further growth of the fluence of heating radiation led to entire dominance of the plasma emission over the harmonic generation (bottom raw image). At these conditions of plasma formation, the large amount of free electrons appearing during ionization of various species existing in the LPP leads to a significant PMM between the interacting waves of driving and harmonic pulses. The femtosecond pulses propagating through such LPPs do not cause the conditions of the accumulation of generating harmonics. The distance (i.e. coherence length) during which the macroprocesses play a decisive role in HHG becomes too short, which leads to the cancellation of harmonic emission during propagation through each next similar distance corresponding to the coherence length of higher order harmonics. These experiments were carried out using the SCP (806 nm) of plasma when only odd harmonics appeared in the XUV spectra. The harmonics from ablated Ag NPs were extended up to the H45 at optimal conditions of ablation (0.5 J cm^{-2}) and fully disappeared at overexcitation of the NP target (1.5 J cm^{-2}).

Most of the harmonic experiments using NIR pulses were carried out by applying the TCP of LPP (1310 + 655 nm). The application of the double-beam configuration to pump the extended plasma, apart from the abovementioned reasons, is related to the small energy of the driving NIR signal pulses ($\leq$1 mJ). The $I_H \infty \lambda^{-5}$ rule (I_H is the harmonic intensity and λ is the driving field wavelength) led to a significant decrease of harmonic yield in the case of the longer wavelength sources compared with the shorter wavelength (806 nm) pump and did not allow the observation of strong harmonics from the 1310 nm pulses. Because of this the second-harmonic generation of signal pulse was used to apply the TCP scheme (NIR + H2) for plasma HHG. The difference between TCPs and SCPs of the plasma produced on the surface of bulk silver is clearly seen in Figure 7.11a. The SCP led to generation of only a few odd harmonics (H17–H23) in the longer wavelength range of XUV (thick curve), while the TCP (1310 + 655 nm) allowed the generation of much stronger and denser harmonic spectrum (thin curve).

The formation of the best conditions for HHG in the analyzed plasmas allowed the generation of sum and difference harmonics during the three-color pump of bulk Ag plasma using signal and idler waves of OPA and 806 nm wave (Figure 7.11b, upper raw image, 1605 + 1607 + 806 nm), as well as the three-color pump of Ag NP plasma using signal, H2, and idler waves (bottom panel, 1310 + 655 + 2075 nm). These processes are extremely sensitive to the phase-matching conditions between a few interacting waves. The achievement of these conditions requires the definition of the optimal ablation and formation of the plasma containing some components, which can enhance the harmonic yield.

In these studies, the moderate fluence (0.6 J cm^{-2}) of 806 nm, 370 ps pulses was used, which corresponded to the conditions of efficient HHG in the plasmas produced during ablation of bulk and NP targets. The TEM

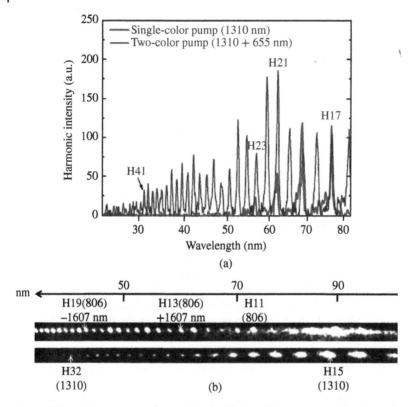

Figure 7.11 (a) Harmonic spectra at optimal ablation of bulk Ag using single-color pump (1310 nm, thick curve) and two-color pump (1310 + 655 nm, thin curve) of extended plasma. (b) Raw images of harmonic spectra in the case of the three-color pump of bulk Ag plasma (upper panel) and Ag NP plasma (bottom panel, see text). *Source*: Ganeev 2017 [38]. Copyright 2017. Reproduced with permission from World Scientific Publishing Co Pvt Ltd.

measurements of original silver NPs showed that the maximum sizes of these particles corresponded to those provided by the manufacturer. However, the NPs had heterogeneous size distribution centered at ~20 and ~5 nm. The ablation of NPs was compared with the ablation of bulk silver. The different fluencies of heating pulses caused the change of the pattern of deposition. At low fluencies on the bulk Ag surface (0.2 J cm^{-2}), the deposition represented a thin layer of the featureless silver film, which did not contain the NPs. The growth of fluence (0.4 J cm^{-2}) caused the appearance of small NPs with the sizes varied between 2 and 7 nm and centered at 3 nm (Figure 7.12a, upper panel). Further growth of the fluence (0.7 J cm^{-2}) led to the appearance of large NPs (>20 nm) and broadening of the size distribution of NPs (bottom panel). The average size of NPs increased up to ~7 nm. The ablation of Ag NPs at these three conditions caused a similar effect. At 0.4 J cm^{-2}, the ~15 nm

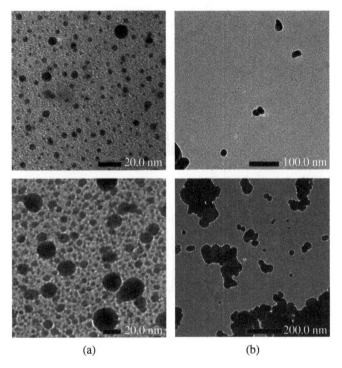

(a)   (b)

Figure 7.12 TEM images of deposited material during ablation of (a) bulk silver and (b) silver NPs at different fluencies of heating pulses (0.4 and 0.7 J cm^{-2} in the case of top and bottom panels, respectively). The lengths of black markers correspond to (a, top) 20 nm, (b, top) 100 nm, (a, bottom) 20 nm, and (b, bottom) 200 nm. *Source*: Ganeev 2017 [38]. Copyright 2017. Reproduced with permission from World Scientific Publishing Co Pvt Ltd.

NPs dominated in the deposited surfaces (Figure 7.12b, upper panel), while at higher fluence (0.7 J cm^{-2}), the aggregates of larger sized NPs were frequently seen in the TEM images (bottom panel).

The content of the plasmas produced during ablation of NPs and bulk silver was analyzed using the ToFMS. The mass spectra of ablated bulk Ag are presented in Figure 7.13a. The 3 ns pulses used for ablation during measurements of mass spectra were focused on the target surfaces to maintain the same fluence of heating radiation, which was used in the studies for the observation of efficient harmonic generation along a longer wavelength range of XUV. These mass spectra were dominated with the singly charged Ag$_1$ ions (upper curve), while containing some amount of singly charged Ag$_2$ molecules. The growth of heating pulse fluence above the optimal value (0.4–0.6 J cm^{-2}) led to significant changes in the mass spectra. At 1.3 J cm^{-2}, the presence of doubly charged Ag$_1$ ions became significantly larger (bottom curve), while some additional low-mass species appeared in the spectra. The Ag$_2$ molecules at these

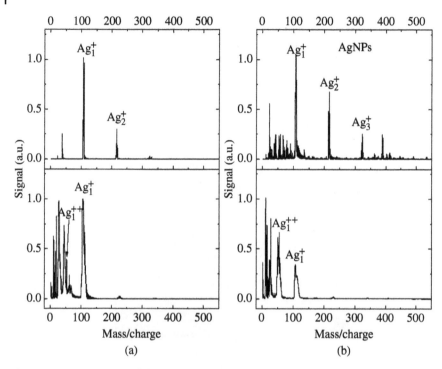

Figure 7.13 Mass spectra of ablated (a) bulk silver and (b) silver NPs at different fluencies of heating pulses (see text). *Source*: Ganeev 2017 [38]. Copyright 2017. Reproduced with permission from World Scientific Publishing Co Pvt Ltd.

conditions nearly disappeared in the mass spectra. We remind that at these conditions of plasma formation, the harmonic generation was significantly suppressed due to large amount of free electrons.

The mass spectrum of ablated Ag NPs at $0.6\,J\,cm^{-2}$ contained the low-mass molecules (Ag_1–Ag_3; Figure 7.13b, upper curve) together with some low-mass and high-mass components. The growth of heating fluence ($1.2\,J\,cm^{-2}$) caused significant changes in the mass spectra (bottom curve). Similar to bulk Ag, the dominant components became the singly charged and doubly charged Ag_1 without the presence of larger clusters.

7.3.4 Discussion

These studies were focused on the definition of the physical reasons of why application of silver plasma demonstrates it as an effective medium for the HHG. This medium has already been well explored in numerous publications; in particular, several studies were devoted to HHG in plasma plumes of silver or silver NPs [40–45]. Some studies explored HHG in silver plasmas under different conditions, like HHG from plasma multijets [13, 24] and HHG by two-color laser pulses in plasma [46]. Thus, the interest to the silver-contained plasmas is quite high.

The silver-containing plasma distinguishes from other metal plasmas. It allows generation of the strongest harmonics in the shorter wavelength region of XUV [40]. This prevalence of silver over other metals was not clearly explained in previous HHG studies, since no analysis of plasma components arriving in the area of interaction with femtosecond pulses was performed. In the following, we discuss the results of HHG, TEM, and ToFMS studies of the ablated species in the case of bulk Ag and Ag NP targets.

No measurements of the absolute value of conversion efficiency of the harmonics in the plateau-like range (20–50 orders) were carried out in those studies. Note that similar experiments using the narrow silver plasma have demonstrated the $\sim 10^{-5}$ conversion efficiency for those harmonics [41]. In discussed studies, the ablation of NP-containing targets by subnanosecond pulses showed the advanced properties of such plasmas. The application of SCPs and TCPs at these conditions resulted in strong odd and even harmonic generation. The free electrons already existing in plasma and appearing during ionization by driving pulses may decrease the coherence length of higher order harmonics. This process stops the accumulation of harmonic yield along the propagation of extended plasma. The density characteristics of the LPP at the delays of 25–45 ns were estimated using the hydrodynamic code HYADES [47]. For the heating pulse intensity of $2 \times 10^9\,\mathrm{W\,cm^{-2}}$ (i.e. at a fluence of $0.8\,\mathrm{J\,cm^{-2}}$), the ionization level and the ion density of the plasma produced on the bulk silver were estimated to be 0.3 and $2 \times 10^{17}\,\mathrm{cm^{-3}}$, respectively. Thus, the influence of the free electrons appearing during plasma formation on the HHG conversion efficiency could not be underestimated.

The increase of HHG conversion efficiency in previous studies of harmonic generation in the NP-contained LPPs was attributed to the change of the cross-section of these species compared with smaller sized particles. The cross-section of recombination of the accelerated electron with the parent particle in the case of NPs is higher compared with atoms. Still debated, this enhancement of harmonics was frequently observed in the ablation experiments using NP targets. The influence of the delay between pump and fundamental pulses on the efficiency of harmonic generation in bulk Ag and Ag NP plasmas was analyzed in the discussed studies. Initially, with a growth of the delay (5–25 ns), the intensity of harmonic radiation was increased. Further growth of the delay led to saturation (at 30–60 ns delay) and gradual decrease of harmonic intensity (at 100 ns delay). The optimal delay depends on the target material. In the case of both the used silver-contained plasmas, the maximal harmonic yield was observed at 45 ns. No opportunity was to analyze the HHG at much longer delays (i.e. of the order of a few tens of microseconds) to observe the influence of large NPs on the harmonic yield, as was reported in recent studies of the HHG in the ablated ZnS [35].

In the following, we address the time dynamics of the movement of the small and large agglomerates of atoms during laser ablation. To better understand the physics behind intense HHG from LPP, the first step would be the identification of the silver species that are responsible for the HHG. This is especially important in the case of NP-ablated plasmas where the composition of the plume is complex, typically involving not just NPs, but also large molecules, clusters, atoms, and ions. As expected from the three-step model of HHG [48], clusters can increase the efficiency of HHG in laser plasmas due to the larger cross-section of recombination of the accelerated electron with the parent particle. The laser pulses can be used to influence the outcome of ablation and ionization of nanomaterials either by acting on the laser/material interaction dynamics [49, 50] or by controlling the phase state of the ablation plasma [51, 52]. At these conditions, the separation of small molecules containing a few atoms of silver from the whole NP may lead to the appearance of the former species above the ablating surface much earlier compared with heavy NPs, which has been shown in the present mass spectroscopy studies of the ablated NPs.

The expansion of laser-generated NP plumes in vacuum was studied in Ref. [34]. The time-resolved CCD images have distinguished two features of component expansion generated by plasma and NP plumes separated in time. These images showed that the emission from atomic plume persisted at ~400 ns, while the NP-contained plumes persisted for a much longer time. The estimations of expansion velocity showed that the NP plume moved at a significantly lower velocity compared to the plasma plume dominantly containing single particles and the expansion velocities differed by ~25 times. One can assume the absence of large (3–20 nm) NPs in the area of ~0.3 mm above the ablated surface shortly (~30–50 ns) after the beginning of ablation. The only multiparticle species here were the few-atom-containing species. Correspondingly, their presence and participation in light–matter interaction allowed the achievement of the high yield of harmonics.

Earlier, it was found that there are three different population species with specific velocities in the expanding plasma [33]. The ion population is the fastest $(8 \times 10^4 \text{ ms}^{-1})$ and is attributed to Coulomb explosion. A neutral population follows adiabatic expansion with a similar or smaller velocity. Finally, the NP population has the slowest velocity. Depending on the sizes of NPs, there is a wide range of the velocities of these species. Lightest clusters (i.e. those containing a small amount of atoms) possess only a few times slower velocity compared with the single atoms and ions. Contrary to them, large NPs, which contain 10^4–10^6 particles, arrive in the area of interaction (~0.3 mm above the target surface) a few tens of microseconds from the beginning of laser ablation. Thus, only small clusters can be involved in the process of HHG, which demonstrated a notable enhancement of harmonic yield. In the meantime, large NPs do not participate in the process of HHG, but rather become the source of the formation of small clusters. The partial disintegration of large NPs during heating, melting, and evaporation causes the formation of the notably tiny clusters containing small amount of atoms, which became the main emitters of harmonics.

One can assume that the reported results may not be crucial for direct demonstration of the role of small molecules or small clusters for achieving the enhancement of HHG in silver plasmas. To prove that, a systematic investigation of the role of sample sizes in HHG yield should be studied at the same experimental conditions.

Furthermore, the results obtained in different targets containing dispersion of NPs with different average sizes and different cluster densities should be compared. Even in that case, the direct confirmation of the crucial role of small molecules or small clusters in the enhancement of harmonics should be followed with simultaneous experiments including both HHG and ToFMS in the same chamber and at the conditions exactly corresponding to the highest conversion efficiency of harmonics. Only in that case, one may draw a conclusive picture of the exceptional role of a few atomic species in the HHG in silver plasma plumes, especially compared with large NPs. The complexity of such studies did not allow to carry out such experiments in a single set of measurements. Moreover, one may understand that both the analysis of NPs deposited on substrates near the plume and the mass spectra reported in this article does not provide enough information about phenomena occurring during the laser–plasma interaction, since they can only attest the final state of the debris after this interaction occurred, as well as characterize the presence of different species in the plasma, which is not exactly similar to the one used in the HHG experiments. Anyway, the qualitative coincidence of the highest conversion efficiency and appearance of small molecules may draw the conclusion on the certain influence of these components of plasma on overall yield of harmonics.

ToFMS diagnostics was applied up to 10 000 mass/charge units. It was sufficient for the detection of clusters containing up to 100 atoms of silver. The 5 nm clusters contain ~27 000 atoms. Obviously, it was not possible to see such a large species using the available ToFMS equipment. However, there is no need to register such large NPs, since, as was underlined, they do not participate in HHG due to their absence in the plasma plume at the moment of femtosecond pulse propagation. Only a few-atoms containing species can be involved in the HHG due to abovementioned reasons. TEM data just indicate the presence of large NPs in the plasmas during the entire stage of plasma spreading out of target surface. These data cannot help in indicating the involvement of small-sized species in the process of HHG.

The density of ablated material was estimated to be ~6×10^{17} cm^{-3} in the case of bulk silver. One can assume that the ionization state of plasma at the moment of HHG was similar to the one defined using the hydrodynamic code HYADES, since during the first few tens of nanoseconds no significant changes in the charge state could be expected at the used conditions of ablation. The increased conversion to harmonics cannot be explained by the increased total density of the ablated material in the case of NP-coated targets because in that case a weaker ablation was used. The ablation of NP-contained target could be maintained at smaller intensity of heating pulses compared with the case of bulk silver. Thus, the optimal ablation conditions, at which the highest harmonic yield was obtained in the case of NP-coated target, were achieved at rather weaker intensity of heating pulse.

The velocity $\sim 4 \times 10^3 \, \text{ms}^{-1}$ corresponds to the fast components responsible for the optimal harmonic conversion, and not the large clusters. ToFMS diagnostics have shown Ag_2, and in some cases Ag_3 molecules, and no larger clusters. Actually, they are two- and three-atomic species, which can hardly be called low-mass clusters. So, one can admit that these molecules could be the sources of increased conversion. Some more diagnostics need to be done to better understand the role of different plasma species in the HHG. Especially in the case of proving the statements about the effect of clusters (or molecules), the density of the ablated material must be measured at the time of the generation of harmonics, either, for example, by interferometry or by light scattering. Even when knowing the bulk density, the size of the nanoclusters will be difficult to estimate. It has to be cleared as well, whether the interaction occurs in plasmas or in a cloud containing neutral atoms, molecules, and fragments. Further studies are needed to quantitatively prove the assumptions drawn from the coincidence of two effects, i.e. the presence of small molecules of silver and the enhancement of harmonic yield.

7.4 Controlling Single Harmonic Enhancement in Laser-Produced Plasmas

7.4.1 On the Method of Harmonic Enhancement

Generation of coherent short-wavelength radiation is the attractive field of optics allowing various scientific and practical applications. The most advanced method here is the HHG of laser radiation that has long been considered a promising tool to retrieve the structural and spectral information through analysis of the harmonics [53]. The role of atomic resonances in increasing the laser radiation conversion efficiency toward a shorter wavelength region was actively discussed in the framework of perturbation theory at the early stages of the studies of low-order harmonic generation [54]. Resonance enhancement introduces an additional possibility of increasing the conversion efficiency of a specific harmonic order by more than 1 order of magnitude. The resonance-induced enhancement of the harmonics of ultrashort pulses is one of the attractive features of coherent radiation frequency conversion in laser-produced plasmas [15], with the maximal enhancement factor of a single harmonic approaching 80 in the case of indium plasma. The use of LPP allowed the analysis of ionic transitions with high oscillator strengths.

Notice that no experiments with gases showing the resonance enhancement of single harmonics have been reported so far. The only study where the partial enhancement of a narrow component of harmonic was demonstrated during HHG in Ar gas was related to the influence of Fano resonances [55]. Why have resonant harmonics not been observed in gas media? The answer to this fundamental question is related to the basic principles of the role of ionic resonances in the enhancement of harmonics. Various factors prevented the observation of resonance enhancement. Among them are the self-absorption near the ionic transitions, which emit strong radiation in the XUV range, and the weak oscillator strength of these transitions. Furthermore, it is obvious that the choice of gases, mostly limited to a few noble gases, is dramatically narrow compared with that of solids, which includes almost all the periodic table predominantly containing the solid elements. Consequently, the probability of finding the appropriate resonance with high oscillator strength, which matches with some harmonics, is significantly smaller in the case of HHG in gases compared with HHG in plasmas. These conditions are rarely fulfilled, and if fulfilled they most probably occur in media produced during the ablation of thousands of solids rather than in a few gases. That is why the analysis of this phenomenon in plasma media has more chances of success.

In this section, we analyze a few plasma samples, which showed numerous cases of the coincidence of harmonics with some emission lines of corresponding ions [56]. We show that ionic transitions demonstrating high oscillator strengths are responsible for enhanced emission of a single order or group of harmonics. Meanwhile, plasma emission plays a rather destructive role during HHG in the plasma plumes.

7.4.2 Experimental Conditions for Observation of the Control of Harmonic Enhancement

The experimental setup comprised the Ti:sapphire laser (810 nm), traveling-wave OPA, and HHG scheme using the propagation of amplified signal radiation from OPA or 810 nm radiation through the extended LPP. Part of amplified uncompressed radiation (370 ps pulse duration, 5 mJ pulse energy) was separated from a whole beam and used as a heating pulse for homogeneous extended plasma formation using the 200 mm focal length cylindrical lens installed in front of the extended targets placed in the vacuum chamber.

The 5 mm long Sb, Sn, Mn, Zn, Se, In, Cr, Ag, Au, Mg, Cd, and graphite samples were used as the targets for ablation and formation of extended homogeneous plasma.

The compressed radiation of Ti:sapphire laser (pulse energy 8 mJ, pulse duration 64 fs, 10 Hz pulse repetition rate, pulse bandwidth 17 nm, and central wavelength 810 nm) pumped the OPA. Signal pulses from OPA allowed the tuning along the 1200–1600 nm range. Most of the experiments were carried out using the 1 mJ, 70 fs signal pulses or 3 mJ, 64 fs, 810 nm pulses. The intensity of focused 1310 nm pulses inside the LPP was 2×10^{14} W cm^{-2}. The 810 nm pulses were also used as a SCP separately from signal pulses for HHG. The harmonic emission was analyzed using an XUVS.

In most of the discussed studies, the TCP scheme comprising NIR pulses and second harmonic radiation as a second field. To add the second field, the 0.5 mm thick beta barium borate (BBO) crystal (type I) was installed inside the vacuum chamber on the path of focused signal pulse. The conversion efficiency of second harmonic pulses was 20%. The use of second wave is not a trick but rather a necessity to study resonant enhancement of harmonics using NIR radiation. The details of the experimental setup are described elsewhere [57].

7.4.3 Featureless and Resonance-Enhanced Harmonic Distributions

In this subsection, the raw images of plasma and harmonic spectra from different LPPs are presented. The reason in presentation of experimental data as the images appearing on the screen of computer rather than the groups of intensity distributions is caused by better viewing of the obtained data and clear definition of the difference between the so-called featureless group of harmonics and the group of harmonics containing a single enhanced harmonic. We would like to stress that the goal of those studies was the qualitative definition of the influence of plasma emission on the harmonic conversion efficiency in different LPPs rather than the quantitative measurements of the gain of single harmonic in the plasmas where this process could be realized. Also notice that visual analysis of raw images of harmonic and plasma emission allows distinguishing these sources of radiation by their divergences. The vertical dimensions of the raw images of plasma emission, which characterize its divergence, were a few times larger compared with those of harmonics.

The saturated images were intentionally chosen to present the spectra for better viewing. Note that for the line-outs of the HHG spectra the unsaturated images were used. The x-axes are shown in the figure on the basis of the calibration of the XUVS for better viewing of the distribution of harmonics along the short-wavelength region. The HHG spectrometer was calibrated using the plasma emission from the used ablated species, as well as other ablating targets.

The first set of studies was carried out in the plasmas where no resonance enhancement was observed. Figure 7.14a shows the plasma and harmonic spectra obtained during ablation of silver and gold targets. The plasma emission spectra were observed at the conditions of ablation at which the HHG could be realized at unfavorable conditions. These plasma spectra, as well as all other emission spectra obtained during ablation of targets, were registered at the conditions when femtosecond pulses did not propagate through the LPPs. Meanwhile, the HHG was performed at the conditions of optimal LPP formation. The term "optimal LPP formation" refers to maximal harmonic yield during propagation of femtosecond pulses through such plasmas. Both Ag and Au plasmas are the attractive media for extended harmonic distribution. The harmonics up to H63 and H57 (not shown in Figure 7.14a) were generated in the silver and gold plasmas during propagation of 810 nm, 64 fs pulses through these LPPs. These distributions did not reveal any enhancement of some specific harmonic order but rather represent the plateau-like shapes when the intensities of harmonics were approximately the same along the whole spectral range of generation with some gradual decrease up to the cutoff orders. One can see that although H11 and H13 coincided with some emission lines of Ag and Au ions, no enhancement of these harmonics was observed.

A similar pattern was obtained in the case of Mg (Figure 7.14b) and graphite (Figure 7.14c) plasmas. One can see that at least H19, H27, and H37 of the SCP (1340 nm, 70 fs) propagating through the magnesium plasma coincided with the plasma emission lines. However, neither enhancement nor decrease of these harmonics was observed in the case of HHG in this plasma, but rather gradual decrease of each next order of harmonics compared with the previous one.

The plasma emission from ablated graphite contained strong emission lines of carbon (upper panel of Figure 7.14c). The propagation of the TCP (1340 + 670 nm) through such plasma allowed generation of approximately equal odd and even harmonics of 1340 nm radiation up to the H43 (middle panel). One can see that the coincidence of a few harmonics with these emission lines did not cause the change of their intensity relative to neighboring orders. Moreover, as was mentioned previously, a decrease of heating pulse energy on the target surface allowed formation of optimal graphite plasma resulting in noticeable enhancement of all harmonics (bottom panel; compare with the middle panel). Again, as in previous cases (Figure 7.14a,b), the featureless harmonic spectrum, without the enhancement of a single harmonic, was obtained.

Most of the plasmas produced on the targets representing the solid elements of periodic table allowed generation of harmonic spectra similar to those shown in Figure 7.14. Meanwhile, a few solid elements showed outstanding features being used for LPP formation and HHG. In Figure 7.15, three samples of harmonic spectra demonstrating the enhanced single harmonic, or group of harmonics, in different ranges of XUV using the selenium, chromium, and tin LPPs are shown.

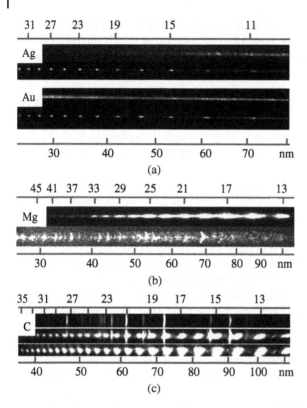

Figure 7.14 Plasma and nonresonant harmonic spectra. (a) Raw images of plasma (upper panels) and harmonic (bottom panels) spectra obtained during ablation of Ag and Au targets and 810 nm femtosecond pulse propagation. Harmonic orders and wavelengths in this and most of the other figures are shown on the upper and bottom axes, respectively. (b) Raw images of harmonic (upper panel) and plasma (bottom panel) spectra obtained during ablation of Mg target and 1340 nm femtosecond pulse propagation. (c) Raw images of plasma (upper panel), plasma + harmonic (middle panel), and harmonic (bottom panel) spectra obtained during ablation of graphite target and 1340 + 670 nm femtosecond pulse propagation. *Source*: Ganeev 2017 [56]. Reproduced with permission from American Institute of Physics.

The SCP of Se plasma using 810 nm radiation allowed odd harmonics generation up to H27 using the optimally formed LPP. The growth of heating pulse fluence on the surface of ablating target led to the appearance of plasma emission in the range of H10–H15 and the decrease of a whole yield of harmonics (Figure 7.15, upper panel). Meanwhile, this overexcitation of target caused the growth of population of some ionic states, which allowed the observation of an enhanced single harmonic (H35) far from the whole set of harmonics. This harmonic was visibly stronger than the lower order ones (H23–H33), contrary to commonly observable patterns of lower intensities for

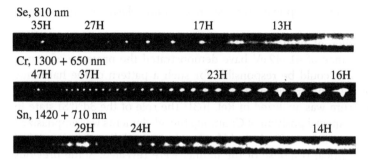

Se, 810 nm
   35H   27H   17H   13H

Cr, 1300 + 650 nm
   47H   37H   23H   16H

Sn, 1420 + 710 nm
        29H   24H   14H

Figure 7.15 Samples of resonance-induced enhancement of single harmonics in various plasmas. Upper, middle, and bottom panels correspond to the harmonic spectra generated in selenium, chromium, and tin plasmas using 810 nm, 1300 + 650 nm, and 1420 + 710 nm pumps, respectively. One can see the enhanced H35 in the case of selenium plasma and 806 nm pump. In the case of chromium plasma, harmonics surrounded H47 were notably stronger compared with those in the range of H41–H45 in the case of 1300 + 650 nm pump. Similarly, tin plasma allowed observation of the enhanced harmonics in the vicinity of H29 of 1420 + 710 nm pump. *Source:* Ganeev 2017 [56]. Reproduced with permission from American Institute of Physics.

the higher orders compared with the lower order ones. In fact, the neighboring harmonics around H35 were barely seen in this spectrum.

The HHG spectrum from Cr plasma also showed the gradual decrease of harmonic yield up to H42 using TCP (1300 + 650 nm). At higher excitation of target, a group of harmonics centered around H47 was observed, apart from the set of lower order harmonics (middle panel).

The third sample of unusual distribution of harmonic spectra was observed during experiments with tin plasma (bottom panel). The TCP (1420 + 710 nm) of overexcited plasma caused the appearance of both the mixture of harmonic and plasma emission in the longer wavelength range of XUV and the group of enhanced harmonics (H31, H30, and especially H29). Notice the absence of H25–H27 in this spectrum.

To address these results, the comparison of harmonic spectra with the spectroscopic analysis of the properties of specific ionic transitions of Se, Cr, and Sn was carried out. The resonance-enhanced single harmonic observed in selenium plasma had a shortest wavelength (H35 of 802 nm pump, ~22.9 nm) reported so far among other resonance enhanced processes [58]. The doubly ionized selenium atom has the ground configuration $4s^2 4p^2$ and the excited configurations $4s^4 p^3$, $4s^2 4p''d$ ($n > 4$), and $4s^2 4p''s$ ($n > 5$). The photon energy studied in previous spectroscopic experiments with Se plasma ranged between 18.0 and 31.0 eV [59, 60]. However, they also reported the strong third-order line that corresponded to the 54.62 eV energy (~22.7 nm), without the deliberation of the origin of this transition. Note that this transition is close to the enhanced H35 in the discussed experiments. Probably, this transition can influence the nonlinear optical response of selenium plasma in this spectral region.

Previous studies of photoabsorption and photoionization spectra of Cr plasma in the range of 41–42 eV have demonstrated the presence of strong transitions, which could be responsible for such a pattern of the harmonic spectrum [61–63]. In particular, the region of the "giant" 3p–3d resonance of the Cr II spectrum was analyzed in Ref. [63]. The role of the $3d^5(^6S)$ state in determining the special position of Cr among the 3d elements was emphasized and the strong transitions, which could both enhance and diminish the optical and nonlinear optical response of the plume, were revealed. Some previous studies of chromium plasma revealed the enhancement of single harmonics in the case of different sources of driving pulses. Particularly, H29 (27.6 nm) of 800 nm pump was noticeably stronger than other harmonics.

The strong Sn II ion has been shown to possess a strong transition $4d^{10}5s^25p^2\,P_{3/2}$–$4d^95s^25p^2(1D)^2\,D_{5/2}$ at the wavelength of 47.20 nm. The *gf* value of this transition has been calculated to be 1.52, and this value is five times larger than other transitions from the ground state of Sn II. Therefore, the enhancement of the H29 (48.9 nm) using the 1420 + 710 nm wavelength pump can be explained as being due to optimal closeness to the above transition. The resonance enhancement of single harmonic (H17 of 795 nm laser, 46.8 nm) has previously been observed and attributed to the influence of the same transition.

Those three samples of harmonic spectra pointed out the role of ablation conditions in the formation of the LPP allowing such an unusual distribution of harmonics appearing in the shorter wavelength range of XUV. During the last decade, this phenomenon was observed in the case of some ionic transitions, the *gf*s of which were significantly larger compared with similar parameters of other neighboring transitions in different spectral ranges [64, 65]. Meanwhile, the role of plasma emission lines was yet to be examined. In the following, we discuss the comparative studies of plasma and harmonic spectra at the conditions of coincidence or mismatch of plasma and harmonic wavelengths.

7.4.4 Comparison of Plasma and Harmonic Spectra in the LPPs Allowing Generation of Resonantly Enhanced Harmonics

To analyze the role of the ionic transitions responsible for plasma emission in the enhancement of single harmonic, one has to use a tunable source of femtosecond pulses. The pump wavelength was tuned using OPA, which allowed the variation of harmonic wavelengths in the vicinity of strong emission lines in the XUV range using Zn, Sb, Cd, and In LPPs. The fixed wavelength of pump radiation was also used to analyze the plasma and harmonic emission from Mn ablation. The data on the plasma emission of various elements were taken from the NIST Atomic Spectra Database [66].

7.4.4.1 Zinc Plasma

The upper panel of Figure 7.16a shows the raw image of zinc plasma emission where a few strong emission lines in the range of 65–90 nm dominated the XUV spectra. This spectrum was obtained at the unfavorable conditions for harmonic generation (i.e. at ∼2 J cm^{-2} fluence of heating pulse), similar to the LPP spectra shown in Figure 7.14. At optimal conditions of zinc ablation, when maximally efficient HHG was achieved, a few plasma emission lines were still seen among the harmonics. Particularly, strong 67.8 nm emission line of Zn III attributed to $3d^{10}$–$3d^9 4p$ transition appeared between H19 and H20 of 1320 + 660 nm pump (second panel). The same can be said about 75.5 nm emission line. The harmonic spectrum was extended up to H36. The H15 and H17, which coincided with 88.1 and 76.7 nm emission lines, appeared to be stronger than other neighboring harmonics. It was difficult to define whether this difference in harmonic intensities can be attributed to the influence of emission line or to the naturally stronger odd harmonics compared to the even ones.

To define the role of plasma emission lines, one can decrease the fluence of heating pulse and tune the harmonics (third panel, 1360 + 680 nm pump). In that case, only 67.8 nm line remained in the spectrum, while other plasma emission lines almost disappeared. The odd and even harmonics (at least up to H22) showed a gradual decrease of intensity for each next harmonic order, without any features indicating the resonance enhancement. The whole decrease of conversion efficiency compared with 1320 + 660 nm pump was caused by the higher energy of latter TCP compared with a previous pump. This decrease of HHG conversion efficiency is further seen in the case of 1380 + 690 nm pump (fourth panel), which also confirmed the assumption about the negligible role of the ionic transitions responsible for plasma emission in the HHG.

The H17 of 1320 + 660 nm pump could be partially enhanced due to different processes. The involvement of 76.7 nm transition of Zn II on the variations of harmonic yield near this transition was examined in Ref. [67]. The width of the resonantly enhanced H17 in those studies was close to that of other harmonics, both in calculated and experimental results, similar to discussed data. Another case was observed in the case of 67.8 nm line. The enhancement region is narrower than the harmonic line width (third panel), so the shape of H20 is close to the one of surrounding harmonics but with a narrow maximum due to the possible resonant enhancement.

Similar observation was reported using 775 nm, 4 fs probe pulses [68]. The observations of the strong emission of 18.3 eV transition (67.8 nm, $3d^{10}$–$3d^9 4p$) of Zn III at the conditions of plasma excitation by a few cycles of broadband pulses centered at 775 nm were attributed to the enhancement of part of the 11th harmonic of this radiation, although the wavelength of this harmonic (70 nm) did not exactly match the wavelength of $3d^{10}$–$3d^9 4p$ transition, which has an oscillator strength significantly greater than other lines in this spectral region [69]. In those studies, the narrowband-enhanced emission was similar

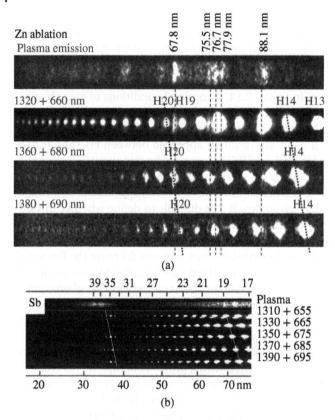

(a)

(b)

Figure 7.16 (a) Raw images of plasma (upper panel) and harmonic (three other panels) spectra obtained during ablation of zinc. Harmonic spectra were tuned by changing the wavelength of driving NIR radiation and its second harmonic from 1320 + 660 nm (second panel from the top) to 1360 + 680 nm (third panel) and 1380 + 690 nm (fourth panel). Dashed lines show the positions of strong plasma emission lines in all four spectra. Dotted lines show the tuning of H14 and H20 at different pumps of plasma. The intensity of H20 remained approximately the same during tuning through the strong resonance transition (~67.8 nm), while H17 was stronger in the case of 1320 + 660 nm pump. (b) Raw images of plasma (upper panel) and harmonic (five other panels) spectra obtained during ablation of antimony. Harmonic spectra were tuned by changing the wavelength of driving NIR radiation and its second harmonic from 1310 + 665 nm (second panel from the top) to 1390 + 695 nm (sixth panel) with a step of 20 nm. Solid lines show the tuning of H36 and H19 at different pumps of plasma. The harmonic orders shown on the top of images correspond to those generated using the 1310 + 655 nm pump. Most of the harmonics remain the same during variation of pump wavelength, excluding those in the region of 37.5 nm. *Source:* Ganeev 2017 [56]. Reproduced with permission from American Institute of Physics.

to that observed here and could also be attributed to the influence of resonance and propagation processes.

7.4.4.2 Antimony Plasma

Another example demonstrating the enhancement of single harmonic by tuning through the ionic transition possessing high gf rather than strong plasma emissions was demonstrated in the case of antimony plasma. This plasma has two strong ionic transitions. The XUV spectra of antimony plasma have been analyzed in Ref. [70]. The oscillator strengths of $4d^{10}5s^22p^3P_2-4d^95s^25p^{33}D_3$ and $4d^{10}5s^22p^1\ D_2-4d^95s^25p^{33}F_3$ transitions have been calculated to be 1.36 and 1.63, respectively, which were a few times larger than those of the neighboring transitions. These strong Sb II transitions ($4d^{10}5s^22p^3P_2-4d^95s^25p^{33}D_3$ and $4d^{10}5s^22p^1\ D_2-4d^95s^25p^{33}F_3$) at the wavelengths of 37.82 and 37.55 nm could be responsible for the enhancement of harmonics in the vicinity of these transitions.

The upper panel of Figure 7.16b shows two groups of plasma emission comprising the continuum and emission lines of antimony. Notice that there are no peculiar emission lines in the vicinity of 37–38 nm. Strongest emission line is seen only at 32.5 nm, quite far from the spectral range of expected enhancement of the nonlinear optical processes.

The application of TCP allowed a significant enhancement of harmonic yield compared with SCP. TCP was gradually tuned between 1310 + 655 nm and 1390 + 695 nm with a step of NIR pulses of 20 nm (second to sixth panels). The lower order harmonics remained approximately the same during this procedure. The harmonic orders shown in the upper axis are attributed to those in the second panel. The white solid line at the right side of this picture shows the tuning of H19. No significant changes in intensities of this harmonic are seen while tuning between the two plasma emission lines shown in the upper panel. The second group of harmonics appears in the 36–38 nm region. The white solid line at the left side of picture shows the tuning of H36. One can see a significant change of the intensity of this harmonic, with maximal conversion efficiency at ~37.3 nm. Probably, the shorter wavelength side of the two abovementioned resonances is responsible for better enhancement of the harmonics tuning in this region. The reasons for this peculiarity could be related to the propagation effect, when the phase-matching conditions became better for the system possessing a lower difference between the wave numbers of pump and harmonic emissions due to anomalous dispersion in this side of resonances.

These lines are not seen in the plasma emission spectrum (see the left group of lines and continuum appearing in the 28–36 nm range on the first panel). The efficiency of each neighboring harmonics in this region depended on the closeness with the $4d^{10}5s^22p^3P_2-4d^95s^25p^{33}D_3$ and $4d^{10}5s^22p^1\ D_2-4d^95s^25p^{33}F_3$ transitions of Sb II ions.

7.4.4.3 Cadmium Plasma

The relative intensities of the H15 generated in Cd plasma were tuned by changing the SCP from 1360 to 1320 nm, as shown in Figure 7.17a (two bottom panels). One can clearly see the relatively equal intensities of other harmonics in the case of 1320 and 1360 nm pumps. Meanwhile, the tuning of pump and harmonic wavelengths toward the longer wavelength region significantly changed the pattern of H15. H15 ($\lambda = 90.7$ nm) using 1360 nm pump became significantly stronger than the H15 obtained using 1320 nm radiation. The emission lines from Cd plasma (upper panel) were far from the region of H15. Thus, the influence of the transitions responsible for plasma emission was insignificant from the point of view of variations of the relative intensities of this harmonic.

7.4.4.4 Indium Plasma

This plasma has been analyzed in previous sections. Strongest enhancement of single harmonic was observed in the case of In LPP (Figure 7.17b, upper

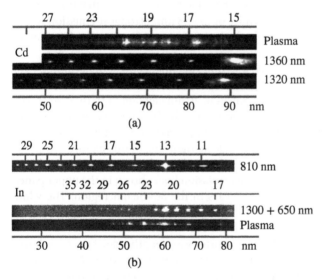

Figure 7.17 (a) Raw images of plasma (upper panel) and harmonic (two bottom panels) spectra obtained during ablation of cadmium. Harmonic spectra were tuned by changing the wavelength of driving NIR radiation from 1320 nm (third panel from the top) to 1360 nm (second panel). The harmonic orders shown on the top of images correspond to those generated using the 1360 nm pump. (b) Raw images of harmonic (two upper panels) and plasma (bottom panel) spectra obtained during ablation of indium. One can see the strong H13 in the case of 810 nm pump (upper panel). The harmonics from 1300 + 650 nm pump were also maximally enhanced in the same spectral region (61.5 nm, middle panel). The harmonic orders shown on the top of upper and middle images correspond to those generated using the 810 nm and 1300 + 650 nm pumps, respectively. *Source:* Ganeev 2017 [56]. Reproduced with permission from American Institute of Physics.

panel). H13 of 810 nm SCP was significantly stronger than other harmonics. The analysis of relative intensities showed that H13 was ~20 times higher compared with neighboring orders.

This harmonic generation process was carried out at a fixed wavelength (810 nm) of Ti:sapphire laser and could not be optimized from the point of view of further enhancement of H13. This optimization assumes better matching of the H13 wavelength with the wavelength of the transition of In II responsible for such resonance-induced enhancement. The difference between the wavelengths of H11, H13, and H15 was large enough, so the ionic transition affected only single harmonic order (H13) and did not influence two other neighboring harmonics. Another situation occurs in the case when the wavelength difference between the harmonics becomes smaller. To do so, the 1300 + 650 nm pump was used, which allowed generation of equal odd and even harmonics in most of the plasmas (see, for example, Figures 7.14c and 7.15, middle panel). It was defined in the case of In LPP that application of this TCP leads to a significant deviation from the featureless homogeneous harmonic distribution. The middle panel of Figure 7.17b shows a noticeable variation of harmonics in the region of 58–77 nm. The harmonic orders (from right to left) in that case start from the H16 (extremely weak emission) and then, their intensity increases from H17 to H21, which is unusual for the HHG process. Strongest among them was the H21 of 1300 nm radiation ($\lambda = 61.9$ nm) almost coinciding with the H13 of 810 nm radiation ($\lambda = 62.3$ nm). Occasionally, there was the emission line ($\lambda = 60.9$ nm, see the bottom panel) in the plasma spectrum almost coinciding with the area of maximal enhancement. Notice that at weaker ablation (i.e. at a fluence of heating pulse of ~1.3 J cm^{-2}), when most of the plasma emission lines disappear, the enhancement of harmonics in this region was even stronger compared with the above-described case (fluence of 1.6 J cm^{-2}). Meanwhile, other plasma lines shown in the bottom line did not affect the harmonic distribution using either 810 nm or 1300 + 650 nm pumps.

7.4.4.5 Manganese Plasma

The final example of harmonic enhancement is shown in Figure 7.18. The Mn plasma was emitted in a broad short-wavelength region (10–35 nm, upper panel) during overexcitation of target by 370 ps pulses (i.e. at a fluence of 1.9 J cm^{-2}). The harmonics at weaker ablation (1.2 J cm^{-2}) were limited by H29. The growth of heating pulse fluence led to the appearance of the extended group of harmonics starting from H33 and extended up to H83 (bottom panel). There was no relation between the emission lines and enhanced H33 and following harmonics. One can assume better conditions of enhancement for the H33 in the vicinity of the giant 3p–3d resonances of Mn around the 51–52 eV range, where the metastable states of manganese are located [71]. Meanwhile, there are no relations between the observed plasma emission

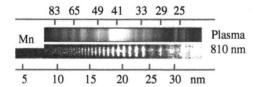

Figure 7.18 Raw images of plasma (upper panel) and harmonic (bottom panel) spectra obtained during ablation of manganese. The intensity of H33 of the 810 nm pump was notably stronger than lower order harmonics. *Source*: Ganeev 2017 [56]. Reproduced with permission from American Institute of Physics.

lines (upper panel) and the harmonic enhancement in the region of ~24 nm (bottom panel).

7.4.5 Basics of Alternative Model of Enhancement

As one can see from previous chapters, plenty of theoretical studies were reported on the possibility of the resonance enhancement of harmonics to determine the availability of this process in HHG experiments. In the following we mention another model of resonance enhancement. This model is based on the multielectron resonance recombination. According to this model, the accelerated electrons while returning to the parent particle can transfer the energy to the electron, or a system of electrons, during the inelastic scattering in such a way that those electrons become excited on the resonant levels, the structure of which is determined by the form of ionic potential. The excited resonance levels are independent of the Stark shift, because these levels are not initially populated. Therefore, these resonances can be reasonably calculated. This model of multielectron resonance HHG model is a generalization of the approach of a four-step model.

7.5 Comparison of Micro- and Macroprocesses During the High-Order Harmonic Generation in Laser-Produced Plasma

7.5.1 Basic Principles of Comparison

The analysis presented in the previous sections of this chapter leads to the anticipation in the advantageous applications of a few methods of harmonic amendment in a single set of experiments. The particular interest here is related with the comparison of the collective mechanisms of accumulation of the nonlinear optical response of medium (so-called macroprocesses) and the mechanisms related with the individual properties of single emitters (so-called microprocesses). Those mechanisms include the fulfillment of the conditions of QPM of the driving and harmonic waves along the whole medium and the conditions of the coincidence of the individual harmonic in the plateau region and the ionic transition possessing strong oscillator strength. Both these processes could be further amended through the implementation of TCP technique in the NIR range.

Although the QPM processes were demonstrated in the gaseous media, no significant resonance-induced enhancement of single harmonic was reported using gas HHG approach. On the contrary, LPP have proven to be the effective media for these two (macro- and micro) processes. On one hand, QPM in LPP has recently been reported in the case of silver plasma [24], though other ablated materials have also proven to be suitable for these purposes [13]. On the other hand, resonance enhancement of single harmonic has been frequently observed during plasma HHG. As it was shown in previous chapters, In, Sn, Cr, Mn, and a few other, predominantly semiconductor, targets have shown the attractiveness for ablation and further generation of enhanced single harmonic during propagation of the ultrashort pulses through the LPP.

Particularly, indium plasma seems to be a good choice to compare the relative influence of micro- and macroprocesses in a single set of experiments. It has demonstrated the strong enhancement of single 13th harmonic (H13) of Ti:sapphire laser. At the same time, this plasma allows generation of the extended set of harmonics up to the short-wavelength region (H43 of 800 nm radiation, ~18.6 nm), which is the requirement for observation of the group of enhanced harmonics due to QPM. In this section, we analyze the advances in using collective processes of enhancement together with the growth of single harmonic induced by the individual properties of emitters for the enhancement of harmonic yield by different means [72]. The indium plasma was used to show the growth of a group of harmonics around the maximally enhanced 30s orders. The tuning of maximally enhanced harmonic was carried out using the variation of the conditions of plasma formation. Simultaneously, the wavelength of driving pulse was adjusted to generate the enhanced H21 of the TCP (1310 to 655 nm) using OPA, which was matched with the abovementioned transition of In II. We discuss the comparison of these two principally different processes of harmonic enhancement and show the advantages of the joint application of as much as possible methods of HHG amendment (QPM-induced enhancement, resonance-induced enhancement, and TCP-induced enhancement) in a single experiment.

The experimental setup consisted of three parts: (i) Ti:sapphire laser, (ii) traveling-wave OPA of white-light continuum, and (iii) setup for generation of high-order harmonics using propagation of amplified signal pulse of parametric amplifier through the extended imperforated and perforated LPP. The principles of the configuration of those experiments were almost similar to those described in this and previous chapters. The size of the In target where the ablation occurred was 5 mm. To create multijet plasmas, the MSM was used. The size of the slits was 0.3 mm with the distance between them of 0.3 mm. The MSM was installed between the focusing cylindrical lens and target to divide the continuous 5 mm long plasma in eight 0.3 mm long plasma jets with ~0.3 mm separation. The number of plasma jets was increased by tilting the MSM. Particularly, tilting the mask at 45° allowed the formation of 11 jets.

7.5.2 Results of Comparative Experiments

In the following, we analyze the QPM of the groups of harmonics using the NIR pulses from OPA. As already mentioned, the principles of the QPM in LPP have recently been demonstrated using the 800 nm lasers. Similar or even larger enhancement of the group of harmonics could be achieved in the case of NIR pump sources, alongside the other methods of HHG amendment.

Tuning of driving radiation allowed defining the maximal enhancement of single harmonic in the vicinity of 62 nm. Propagation of 1300 nm radiation and its second harmonic through the extended imperforated plasma led to generation of a few strong odd and even harmonics, with the maximum enhancement in the case of H21 (Figure 7.19a, thick curve). The spectrum of harmonics is quite unusual, since it contains a group of strong harmonics in the vicinity of the ionic transition of indium possessing large gf, contrary to earlier reported generation of single enhanced harmonic using the 800 nm class lasers [10, 15]. Lower order harmonics (i.e. those below H21) become weaker with the growth of XUV wavelength, contrary to commonly reported pattern of the plateau-like shape of harmonics demonstrating the gradual decrease of harmonic intensity toward the shorter wavelength region. The latter behavior was observed in other plasmas used for comparison with the indium plasma. Particularly, harmonic generation using TCP of silver plasma showed almost plateau-like distribution of harmonics along the broad range of XUV (compare the insets in Figure 7.19a showing the raw images of harmonic spectra in the case of indium (upper panel) and silver (bottom panel) plasmas).

Upper inset of the harmonic spectrum image from indium plasma shown in Figure 7.19a contains the saturated harmonics (particularly H21 and H20). This image is presented in order to clearly underline the difference in the harmonic distribution produced from indium plasma with regard to the plasma generated on the silver target. Notice that all line-outs were taken from the collected spectra where no saturation of registrar was observed. The raw image of the spectrum in the case of indium plasma showed that the enhancement of harmonics occurred at the exact resonance conditions (see maximally enhanced H21) and in the vicinity of this region (H16–H22). Higher orders of harmonics demonstrated almost plateau-like shape starting from the H23 until the cutoff region (H39), similar to the homogeneous spectrum from silver plasma.

The purpose in showing the raw images in Figure 7.19a is to acquaint the reader with real collection data and visually demonstrate the appearance of

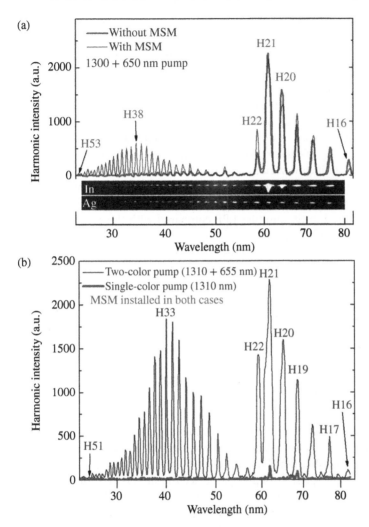

Figure 7.19 (a) Application of imperforated (thick curve) and multijet (thin curve) indium plasmas in the case of two-color (1300 + 650 nm) pump. Inset shows two raw images of harmonic spectra obtained in the cases of the two-color pump of expended imperforated indium plasma (upper panel) and silver plasma (bottom panel). (b) Comparative spectra from In multijet plasma in the case of single-color (1310 nm, thick curve) and two-color (1310 + 655 nm, thin curve) pumps. *Source*: Ganeev 2016 [72]. Reproduced with permission from American Institute of Physics.

the separated group of enhanced harmonics in the case of indium plasma. The spectra were intentionally chosen to present as the saturated images for better viewing. The x-axis is shown in the figure on the basis of the calibration of XUVS. The HHG spectrometer was calibrated using the plasma emission from the used ablated species, as well as other ablating targets.

These two insets showed the harmonic spectra generated using the TCP of the homogeneous 5 mm long indium and silver plasmas produced using the fluence of heating pulse of $F \sim 1.0\,\text{J cm}^{-2}$. Once an extended indium plasma was separated onto a group of eight jets by using the MSM placed in front of ablating indium target, a significant variation of harmonic distribution was observed (Figure 7.19a, thin curve). A group of harmonics centered near the H38 was notably enhanced compared with the lower orders. Furthermore, 50× growth of the H38 generated in the multijet LPP compared with the same harmonic generated in the extended plasma was achieved. Similar pattern was observed in the case of silver plasma. The analysis of QPM in silver plasma has earlier been presented in Ref. [24] (see also Section 7.2). Those studies were carried out using the SCP (800 nm). In the discussed studies, the order of QPM-enhanced harmonics became larger compared with the case of using 800 nm pump due to lesser dispersion of plasma in the used NIR region. In the case of NIR pulses (1300 nm), the maximally enhanced harmonic order (q_{qpm}) corresponded to the relation $q_{\text{qpm}} \approx 1 \times 10^{18}/(l_{\text{jet}} \times N_{\text{e}})$, where l_{jet} and N_{e} are the length of single plasma jet in multijet plume (measured in millimeters) and electron density of plasma (measured in cubic centimeters), respectively. Remember that the size of each jet was $\sim 0.3\,\text{mm}$.

One can see that lower order harmonics (H16–H26) in these two cases were approximately similar to each other. Indeed, the only difference between these two conditions of plasma formation was the decreased length of the LPP, since the MSM allowed the ablation of the half of a whole length of target. The saturation of harmonics at large sizes of plasma caused the similar harmonic yield for both 5 and 2.5 mm long plasmas. The meaning of "saturation" in that case is as follows. The dependence of harmonic yield on the length of extended plasma for different harmonic orders was analyzed during those studies. Particularly, the $I_{21\text{H}}(l_{\text{plasma}})$ dependence showed the quadratic growth up to $l_{\text{plasma}} \approx 1.5\,\text{mm}$, and then this dependence had the slope of ~ 0.5 and some instability.

The lower order harmonic spectra generated in the case of multijet and extended plasmas are similar to each other (Figure 7.19a). However, it is not correct to compare these two spectra from the point of view of the variation of harmonic yield as a function of the medium length, though the whole lengths of plasma media used in the cases of homogeneous and perforated plasmas were 5 and 2.5 mm. The small slope of $I_{\text{H}}(l_{\text{plasma}})$ for the lengths exceeding 1.5 mm (~ 0.5) and additional mechanisms of HHG in the case of multijet plasma can affect the yield of lower order harmonics.

At the used fluence of heating pulse $(1\,J\,cm^{-2})$, the plasma density $(3 \times 10^{17}\,cm^{-3})$ was low enough to exclude the role of absorption processes in the studied spectral region. Thus, the saturation, i.e. decrease of the growth of harmonic yield, was attributed to the propagation effect. One can expect a decrease of harmonic yield for larger lengths of medium. Such case could be realized for the on-axis components of harmonic spectra, while off-axis components still became enhanced. Note that those experiments were carried out at the conditions of collection of the whole spectrum of harmonics using the cylindrical mirror inside the XUVS. However, once the flat mirror was used instead of the cylindrical one, the off- and on-axis distribution of harmonics clearly showed a difference in the harmonic yields for the central and outer parts of the driving beams.

Another situation occurred in the case of higher order harmonics. While in the case of imperforated plasma, the intensity of harmonics starting from the 25th order gradually decreased down to the cutoff harmonic (~H40, thick curve), the perforated plasma created the preferable conditions for the harmonics centered round the 38th order. The phase-matching conditions of the group of harmonics around H38 became significantly improved, which led to the notable enhancement compared with the imperforated plasma. The harmonic cutoff was also extended up to the H52.

The maximally enhanced harmonic order at the conditions of QPM depended on the fluence of heating pulse. It is obvious that at higher fluence the electron density of plasma became larger and vice versa. Correspondingly, the q_{qpm} at higher fluencies was shifted toward the lower orders. Thin curve of Figure 7.19b shows the group of enhanced harmonics centered near the H33. This spectrum was obtained at larger fluence compared with the case presented in the upper panel of Figure 7.19a (1.2 and $1\,J\,cm^{-2}$, respectively). Different wavelengths of NIR pumps in these two cases (1310 and 1300 nm) did not crucially influence the QPM conditions compared with different amount of emitters and free electrons. One can see that the QPM-induced enhancement of a group of harmonics became almost equal to the resonance-induced enhancement of single harmonic (compare H33 and H21). Moreover, the whole enhancement of this group of harmonics (between H24 and H48) exceeds the one of harmonics centered around H21.

Once the BBO crystal was removed from the path of NIR radiation, the harmonic spectrum from multijet plasma became drastically weaker compared with the TCP (Figure 7.19b, thick curve). In the case of SCP, the modulation of indium plasma did not improve the harmonic generation efficiency. Only two harmonics were generated (H19 and H21) similar to the case of extended imperforated plasma, which were far from the spectral range (~35–45 nm) where the QPM-induced enhancement was expected. Correspondingly, no harmonics were observed in the spectral region where the QPM conditions for the TCP were fulfilled. Furthermore, one can see that the resonance enhancement of H21 was significantly decreased in the case of SCP. The ratio of enhanced H21 in the case of TCPs and SCPs was 18. Thus, the TCP of multijet plasma allowed both significant QPM-induced growth of the group of harmonics and resonance-induced growth of the single harmonic and of a few neighboring orders.

The efficiency of the QPM concept for the plasma harmonics depends on many factors. Figure 7.20 shows the set of harmonic spectra at the conditions of the application of multijet indium plasma plumes and tunable NIR radiation. The tuning of NIR pulses along the 1300–1340 nm range led to variation of the

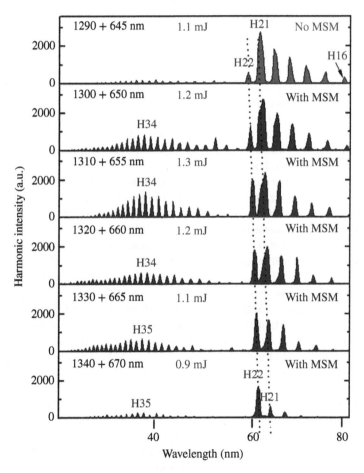

Figure 7.20 Dependences of the QPM-enhanced harmonics generated in the multijet indium plasma on the tuning of the driving NIR pulses. Upper panel shows the harmonic spectrum from imperforated indium plasma. Each panel contains the wavelengths of pumps and the pulse energies. *Source*: Ganeev 2016 [72]. Reproduced with permission from American Institute of Physics.

intensity of tunable XUV radiation due to the change of the conversion efficiency and amplification for different wavelengths of OPA. Correspondingly, the conversion efficiency of this radiation and its second harmonic toward the odd and even harmonics was changed along the XUV region. The variation of the resonantly enhanced harmonic generated in multijet plasma can be seen by comparing the lower order harmonics (H16–H22) of the five wavelengths of NIR radiation (1300, 1310, 1320, 1330, and 1340 nm). One can clearly see the redistribution of the relative intensities of the 21st and 22nd harmonics tuned along the strong resonance transition of In II (62.24 nm). In the meantime, the QPM-enhanced harmonics did not tune along the spectral range, while their intensities followed the energies of the NIR pumps. Strongest enhancement of the QPM harmonics was observed in the case of 1310 + 655 nm pump (third panel from the top; compare this spectrum with the upper panel obtained in the case of imperforated 5 mm long plasma pumped by 1290 + 645 nm pump). The maximum enhancement was maintained for almost the same order of harmonics (H34 or H35), which points out the independence of the electron density, which defines the q_{qpm}, on the variation of driving pulse intensity. The enhancement of the groups of enhanced harmonics decreased in the case of weaker pump.

Two enhancement mechanisms (resonance- and QPM-induced growth of harmonic yield) were tried to combine for the group of harmonic and the single harmonic belonging to this group. In the used particular case (indium plasma; NIR pulses; enhanced radiation near 61 nm, which corresponds to approximately H20 or H21 depending on the pump wavelength; separation of the continuous 5 mm long plasma into eight 0.3 mm long plasma jets with ~0.3 mm separation), there is a difference in the wavelengths of resonance-enhanced harmonics (60–65 nm) and QPM-enhanced harmonics (28–45 nm, Figure 7.19). The first spectral region could not be changed, since it related with the microprocess attributed to the peculiarities of indium ion. The second region is defined by the propagation process described by the relation for the maximally enhanced harmonic (see Section 7.2).

In accordance with this relation, there are two options to move q_{qpm} toward the longer wavelength region. One can either increase the sizes of single jet by increasing the slit sizes of the mask or increase the electron density by increasing the plasma density. Notice that the used intensity of driving pulse ionizes almost every particle within the focal region. The first option was tried by using the MSM containing the 0.8 mm wide slits. There was the shift of q_{qpm} toward the lower wavelengths. However, the enhancement of the group of harmonics in the case of three 0.8 mm long jets was not as strong as the one obtained during application of eight 0.3 mm long plasma jets, probably due to the smaller amount of separated emitting media. Notice that the yield of phase-matched harmonics quadratically increases with the growth of the number of emitting jets. Second option requires stronger excitation of target and correspondingly larger plasma density. The growth of plasma density causes larger amount of the free electrons appearing through either initial stronger ablation or tunnel ionization of larger amount of plasma particles. However, there is another impeding process, strong incoherent plasma emission in the same XUV region at larger fluence of heating pulse, which decreases the usefulness of generated harmonics.

To realize the dual enhancement of harmonics at the same spectral region, one should use the ions allowing the resonance enhancement of shorter wavelength harmonics. In that case, one can achieve the conditions of mutual appearance of two enhancement processes in the same spectral region, while using large amount of jets. This coincidence of two processes was observed in the tin (see below), chromium, and manganese plasmas, though the enhancement of single harmonics in those plasmas was less pronounced with regard to the indium plasma.

The installation of MSM on the path of heating pulse for multijet plasma formation leads to the enhancement of higher order harmonics, only in the case when the phase-matching conditions of the harmonic generation in the extended imperforated plasma became worsened due to the presence of large amount of the free electrons strongly affecting the refractive index of plasma. Once the conditions of extended plasma formation were maintained in a proper way (i.e. the plasma contained an insignificant amount of electrons affecting phase-matching conditions), the installation of the mask on the path of heating pulse led to a decrease of plasma length followed with the decrease of a whole conversion efficiency.

The example of such gradual decrease of harmonic yield along the whole spectral range of generation is shown in Figure 7.21a. Upper panel shows the harmonic spectrum produced in the 5 mm long imperforated chromium plasma using the $1310 + 655$ nm pump. The harmonic generation was obtained at the conditions of Cr target ablation, which did not spoil the phase matching between the interacting waves. Those conditions refer to the availability of further growth of conversion efficiency in the case of longer plasma formation (i.e. at the plasma lengths l_{plasma} exceeding 5 mm). The absence of saturation in the $I_H \infty (l_{plasma})^2$ dependence points out the insignificant influence of various detrimental factors, such as plasma dispersion, which can destroy the coherent accumulation of the generating XUV photons. The modulation of such plasma using MSM, which causes a twofold decrease of plasma length, leads only to the worsening of HHG conversion conditions due to decrease of the interaction length. This assumption was proven in the case of the modulation of chromium plasma (see bottom panel). The shape of the envelope of harmonic distribution remained same as in the case of imperforated plasma (upper panel), with the harmonic yield for all orders decreasing with a factor of ~1.5–2.5 (especially, for the longer wavelength part of the spectrum).

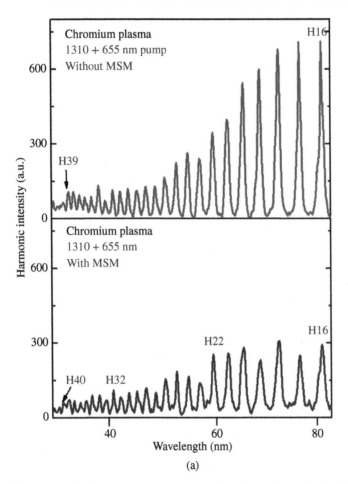

Figure 7.21 (a) Harmonic spectra generated from the low-ionized chromium plasma using two-color pump. Upper panel was obtained in the case of imperforated 5 mm long plasma, while bottom panel was obtained in the case of eight-jet plasma. No QPM effect was observed in that case due to small density of free electrons appeared during laser ablation and tunnel ionization. (b) Harmonic spectra generated in the imperforated (thick curve) and multijet (thin curve) tin plasmas. *Source:* Ganeev 2016 [72]. Reproduced with permission from American Institute of Physics.

Another example of the nonoptimal use of multijet plasma is shown in Figure 7.21b. The application of imperforated tin plasma (thick curve) showed a gradual decrease of harmonic efficiency followed with the enhancement of a group of harmonics in the region of the ionic transition of Sn II possessing large oscillator strength. The enhanced harmonic (H29) almost coincided with the $4d^{10}5s^25p$ $^2P_{3/2}$–$4d^95s^25p^2$ ($^1D)^2D_{5/2}$ transition of the Sn II ion.

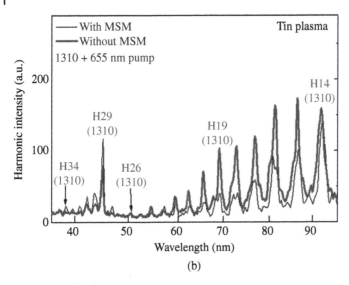

Figure 7.21 (*Continued*)

The studies using the tunable NIR pulses and their second harmonics showed a fine-tuning of the resonance-enhanced harmonic and change of the order of this harmonic. The maximally enhanced harmonics, for which both micro- and macroprocesses were optimized to generate highest photon yield, were changed from H27 in the case of 1290 + 645 nm pump to H31 in the case of 1450 + 725 nm pump. In all these cases, the preceding harmonics were suppressed compared with the resonance-enhanced ones (compare H29 and H27 in the case of 1310 + 655 nm pump, thick curve of Figure 7.21b). The modulation of extended plasma in that case led to a decrease of lower order harmonics and the growth of the harmonics laying around the H30, which was enhanced with a factor of 2.

Such small enhancement compared with the case of indium plasma was caused by nonoptimal excitation of target during laser ablation. In other words, once one excites and ablates the target at the conditions when plasma concentration increases together with plasma emission, the advantages of the former process became less noticeable compared with the negative influence of strong incoherent plasma emission (from the point of view of further applications of generated coherent radiation).

To justify the usefulness of the harmonics generated using QPM and resonance techniques in indium plasma for various applications, one has to present information not only about relative efficiencies when the different parameters change but also absolute efficiencies of the HHG process and its comparison with coherent XUV radiation generated using other techniques. The comparative studies of gas- and plasma-based HHG have already been carried out in a few laboratories [73–75]. It has been shown that some plasma species, particularly carbon plasma, showed stronger harmonic yield, at least for the lower orders, compared with the argon gas harmonics. One can antic-ipate that resonantly enhanced lower order harmonic from indium plasma also would be strong enough to compete with the gas-based harmonics. The resonantly enhanced H21 and QPM-enhanced H33 from the In plasma were compared with the harmonics generated in the graphite ablation and it found that the former harmonics were as strong as the harmonics generated in the carbon-contained plasma plumes. From this comparative study and previous abovementioned reports, one can conclude about the usefulness in application of these enhanced harmonics in the experiments, similar to the gas harmonics.

The absolute value of harmonic energy generated from indium plasma was measured by using the method described in Ref. [40]. By using this calibra-tion method, the energy of the QPM-enhanced H33 generated from the indium plasma was measured to be 0.08 μJ. One can assume the 8×10^{-5} conversion efficiency of this harmonic taking into account the 1 mJ energy of converted 1310 nm pulses.

7.5.3 Discussion of Comparative Experiments

There are a few issues, which should be addressed for better understanding of the processes occurring during HHG in multijet plasmas. Particularly, what could cause saturation when absorption by the plasma itself is excluded and QPM is centered in the vicinity of H38? In the following we address the mean-ing of saturation with regard to the analyzed processes. There are two options, which may lead to the observation of saturation of the QPM-enhanced har-monics. The first option is related with the limits of coherent adding of the harmonic yield along the whole set of jets. In the experiments, the formation of phase-matching conditions along the single jet was again maintained in the following jet leading to the significant growth of the intensity of harmonics. To prove this assumption and the role of QPM in the observed peculiarities of har-monic spectra, the intensity (I_H) of harmonics was studied as a function of the number of plasma jets (n) in the case of ablation of target. The heating zones on the target surface were shielded step-by-step to create different number of plasma jets (from single-jet to eight-jet formations).

The anticipated featureless shape of the high-order harmonic spectra from the single 0.3 mm long plasma jet was similar to those observed in the case of 5 mm long indium plasma (see insets in Figure 7.19a and upper panel of Figure 7.20). With the addition of each next jet, the spectral envelope was drastically changed, with the 38th harmonic intensity in the case of three-jet configuration becoming almost eight times stronger compared with the case of single-jet plasma. One can anticipate the n^2 growth of harmonic yield for the n-jet configuration compared with the single jet once the PMM becomes suppressed. These conditions allow expecting the growth factor of 9 in the case of three-jet medium, which was close to the experimentally measured enhancement factor (~8×). At larger amount of jets, the maximum enhancement factor in the QPM region deviated from the $I_H \propto n^2$ dependence, probably due to unequal properties of the jets, which can arise from the heterogeneous excitation of the extended target. Another reason of the deviation from the quadratic rule could be the growing influence of the driving pulse on the variation of phase relations in the case of a larger amount of contributing zones. Particularly, in the case of eight separated zones, the enhancement of harmonics with a factor of ~25× was observed, contrary to the anticipated 64× enhancement. In this connection, the saturation of QPM was caused by abovementioned processes.

Second option is related with the method of spectrum collection. The harmonic emission was focused inside the XUVS along the vertical axis using the gold-coated cylindrical mirror placed at 87° with regard to the interacting radiation. In that case, the raw images of harmonic spectra represented the series of the "dots" corresponding to the distribution of harmonics along the XUV region. Two insets of Figure 7.19a show the examples of such raw images of the harmonic spectra captured by CCD camera.

While improving the visibility and fluence of generated XUV radiation, this method of harmonic images collection did not allow the analysis of the contribution of different parts of driving beam on the spatial distribution of harmonics. To visualize the spatial shapes of harmonics along the vertical axis, one has to use the plane mirror instead of the cylindrically focused one. In that case, it was observed that the on-axis parts of harmonics were significantly suppressed compared with the off-axis components of the same orders.

The origin of this spatial shape of high-order harmonics is related with the PMM in the axial region of propagation of the driving beam. Strong intensity of laser radiation led to the appearance of additional tunneled electrons during ionization by the driving pulses. Less intensity of the driving beam on the wings of spatial distribution caused weaker influence of the abovementioned processes on the HHG, which led to the appearance of the higher order harmonics dominantly in the off-axis area. The installation of the MSM with 0.3 mm slits caused a dramatic redistribution of harmonics along the spatial distribution of driving radiation. The spectrum showed a significant growth of harmonic emission along the on-axis region.

The divergence of the enhanced radiation was a few times smaller compared with the case of the HHG in the imperforated plasma. The maximally enhanced harmonic (H38) in the on-axis area was significantly stronger compared with the same harmonic order observed in the case of imperforated plasma. The harmonic emission was concentrated solely on the on-axis area. The meaning of the saturation of QPM in that case was related with the redistribution of this process along the spatial shape of converting beam. The conditions of QPM were far from optimal for different parts of driving beam.

The next issue is related with the necessity in formation of appropriate conditions for the HHG and QPM. Figure 7.21b shows the decreased low-order harmonics from tin plasma for the spectrum with the MSM, as expected for shorter interaction lengths below the coherence length. In contrast, there is an increased signal for a few harmonics around the resonance. This result points out the relation of the decrease of lower order harmonics in the case of using the MSM with the twofold decrease of the whole length of plasma. Meanwhile, the higher order harmonics (between H28 and H34) are enhanced due to formation of the QPM conditions. However, there are different ranges of these conditions. At ideal conditions, when the coherence lengths of H29 or H30 exactly match with the sizes of jets, one can expect a significant gain of those harmonics. However, once the harmonic yield does not coherently add along the whole set of jets, partially due to imperfect distribution of atoms/ions and free electrons in each jet, the QPM conditions became deteriorated to some extent. That is why the observed enhancement factor was only ∼2×. However, even at these "non-optimal" conditions of plasma formation, it was possible to observe the multiparticle-induced growth of the harmonic, which was notably suppressed by the single-particle-induced process (compare H28 in these two cases).

References

1 Teubner, U. and Gibbon, P. (2009). *Rev. Mod. Phys.* 81: 445.

2 Zhang, G.P. (2007). *Int. J. Mod. Phys. B* 21: 5167.

3 Ganeev, R.A. (2014). *J. Opt. Soc. Am. B* 31: 2221.

4 Donnelly, T.D., Ditmire, T., Neuman, K. et al. (1996). *Phys. Rev. Lett.* 76: 2472.

5 Ganeev, R.A., Baba, M., Suzuki, M., and Kuroda, H. (2014). *J. Phys. B* 47: 135401.

6 Kim, I.J., Kim, C.M., Kim, H.T. et al. (2005). *Phys. Rev. Lett.* 94: 243901.

7 Ganeev, R.A., Singhal, H., Naik, P.A. et al. (2010). *Phys. Rev. A* 82: 053831.

8 Zhang, X., Lytle, A.L., Popmintchev, T. et al. (2007). *Nat. Phys.* 3: 270.

9 Winterfeldt, C., Spielmann, C., and Gerber, G. (2008). *Rev. Mod. Phys.* 80: 117.

10 Ganeev, R.A., Strelkov, V.V., Hutchison, C. et al. (2012). *Phys. Rev. A* 85: 023832.

11 Pirri, A., Corsi, C., and Bellini, M. (2008). *Phys. Rev. A* 78: 011801.

12 Bahabad, A.B., Murnane, M.M., and Kapteyn, H.C. (2010). *Nat. Photonics* 4: 570.

13 Ganeev, R.A., Suzuki, M., and Kuroda, H. (2014). *Phys. Rev. A* 89: 033821.

14 Constant, E., Garzella, D., Breger, P. et al. (1999). *Phys. Rev. Lett.* 82: 1668.

15 Ganeev, R.A., Suzuki, M., Baba, M. et al. (2006). *Opt. Lett.* 31: 1699.

16 Madsen, C.B. and Madsen, L.B. (2006). *Phys. Rev. A* 74: 023403.

17 Lan, P., Takahashi, E.J., and Midorikawa, K. (2010). *Phys. Rev. A* 81: 061802.

18 Hutchison, C., Ganeev, R.A., Castillejo, M. et al. (2013). *Phys. Chem. Chem. Phys.* 15: 12308.

19 Ganeev, R.A., Suzuki, M., Redkin, P.V., and Kuroda, H. (2014). *J. Nonlinear Opt. Phys. Mater.* 23: 1450013.

20 Ganeev, R.A., Suzuki, M., and Kuroda, H. (2014). *J. Phys. B* 47: 105401.

21 Seres, J., Yakovlev, V.S., Seres, E. et al. (2007). *Nat. Phys.* 3: 878.

22 Tosa, V., Yakovlev, V.S., and Krausz, F. (2008). *New J. Phys.* 10: 025016.

23 Zheng, L., Chen, X., Tang, S., and Li, R. (2007). *Opt. Express* 15: 17985.

24 Ganeev, R.A., Tosa, V., Kovács, K. et al. (2015). *Phys. Rev. A* 91: 043823.

25 Tosa, V., Takahashi, E., Nabekawa, Y., and Midorikawa, K. (2003). *Phys. Rev. A* 67: 063817.

26 Ammosov, M.V., Delone, N.B., and Krainov, V.P. (1986). *Sov. Phys. JETP* 64: 1191.

27 Lewenstein, M., Balcou, P., Ivanov, M.Y. et al. (1994). *Phys. Rev. A* 49: 2117.

28 Heyl, C.M., Güdde, J., Höfer, U., and L'Huillier, A. (2011). *Phys. Rev. Lett.* 107: 033903.

29 Kovacs, K., Balogh, E., Hebling, J. et al. (2012). *Phys. Rev. Lett.* 108: 193903.

30 Singhal, H., Arora, V., Rao, B.S. et al. (2009). *Phys. Rev. A* 79: 023807.

31 Ganeev, R.A., Suzuki, M., and Kuroda, H. (2014). *Phys. Plasmas* 21: 053503.

32 Singhal, H., Naik, P.A., Kumar, M. et al. (2014). *J. Appl. Phys.* 115: 033104.

33 Albert, O., Roger, S., Glinec, Y. et al. (2003). *Appl. Phys. A: Mater. Sci. Process.* 76: 319.

34 Harilal, S.S., Farid, N., Hassanein, A., and Kozhevin, V.M. (2013). *J. Appl. Phys.* 114: 203302.

35 Oujja, M., Lopez-Quintas, I., Benítez-Canete, A. et al. (2017). *Appl. Surf. Sci.* 392: 572.

36 Pertot, Y., Elouga Bom, L.B., Bhardwaj, V.R., and Ozaki, T. (2011). *Appl. Phys. Lett.* 98: 101104.

37 Amoruso, S., Bruzzese, R., Wang, X. et al. (2007). *J. Phys. B: At. Mol. Phys.* 40: 1253.

38 Ganeev, R.A. (2017). *J. Nonlinear Opt. Phys. Mater.* 26: 1750010.

39 Ganeev, R.A., Wang, Z., Lan, P. et al. (2016). *Phys. Rev. A* 93: 043848.

40 Ganeev, R.A., Baba, M., Suzuki, M., and Kuroda, H. (2005). *Phys. Lett. A* 339: 103.

41 Elouga Bom, L.B., Kieffer, J.-C., Ganeev, R.A. et al. (2007). *Phys. Rev. A* 75: 033804.

42 Ganeev, R.A., Singhal, H., Naik, P.A. et al. (2007). *Appl. Phys. B: Lasers Opt.* 87: 243.

43 Ganeev, R.A., Suzuki, M., Baba, M. et al. (2008). *J. Phys. B: At. Mol. Phys.* 41: 045603.

44 Singhal, H., Ganeev, R.A., Naik, P.A. et al. (2010). *Phys. Rev. A* 82: 043821.

45 Andreev, A.V., Ganeev, R.A., Kuroda, H. et al. (2013). *Eur. Phys. J. D* 67: 22.

46 Emelina, A.S., Emelin, M.Y., Ganeev, R.A. et al. (2016). *Opt. Express* 24: 13971.

47 Rubenchik, A.M., Feit, M.D., Perry, M.D., and Larsen, J.T. (1998). *Appl. Surf. Sci.* 129: 193.

48 Corkum, P.B. (1993). *Phys. Rev. Lett.* 71: 1994.

49 Englert, L., Rethfeld, B., Haag, L. et al. (2007). *Opt. Express* 15: 17855.

50 Bourquard, F., Tite, T., Loir, A.-S. et al. (2014). *J. Phys. Chem. C* 118: 4377.

51 Amoruso, S., Bruzzese, R., and Wang, X. (2009). *Appl. Phys. Lett.* 95: 251501.

52 Guillermin, M., Colombier, J.P., Valette, S. et al. (2010). *Phys. Rev. B* 82: 035430.

53 Corkum, P.B. and Krausz, F. (2007). *Nat. Phys.* 3: 381.

54 Reintjes, J.F. (1984). *Nonlinear Optical Parametric Processes in Liquids and Gases*. Orlando: Academic Press.

55 Rothhardt, J., Hadrich, S., Demmler, S. et al. (2014). *Phys. Rev. Lett.* 112: 233002.

56 Ganeev, R.A. (2017). *J. Appl. Phys.* 121: 133108.

57 Ganeev, R.A. (2016). *J. Opt. Soc. Am. B* 33: E93.

58 Ganeev, R.A., Suzuki, M., Yoneya, S., and Kuroda, H. (2015). *J. Appl. Phys.* 117: 023114.

59 Tauheed, A. and Hala, A. (2012). *Phys. Scr.* 85: 025304.

60 Esteves, D.A., Bilodeau, R.C., Sterling, N.C. et al. (2011). *Phys. Rev. A* 84: 013406.

61 McGuinness, C., Martins, M., Wernet, P. et al. (1999). *J. Phys. B: At. Mol. Phys.* 32: L583.

62 McGuinness, C., Martins, M., van Kampen, P. et al. (2000). *J. Phys. B: At. Mol. Phys.* 33: 5077.

63 West, J.B., Hansen, J.E., Kristensen, B. et al. (2003). *J. Phys. B: At. Mol. Phys.* 36: L327.

64 Suzuki, M., Baba, M., Kuroda, H. et al. (2007). *Opt. Express* 15: 1161.

65 Ganeev, R.A., Zheng, J., Wöstmann, M. et al. (2014). *Eur. Phys. J. D* 68: 325.

66 Kramida, A., Ralchenko, Y., Reader, J., and NIST ASD Team (2013). *NIST Atomic Spectra Database (Version. 5.1)*. Gaithersburg, MD: National Institute of Standards and Technology.

67 Ganeev, R.A., Suzuki, M., Yoneya, S. et al. (2016). *J. Phys. B: At. Mol. Phys.* 49: 055402.

68 Ganeev, R.A., Abdelrahman, Z., Frank, F. et al. (2014). *Appl. Phys. Lett.* 104: 021122.

69 Dick, K.A. (1968). *Can. J. Phys.* 46: 1291.

70 D'Arcy, R., Costello, J.T., McGuinness, C., and O'Sullivan, G. (1999). *J. Phys. B: At. Mol. Phys.* 32: 4859.

71 Kjeldsen, H., Folkmann, F., Kristensen, B. et al. (2004). *J. Phys. B: At. Mol. Phys.* 37: 1321.

72 Ganeev, R.A. (2016). *J. Appl. Phys.* 119: 113104.

73 Ganeev, R.A., Witting, T., Hutchison, C. et al. (2012). *Phys. Rev. A* 85: 015807.

74 Pertot, Y., Chen, S., Khan, S.D. et al. (2012). *J. Phys. B: At. Mol. Phys.* 45: 074017.

75 Wöstmann, M., Redkin, P.V., Zheng, J. et al. (2015). *Appl. Phys. B: Lasers Opt.* 120: 17.

Summary

In the following, we summarize different issues analyzed in this book. All of the discussed studies were related with resonance enhancement of the high-order harmonics generating in different laser-produced plasmas (LPPs). Most important conclusions arising from the reviewed studies are as follows:

1) Different theoretical approaches describing the resonance enhancement of single harmonic are analyzed. We first discussed the observations of the plasma harmonic spectral modifications using two-color near-infrared (NIR) pulses allowing definition of the excited neutral and ionic species responsible for the enhancement of single harmonics close to the resonances of medium. This approach allows to precisely analyze the role of those transitions by tuning various harmonics through the resonances under consideration at controllable conditions of excitation of neutral and ionic transitions. The relation between the appearance of strong emission lines close to those transitions and the enhancement of specific harmonics was analyzed. Those studies allow further development of the laser-ablation-induced high-order harmonic generation (HHG) spectroscopy of plasma species. The discussed studies also show that the application of tunable NIR pulses may optimize earlier reported resonance enhancement of specific harmonics and identify new areas of harmonic enhancement. Microprocesses and propagation effect are discussed to describe the narrowing of the enhanced emission spectra from plasmas.

2) High-order harmonics generated in LPP can be resonantly enhanced in the energy range, which corresponds to the plasma ion transitions possessing large oscillator strengths. This effect was earlier discovered using the fixed wavelength pump sources (Ti:sapphire laser). In order to optimize these enhancements as well as to study these ionic transitions in more detail, the application of tunable NIR radiation from optical parametric amplifier is proposed. We have analyzed the enhancement of tunable harmonics in the regions of 27, 38, and 47 nm using tin, antimony, and chromium plasmas.

Resonance Enhancement in Laser-Produced Plasmas: Concepts and Applications,
First Edition. Rashid A. Ganeev.
© 2018 John Wiley & Sons, Inc. Published 2018 by John Wiley & Sons, Inc.

3) We discussed the first observation of significant resonance enhancement of a single high-order harmonic in extreme ultraviolet (XUV)
region generated during propagation of femtosecond pulses through the
low-excited indium plasma. A strong 13th harmonic (61.2 nm) from a
Ti:sapphire laser radiation with conversion efficiency of 8×10^{-5} and
output intensity almost two orders of magnitude higher than neighbor
harmonics was demonstrated in those studies. Indium-containing targets
were used for the preparation of laser plasma for the HHG (up to the 39th
order) using the femtosecond pulses with variable chirp. The high-order
harmonics generated from indium plume showed a plateau pattern,
while in the case of plumes prepared on the surfaces of In-containing
semiconductors a steady or even steep decrease of conversion efficiency
for each next harmonic was observed.

4) We have discussed the chirp control of the fundamental laser leading to
a significant variation of harmonic intensity distribution in the plateau
region for some specific ablated targets. Such an approach paves the
way for efficient single-harmonic enhancement in the XUV range using
different plasma sources. Results of the discussed studies have shown
that, by tuning the harmonic wavelength to an ionic transition with strong
oscillator strength, the intensity of harmonic can be drastically enhanced.
One should choose an ion with relatively low ionization, preferably one in
a neutral or singly ionized state, since the experiments have shown that an
increase in the electron density will suppress harmonic generation. The
present method of harmonic intensity enhancement can be applicable
to ablation that includes lowly ionized atoms that have strong transition
rates in the XUV wavelength range.

5) The observation of harmonic generation (up to the 33rd order) after the
propagation of femtosecond laser pulses through low-excited chromium
plasma was discussed. The steep decrease of intensity for low-order harmonics was followed by a plateau pattern with conversion efficiency in the
range of 10^{-7}. A considerable restriction of 27th harmonic generation, as
well as the growth of 29th harmonic, were observed in different focusing
conditions.

6) We analyzed the results of studies on the resonance-induced enhancement of single high-order harmonics generated in laser plasmas using
femtosecond pump lasers with 800 and 400 nm central wavelengths.
For these purposes, the Mn, Cr, Sb, and Sn plasmas were identified as
the appropriate nonlinear media for efficient harmonic generation and
single-harmonic enhancement. Most of the ionic/neutral transitions
responsible for the observed resonance-induced enhancement were
identified, which showed strong oscillator strengths, in accordance with
previous photoabsorption studies of the above plumes. The enhancement of the 13th, 17th, 21st, 29th, and 33rd harmonics from the above

targets was discussed. We also analyzed the observation of harmonic enhancement from some targets in the case of a second harmonic pump laser (400 nm central wavelength). Using Mn plume, the highest harmonic photon energy (52.9 eV) was demonstrated at which single enhancement has been observed (17th order, $\lambda = 23.5$ nm).

7) The spatial coherence measurements of the high-order harmonics generating in LPP plumes for resonant and nonresonant harmonics were discussed. Reasonably high visibilities in the range of ≈ 0.6–0.75 were measured for C, Zn, and In targets for harmonics in the range of 15–25 eV. The maximum visibility obtained is consistent with only a very small departure from full coherence of driving laser beam and thus may not represent a real upper limit for spatial coherency. Those results confirm that high-order harmonic from ablation plasmas can be used for applications requiring high spatial coherence, such as diffraction imaging. The higher visibility for the plasma harmonics compared with an argon gas target in discussed experiments was attributed to a reduced production of free electrons during the femtosecond pulse propagation in the preformed plasmas.

8) It was shown that doubly charged ions at appropriate plasma conditions can be used to increase the coherent photon energies obtained from ablation HHG. Those results reveal some new expectations for further extension of the harmonic cutoff through the generation from doubly charged ions, when the plasma conditions remain optimal for the HHG. Such an approach, together with already demonstrated single-harmonic enhancement in the plateau region, can pave the way for further improvements in conversion efficiency and maximum cutoff energy of the harmonics generating through the interaction of the radiation of moderate intensity with the laser ablation.

9) We analyzed the harmonic spectra from manganese plasmas driven by 3.5 fs pulses, which were dominated by a single enhanced 31st harmonic at around 50 eV. The spectro-temporal experiments reveal the influence of the drive pulse duration on this process. Analysis of the harmonic energy tuning upon the variation of driving laser pulse duration controlled by change of gas pressure in the hollow fiber compression system have shown stabilization of the enhanced harmonic's energy and wavelength, which can be explained by resonance-induced enhancement at 51 eV. The theoretical modeling suggests that the emission could constitute an isolated subfemtosecond pulse. The observed weak carrier envelope phase (CEP) dependence might reduce the requirements for CEP stabilization of the laser. The isolation of a single harmonic order without any filtering could also be useful for various applications where the coherent, short pulse XUV radiation is required, without the losses induced by spectral dispersion or filtering.

10) We analyzed the studies of the high-order harmonics of ultrashort pulses generated in the plasma plumes produced by ablation of various elemental semiconductors (Te, Se, Si, As, Sb, and Ge). These plasma media were distinguished by the generation of the harmonics ranging between the 27th and 51st orders. Application of two-color pump allowed the generation of both odd and even harmonics, with the intensities of latter harmonics stronger than those of former ones. Strong resonance-induced 35th harmonic, with the enhancement factor of 12× compared with the neighboring harmonic orders, obtained in selenium plasma was analyzed. The resonance-induced enhancement of single harmonics was also observed in Ge, Sb, Te, and As plasmas.

11) The GaAs plasma produced by the picosecond pulse on the target surface was studied as the medium for the HHG (up to the 43rd order) using femtosecond pulses with different chirp. The high-order harmonics generated from GaAs plume showed a plateau pattern. A significant enhancement of single harmonic (27th) in the vicinity of 29 nm was achieved. We analyzed the plasma and harmonic spectra of the components of GaAs molecule, which showed that the As ions are responsible for the observed peculiarity of the considerable enhancement of intensity of the 27th harmonic. Additionally, the enhancement of the 27th harmonic at the wavelength of 29.44 nm by using the laser-ablation tellurium plume was discussed. The maximum cutoff wavelength of the HHG from the laser-ablation tellurium plume was found to be 15.58 nm. The intensity of the 27th harmonic was three times higher compared to the neighboring harmonics, and the 27th harmonic energy was estimated to be 12 nJ. Finally, we analyzed the enhancement of the 21st harmonic by using the antimony laser-ablation plume as a nonlinear medium. The energy of this harmonic was measured to be 0.3 μJ, and the conversion efficiency was 2.5×10^{-5}. The intensity of this harmonic 20 times exceeded those of the 23rd and 19th harmonics. The maximum cutoff wavelength of the HHG using antimony laser-ablation plume was 14.45 nm (photon energy: 86 eV).

12) The discussed studies demonstrated the generation of high-order harmonics in ablated plumes using the moderate-class picosecond lasers as well. Carbon-containing plasmas proved to be efficient HHG media using the picosecond Nd:YAG lasers and moderate laser intensities. The important peculiarity of those studies was the observation of enhanced single harmonic in the vacuum ultraviolet range (100–200 nm), which was originated from the resonance-induced growth of the nonlinear optical response of carbon-containing plasmas in the vicinity of strong neutral transitions (C I, $\lambda = 148$ nm and C I, $\lambda = 156$ nm) causing the increase of 7th harmonic ($\lambda = 152$ nm) conversion efficiency. The influence of the Kerr nonlinearities can explain the relative enhancement of the 7th harmonic and its

stabilization over rather broad range of laser intensities. Summarizing, the application of picosecond pulses for plasma HHG, analysis of low-order harmonics from Nd:YAG laser, and resonance enhancement of single harmonic at 152 nm with following theoretical treatment of this process are the three peculiarities, which distinguish the discussed study from other reported research in this field. None of these three topics were treated previously for plasma HHG. Moreover, even comparing with gas HHG, no analogous effects were observed in this spectral range.

13) We also discussed the generation of high-order harmonics from the Pb plasma using the mode-locked picosecond laser. The important peculiarity of those studies was the observation and analysis of the enhanced 11th harmonic (96.7 nm) in the vacuum ultraviolet range, which originated from the resonance-induced growth of the nonlinear optical response of lead plasma at the wavelength of 11th harmonic lying close to the strong ionic transition (Pb II, $\lambda = 96.72$ nm). The conversion efficiency of this harmonic was measured to be 3×10^{-6}. A strong departure from enhanced yield to suppressed state of the 11th harmonic in the case of the addition of He in the area of plasma formation clearly indicated the involvement of the propagation effect, which destroys the favorable conditions for the 11th harmonic generation while remaining the yields of other harmonics less influenced. The clear evidence for the availability to monitor the involvement of resonances in HHG through the modulation of phase mismatch is also presented in the case of carbon plasma. The similarity in the behavior of resonance harmonics in the cases of addition of the gas with positive dispersion in the carbon and lead plasmas is an additional proof of the influence of C and Pb ionic transitions on the nonlinear optical response and particular harmonic yields.

14) The discussed enhancement of harmonic yield in the low-energy XUV range in the case of nanoparticle (NP)-containing plasmas compared with monomer plasmas allows demonstration of the conditions when some additional enhancement was obtained for the single harmonics generated in the former plasmas at the conditions of the closeness of the wavelengths of these harmonics with the ionic transitions. Particularly, we have discussed the enhancement of single (13th, 17th, and 33rd) harmonics in the case of a few NP-containing plasmas (In_2O_3, Sn, and Mn_2O_3, respectively). The comparison of the harmonic spectra obtained using the plasmas produced on the NP and bulk materials of the same origin showed that similar enhancement of single harmonics occurs only in the case of strong excitation of the NP-containing targets. The feature specific to NP-induced HHG is the variation of the spectrum of individual enhanced harmonic. Another peculiarity of NP-induced harmonics is the appearance of strong blue-sided lobes near each lower order harmonic. Those studies showed that, at weak excitation of the targets contained

similar elements of periodic table, the harmonics are mainly produced by ionized monomers from the ablated bulk targets and by neutral NPs from the ablated NP-covered targets. However, with the growth of heating pulse fluence on the target surfaces, the most prominent feature, i.e. a single harmonic enhancement, can be straightforwardly interpreted by the resonances in ions.

15) HHG in fullerenes that exhibit an extended cutoff and enhancement (up to 25 times) of harmonic intensity in the low-energy plateau region compared with those generated from monomer carbon particles was discussed. The enhancement is attributed to the multi-electron dynamics in C_{60}. Further studies on HHG involving isolated C_{60} molecules irradiated with light pulses at different wavelengths and polarizations would provide deeper insight into the mechanisms responsible for the enhancement of high harmonics. Those results along with theoretical calculations suggest that multi-electron effects will vitiate the process of tomographic imaging of molecular orbitals that involves retrieving the structural information of molecule encoded in the high-order harmonic spectrum. Another conclusion of those studies is a demonstration that systems exhibiting collective oscillations at lower energies, such as small metal clusters, could lead to larger enhancements of high-order harmonics due to higher probability of exciting multi-electron dynamics.

16) We analyzed different approaches in the harmonic enhancement and compared them with the resonance approach. Particularly, we discussed the experimental studies of the quasi-phase-matched (QPM) harmonics generated in spatially modulated silver plasma. We analyzed the influence of driving and heating pulse energies on the enhancement of the QPM harmonics and showed the tuning of those enhanced harmonics by tilting the multislit mask placed in front of the Ag target. The results of calculations of the QPM conditions in laser plasmas suggest that the experimental observations could be qualitatively explained by the joint involvement of the two groups of electrons, those appearing during ablation and those coming from tunnel ionization in the formation of the QPM conditions. Another approach is the use of small particles for harmonic enhancement. The conditions of laser ablation of the bulk silver and silver NP powder were varied and analyzed to define various processes to enhance and restrict the harmonic yield during propagation of ultrashort pulses through such LPPs. The discussed studies showed that the enhancement of harmonics during either ablation of NP targets or formation of NPs from ablated bulk targets could be caused by the appearance of small species, either small molecules or small clusters of silver, at the moment of propagation of the laser pulses above the target surface. The easiness in the formation of a few atom-containing molecules during the ablation of the NP media may lead to their stronger influence on the HHG, since

these species appear in the area of ultrashort pulses propagation just at the moment corresponding to the delay between the heating and driving pulses. This assumption was confirmed during subsequent HHG studies, which revealed the correlation of the strong yields of harmonics from the ablation of bulk Ag and Ag NPs and the appearance of a few-atomic silver molecules in the mass spectra. Forming resonance conditions to enhance the nonlinear optical response of the medium may be an alternative to the phase-matching technique previously used for harmonic enhancement. In the meantime, if this process could be combined with the phase-matching effects and/or coherent control of HHG, one will be able to generate the harmonics enhanced by both collective and individual features of the medium. Application of tunable NIR pulses is the additional advantage in this direction since, on one hand, the tuning of driving pulses toward the wavelength the integers of which exactly coincide with the ionic transitions possessing large *gf* allows achieving the maximal resonantly enhanced yield of single harmonic. On the other hand, the use of longer wavelength laser sources allows the extension of harmonic cutoff and correspondingly the observation of QPM effect in the shorter wavelength region. Indium plasma perfectly matched with the above conditions and became a best choice for comparative studies to answer the question as to which (either collective or individual) process allows stronger nonlinear optical response of the medium. Further amendments are achieved once one uses the two-color pump technique for these two mechanisms.

Index

Resonance Enhancement in Laser-Produced Plasmas: Concepts and Applications,
First Edition. Rashid A. Ganeev.
© 2018 John Wiley & Sons, Inc. Published 2018 by John Wiley & Sons, Inc.